Contemporary Ergonomics and Human Factors 2013

Contemporary Ergonomics and Human Factors 2013

Editor

Martin Anderson

Health and Safety Executive, Bootle, UK

CRC Press
Taylor & Francis Group
Boca Raton   London   New York   Leiden

CRC Press is an imprint of the
Taylor & Francis Group, an **informa** business

A BALKEMA BOOK

Institute of
Ergonomics &
Human Factors

Preferred or adopted time headway? A driving simulator study
M. Gouy, C. Diels, N. Reed, A. Stevens & G. Burnett
© Transport Research Laboratory, UK, 2013

Hierarchical System Description (HSD) using MODAF and ISO 26800
M. Tainsh
© Lockheed Martin, UK, 2013

A quick method of assessing situation awareness in air traffic control
J. Nixon & A. Lowrey
© BAE Systems Ltd, UK, 2013

Reproduced with permission of the Controller of her Britannic Majesty's Stationary Office. The views expressed are those of the Author and do not necessarily reflect the views or policy of the Controller or any government department. The Controller, any government and Taylor & Francis Ltd accept no responsibility for the accuracy of any recipe, formula instruction contained within this population.

Typeset by MPS Ltd, Chennai, India
Printed and bound in Great Britain by CPI Group (UK) Ltd, Croydon, CR0 4YY.

Every effort has been made to ensure that the advice and information in this book is true and accurate at the time of going to press. However, neither the publisher nor the authors can accept any legal responsibility or liability for any errors or omissions that may be made. In the case of drug administration, any medical procedure or the use of technical equipment mentioned within this book, you are strongly advised to consult the manufacturer's guidelines.

ISBN 978-1-138-00042-1 (Pbk)
ISBN 978-0-203-74458-1 (eBook)

Contents

Preface

These are the proceedings of the International Conference on Contemporary Ergonomics and Human Factors 2013, held in April 2013 at Robinson College, Cambridge, UK. The conference is a major international event for ergonomists and human factors professionals, and attracts contributions and delegates from around the world. This conference is also the annual conference of the Institute of Ergonomics & Human Factors.

Papers are subject to a peer review process by a panel of reviewers before they are published in Contemporary Ergonomics and Human Factors. Topics covered in this edition include work & wellbeing, design, transport, safety culture, green ergonomics, healthcare, accessibility, human cognition, biomechanics, crowd behaviour and the systems approach.

The 2013 conference also included a symposium on human computer interaction led by the Donald Broadbent Lecture, and a symposium on standards preceded by a Plenary Lecture.

The Institute of Ergonomics & Human Factors is the professional body for ergonomists and human factors specialists based in the United Kingdom. It also attracts members throughout the world and is affiliated to the International Ergonomics Association. It provides recognition of competence of its members through its Professional Register. For further details contact:

Institute of Ergonomics & Human Factors
Elms Court
Elms Grove
Loughborough
Leicestershire
LE11 1RG
UK

Tel: (+44) 1509 234 904
Fax: (+44) 1509 235 666

Email: iehf@ergonomics.org.uk
Webpage: www.ergonomics.org.uk

Annual Conference 2013 Programme Committee

Chair   Professor Sarah Sharples
Editor Martin Anderson
       Dr Wen-Ruey Chang
       Dr Carole Deighton
       Dr David Golightly
       Margaret Hanson
       Dr Sue Hignett
       Murray Sinclair
       Dr Richard So
       Dr Mike Tainsh
       Dr Patrick Waterson

DONALD BROADBENT
LECTURE

INTERACTIONS "IN THE WILD": EXPLORATIONS IN HEALTHCARE

Ann Blandford

UCL Interaction Centre, University College London, UK

The real world is an essential laboratory for understanding people's interactions with technology. However, the real world is complex. Healthcare poses particular challenges; I highlight seven challenges faced by researchers conducting situated studies of healthcare technologies, from gaining access, through engaging with participants, to reporting findings. I draw on experiences of studies in hospitals and homes to propose ways to address these challenges, including engaging clinicians and patients early in study design and using Distributed Cognition to structure observation and analysis.

Introduction

Healthcare is evolving: towards greater reliance on interactive technologies, and more care being delivered outside clinical settings. This places increasing demands on healthcare technologies to be truly usable, by people with limited medical knowledge but whose health depends on effective use of the technology. Available technologies range from devices, such as blood pressure monitors, that can be bought over-the-counter to home haemodialysis systems that require extensive training and clinical oversight. While classical usability studies of these technologies are necessary, they are not sufficient to ensure that products are truly usable and safe in the intended (and maybe even unexpected) contexts of use. For example, poor lighting can render the displays of mobile devices unreadable; audible alarms can wake people up unnecessarily; people discover workarounds that may compromise the safety of devices (O'Connor, 2010); and interruptions may make it difficult for people to reconstruct where they were in a task. To fully assess the appropriateness of the design of existing systems, and identify requirements for next-generation products, it is essential to study use in context. This can be challenging for any product, but particularly so in healthcare: for example, they are typically used by people who are overworked, stressed or unwell; there is great variety in the settings in which they are used; and the human factors researcher typically has limited understanding of the health conditions being treated.

Seven challenges to studying healthcare technologies

In this paper, we consider seven challenges faced by human factors researchers studying the situated use of healthcare technologies, from gaining access through to reporting findings, and outline strategies for addressing those challenges.

Challenge 1: gaining access

As with any situated study, gaining access to relevant study settings is easier if an existing connection has been established. We have found it invaluable to work directly with enthusiastic and informed clinicians, who both facilitate access and also provide a reality check on the design and conduct of studies.

It is essential to navigate a maze of procedures and paperwork to gain the necessary formal approvals to conduct any study. For technology studies, it may not be necessary to gain full ethical clearance; many studies can be classed as service evaluations or audit studies. In our experience there are both geographical variations (between health service trusts) and evolving practices over time, so it is invaluable to engage early and often with local experts on gaining formal approval.

Where to start? Without testing the study protocol, it can be difficult to be sure what will work well, and yet it is not permissible to conduct the study without approval, which requires a detailed description of what is planned. We have sometimes been too cautious in our research plans. Engaging with potential participants and spending time in the proposed study settings builds awareness of what is possible.

Challenge 2: engaging with participants

Few people enter the healthcare profession because they are interested in technology; similarly, few patients and carers are actively interested in the technology they need to use to maintain their health. Indeed, the technology may be perceived as threatening, and studies that focus on safety, and the role of human error within that (e.g. Furniss et al, 2011), particularly so. Furthermore, the benefits to participants of taking part are often indirect and long-term. We have chosen to focus on people's skills and expertise, and on what works well, and to limit the demands on participants (e.g. their time); we had early ideals of engaging participants as co-researchers (e.g. using cultural probes or diaries (Cheverst et al, 2003)), but quickly discovered that our participants had too many other demands on their time and energy to engage in such data gathering.

Challenge 3: informed consent

Informed consent is widely regarded as essential, and yet obtaining formal informed consent from every individual within a typical healthcare setting is disproportionately time-consuming and distracting. It is often necessary to anticipate future possibilities. For example, when observing nurses working with infusion devices on night duty, we obtained consent from all patients who might be visited while they slept, before they settled down for the night.

Challenge 4: being an outsider

Human factors specialists bring a perspective to the design and use of healthcare technology that complements that of the clinicians, patients and carers who have to use it. However, they will always remain outsiders within the clinical setting and also in the homes of patients. They are neither professionals or personal friends. This can make it difficult to know how to act, e.g. when faced by difficult news (e.g. the death of a patient), observing procedures that are not going to plan, or encountering unexpectedly intimate moments. We have found that it is generally more effective to be a person first and a researcher second: to help out where possible, to get involved (e.g. fetching new equipment to replace a faulty device), to empathise and chat, while also being sensitive and withdrawing when circumstances demand it. It may be more difficult to write up the "method" section for such as study, but in practice it results in greater acceptance, greater engagement, and hence a greater depth of understanding of the situation and of technology use.

For the researcher, there may be personal challenges that are less common in other settings: encountering death and suffering and observing clinical procedures. To help deal with these, we have found it important to have peer support in place, as well as more formal counselling services where needed.

Challenge 5: knowing where to look

Most healthcare settings are complex: involving people with various roles and a plethora of equipment, highly reconfigurable, and often with a configuration that is changing dynamically. A particular piece of technology may be used only intermittently. This can make it very difficult to know where to look: for example, whether to accompany a particular nurse, to stay with particular pieces of equipment, or to loiter hopefully, waiting to be told (for example) that an infusion pump is about to be programmed. To respect patient confidentiality, there is limited opportunity to take video recordings or still photographs, so choices have to be made about what to record in real time. We have found Distributed Cognition (Rajkomar and Blandford, 2012) to be a useful perspective to take, particularly when familiarising ourselves with a new research context.

Challenge 6: innovating in a regulated environment

Within most regulatory systems, including Europe and North America, any device that is classified as a medical device has to be subjected to extensive approval processes before being deployed in healthcare, with only a few exceptions. This severely limits the scope for iterative design and testing in context. Our approach to this has been to engage with manufacturers, in order to influence the design of next-generation products. We are encapsulating our findings in personas, scenarios and other resources for manufacturers and other stakeholders, to inform design indirectly.

Challenge 7: knowing your audience

There are many possible audiences for findings on human factors in healthcare: our peers; technology manufacturers; regulators and policy makers; and technology users. Inevitably, each audience has sensitivities and interests. Two particular challenges we have faced in reporting on human error and technology design are, firstly a cultural attitude in healthcare: that only bad doctors and nurses make mistakes; and secondly a regulatory perspective: that technology is designed to be used in particular ways, and that to report on use that is outside that design space is to breach a culture of silence about actual practice. In writing, we have chosen to recognise, and respond directly to, such cultural factors, so as to build more effective dialogues between the different communities involved in healthcare technology research, design and use.

Conclusion

This is inevitably a skim across the surface of challenges and ways of responding to them in healthcare human factors research. At the heart of effective research in healthcare is engagement, building dialogues, and being sensitive to both the limitations and the opportunities provided. The challenges of working in healthcare are common, in many regards, to those faced in other settings, and yet are often more extreme. However, the personal rewards – of working with people who really care about their work, and of the prospect of making a difference to people's quality of life through improved design of healthcare technologies – are also great.

Acknowledgements

I am indebted to my research team, particularly Dominic Furniss, Aisling O'Kane, Atish Rajkomar and MSc students, and to many other colleagues for discussion of the ideas presented here. I have learned, and am still learning, so much! This research is funded by EPSRC EP/G059063/1.

References

Cheverst, K., Clarke, K., Dewsbury, G., Rouncefield, M., Rodden, T., Hughes, J., Crabtree, A. and Hemmings, T. 2003, Designing with Care: Adapting Cultural Probes to Inform Design in Sensitive Settings. *Proc. OzCHI.* 4–13
Furniss, D., Blandford, A. and Mayer, A. 2011, Unremarkable errors: Low-level disturbances in infusion pump use. *Proc. British HCI.* 197–204
O'Connor, L. 2010, *Workarounds in accident and emergency and intensive therapy departments: resilience, creation and consequences.* MSc thesis, UCL
Rajkomar, A. and Blandford, A. 2012, Understanding infusion administration in the ICU through Distributed Cognition. *Journal of biomedical informatics, 45*(3), 580–90

INSTITUTE LECTURE

INSTITUTE OF ERGONOMICS & HUMAN FACTORS 2013 LECTURE

P. John Clarkson

Cambridge Engineering Design Centre, Cambridge, UK

John Clarkson is directly involved in the teaching of design at all levels of the undergraduate course. At PA Consulting Group John gained wide experience of product development with a particular focus on the design of medical equipment and high-integrity systems, where clients required a risk-based systems approach to design to ensure timely delivery of safe systems. His research interests are in the general area of engineering design, particularly the development of design methodologies to address specific design issues, for example, process management, change management, healthcare design and inclusive design. As well as publishing over 450 papers, he has written and edited a number of books on medical equipment design and inclusive design.

KEYNOTE LECTURE

PERILS AND POSSIBILITIES FOR COMMUNICATING RISK AND UNCERTAINTY

D. Spiegelhalter

Winton Professor of the Public Understanding of Risk,
University of Cambridge, UK

David was previously Senior Scientist in the MRC Biostatistics Unit. His background is in medical statistics, particularly the use of Bayesian methods in clinical trials, health technology assessment and drug safety. He led the statistical team in the Bristol Royal Infirmary Inquiry and also gave evidence in the Shipman Inquiry. In his post he leads a small team which attempts to improve the way in which the quantitative aspects of risk and uncertainty are discussed in society. He works closely with the Millennium Mathematics Project in trying to bring risk and uncertainty into education. He gives many presentations to schools and others, advises organisations on risk communication, and is a regular newspaper columnist on current issues. He was elected FRS in 2005 and awarded an OBE in 2006 for services to medical statistics.

PLENARY LECTURES

STANDARDS AS HUMAN FACTORS BEST PRACTICE FOR INDUSTRY

Scott Steedman

Director of Standards, BSI Group, UK

Voluntary consensus standards embody best practice knowledge for business and industry and play a vital role in driving innovation. Leading companies now recognise that attitudes to boardroom behaviour, leadership and risk are all driven by an understanding of the values and principles that underpin the organisation. This paper, supported by a symposium of companion papers and debate, explores the background and opportunity to further integrate human factors into organisations, through a new generation of codes, guidance and standards, drawing on examples of past success and future promise.

Introduction

The integration of human factors thinking in business and industry is vital to the success of today's leading companies. It has become clear that best practice standards form a valuable tool for organisations to demonstrate commitment to better behaviours. The UK has a strong opportunity to show leadership in this area and to shape the form and content of such guidance. Integrating human factors expertise into the standards-making process, both in terms of recognising new issues that might be addressed in this way and in supporting committees already working in established fields, is an important first step.

First generation: Standards for products

As mass production and electricity became commonplace in the early days of the twentieth century it became necessary for industries to agree formats and specifications for product lines. By agreeing, through standards, on a way forward for a new or emerging supply chain, industries were able to invest with confidence in the new market opportunity. One key outcome would be interoperability, including standardising user interfaces. Along the way, issues of safety and reliability became embedded in the world of product standards. This was the first generation of voluntary consensus industry standards. Several examples of product standards exist in the area of ergonomics. One of the earliest was BS 3044 *Anatomical, Physiological and Anthropometric Principles in the Design of Office Chairs and Tables*, written by W F Floyd and D F Roberts. Although long since withdrawn, the key principles from this standard still hold good. Product standards related to the safety and usability of medical devices are considered in the Standards symposium [Vincent].

Second generation: Standards for processes

The second generation of standards had a more organisational dimension. Pioneers of quality management recognised that beyond product lay process, and to achieve further competitive advantage it was necessary to consider the management processes within the organisation. Such standards are now well established, for example: ISO 9001 *Quality Management* and OHSAS 18001 *Occupational Health and Safety*. A process standard familiar to the human factors community is ISO 9241-210: *Human-centred design for interactive systems*. An example of the benefits of using this standard in delivering user centred design is covered in the symposium [Väänänen]. Other symposium papers consider the use of various product and process standards in the engineering process in a safety critical context [Barge] and the case for moving towards more performance-based standards for human and organisational factors such as safety management systems [Miles].

Third generation: Standards for business potential and values

Over the past decade, the importance of behaviour in organisational development and business performance has assumed a much higher priority. The scope for improvements in productivity, talent management, safety or business performance is clearly associated with better understanding of the influence of culture and human factors in the workplace.

New standards, that seek to capture and share best practice, are now being developed by business and industry groups together with their stakeholders. The UK, through its national standards body BSI, is taking a leadership role in developing such standards. An example is BS 11000 *Collaborative business relationships*. This, in two parts, provides a framework and guidance for establishing effective collaborative business relationships between different organisations through embedding the right processes and behaviours. A key example of where this is valuable is in the area of Research and Design. The sorts of partnerships which involve Human Factors specialists are very much within the scope of this work. Another example, BS 8900, Sustainable development, is covered in the symposium [Watkins]. These standards are not of the traditional form, associated with product quality or management process, but represent a third generation of standards suited to the twenty-first century, where people values are at the forefront of all leading organisational practice.

This is a new opportunity for the human factors community to integrate their thinking into business and industry. Shaping the behavioural standards of the future – from boardroom performance to best practice in innovation management – is something we should embrace as the next chapter in organisational achievement.

Opportunity for standards to further embrace human factors

The convergence of standards – as knowledge for business and industry – with human factors is now fully upon us. The principle of standards making by consensus is that the process must be open, fair and transparent to all stakeholders. Third generation standards are aimed at helping people in the workplace deliver their own potential and satisfy their organisation's customers more effectively through better behaviours, set down in the language of values and principles, something which resonates well with the aims of the human factors community. The role of BSI as the national standards body is to provide a level playing field for experts and users to reach consensus on the issues they face. Whether it is in health, safety, financial services, workplace behaviours or wherever, our national standards-making process is ideally suited to supporting the integration of best practice in human factors into organisational development.

References

Barge R, 2013, *Using standards to support Human Factors engineering*, Contemporary Ergonomics and Human Factors 2013, Taylor and Francis

BS 3044: 1958 *Anatomical, Physiological and Anthropometric Principles in the Design of Office Chairs and Tables*, W F Floyd and D F Roberts, British Standards Institution

BS 11000-1: 2010 *Collaborative business relationships. A framework specification*, British Standards Institution

BS 11000-2: 2011 *Collaborative business relationships. Guide to implementing BS 11000-1*, British Standards Institution

BS ISO 9241-210: 2010 *Human-centred design for interactive systems*, British Standards Institution

BS OHSAS 18001: 2007 *Occupational Health and Safety*, British Standards Institution

BS 8900, publication due 2013, *Sustainable development*, British Standards Institution

BS ISO 9001:2008 *Quality Management*, British Standards Institution

Miles, R, *The case for human and organisational factors standards*, Contemporary Ergonomics and Human Factors 2013, Taylor and Francis

Väänänen, T, *Putting the customer first – delivering user-centred design using ISO 9241*, Contemporary Ergonomics and Human Factors 2013, Taylor and Francis

Vincent, C, *Safety and usability of medical devices*, Contemporary Ergonomics and Human Factors 2013, Taylor and Francis

Watkins, M, *Defining and capturing human factors in sustainable development*, Contemporary Ergonomics and Human Factors 2013, Taylor and Francis

TASK, TEAM AND TECHNOLOGY INTEGRATION IN SURGICAL CARE

Ken Catchpole

Director of Surgical Safety and Human Factors Research,
Cedars Sinai Medical Centre, Los Angeles, USA

Teamwork training solutions based on Crew Resource Management have been popular in healthcare as a solution to patient safety problems. In this paper, I argue that team and technology requirements vary considerably with the types of surgery undertaken; and thus, that a more sophisticated model of process improvement in healthcare is required.

Introduction (example of a main heading)

Arguably the two major deployments of human factors expertise in healthcare in the last decade have been the use of direct observation methodologies to understand the complexity, strengths, and weaknesses of healthcare system; and the focus on teamwork and communication as a cause and solution to healthcare safety and performance problems. The methodological model has been perhaps the only way to understand work as performed at the sharp end (Catchpole & Wiegmann 2012), in order to distinguish between "work as imagined" vs "work as performed". Each new direct observation study has revealed new insights into the pressures, and every day problems associated with working in healthcare that illustrate the vast mismatch not only between human abilities and system requirements – but also between what "should" happen and what "really" happens. Indeed, by and large the understanding of safety and performance – and teamwork in particular – has been dominated by simplistic views derived from aviation or other high risk industries (Leonard et al. 2004). Though once valuable to begin to understand the nature of errors in healthcare, it is time that this view should be challenged. Direct observation tells us that he relationship between team, task and process in different is complex.

It is an oft-quoted statistic that 80% (or a similar proportion) of incidents are caused by communication, and the emphasis on aviation models of care has been explicit and extensive. A great many observation studies in the last decade have examined teamwork and process in surgical care; and there have been multiple attempts to demonstrate the value of Crew Resource Management training in a variety of healthcare settings (Awad et al. 2005; Grogan et al. 2004; Young-Xu et al. 2011; McCulloch et al. 2011). There have also been a range of useful studies that have defined the non-technical skills of a variety of healthcare practitioners (Fletcher et al. 2003; Yule et al. 2006). This work has begun to establish a human factors presence in healthcare that hitherto had been omitted from the consideration of most healthcare systems. By and large, however, it has been assumed that teamwork exists independently of the task or the technology being used. To examine the problems behind this assumption, we will focus on four different types of surgical care.

Task, technology and teamwork in cardiac surgery

Cardiac surgery is the classic "high technology, high risk" surgery. In a seminal human factors study in surgery, direct observation of teamwork and process demonstrated that small process problems could accumulate to affect outcome if the surgical teams did not appropriately compensate for those problems (de Leval et al. 2000). Further studies using video recordings were able to examine exactly how these life-threatening events arose (Catchpole et al. 2006). Analyses of these events demonstrate the close coupling between task, team and technology for the management of safe, effective cardiac surgery. The management of perfusion (sufficient oxygen supply to the brain and vital organs) while the heart is being operated is the key component of the success of an operation. The anesthetist, the surgeon and the perfusionist need to co-ordiate their actions and use of equipment to ensure that the patient is kept appropriately stable and the surgical field free of blood to allow the operation to move forward. This can be a delicate balance, and malfunctioning technology, mis-communications or unclear task definition can upset the process at any one time. Examples have been previously published of where and how problems with any one of these components can upset the rhythm of the operation, and can lead to vastly increased risks (Catchpole 2011). As a consequence, in these types of operations, the most frequent disruptions to optimal care are communication/co-ordination issues, staff absences, and equipment problems.

Task, technology and teamwork in orthopedic surgery

Several studies have been conducted in orthopedics. In hip and knee replacement surgery, the key relationship is between the scrub nurse and the surgeon. The anesthesiologist is rarely involved directly in the team as their task – to keep the patient stable – is largely unaffected by the course of the operation. Orthopedic surgical kits utilize 6 to 18 or more trays of jigs, cutting blocks, screws, rods, inserts and other highly technical metalwork in order to measure and configure the replacement joint. In order to keep the process running, the scrub nurse needs to have a well developed understanding of when, and especially how, all the kit will be used. Each manufacturer's method, technology and configuration requirements are different, which means that process and sometimes teamwork will break down if either the surgeon or the scrub nurse are using equipment they are unfamiliar with. Given that space in an operating room can be limited, management of the different trays, which need to be stacked on top of each other, requires considerable physical effort, a detailed knowledge of where every piece can be found in every tray, and good anticipation of and communication about the next phase in the surgical process. As a further observation, this equipment has generally not been optimized from a usability stand point, and allows incorrect configurations, which clearly raises further safety issues. In these operations, the most frequent problems are related to distractions and equipment or workspace management (Catchpole et al. 2007).

Task, technology and teamwork in robotic surgery

We have also been conducting new studies examining robotic surgery. This growing field of surgical endeavor has distinct advantages in offering keyhole (laparoscopic) surgery with a greater range of articulation than would be offered in more traditional laparoscopy, while removing hand tremor effects. Several ports are placed into the abdomen of the patient, which are then docked to the robot. The surgeon then sits at a terminal, several feet away from the operating table, placing his head and hands into an enclosure that allows him to see via 3D camera, and manipulate his devices remotely in several articulatory dimensions. This new technology changes the teamwork and task requirements in several ways. First, being remote from the scrub tech and assistant surgeon who remain at the operating table means that they are no longer directly working alongside the rest of their team. There may be a microphone/speaker system to substitute for direct communication, but this does not wholly substitute. For example, a frequent observation is where the arms of the robot interfere and are interfered with accidentally by the assistant surgeon, who can experience discomfort and minor crushing as the arms can pin him in unsually contorted locations. Furthermore, the need to change instruments – which is a process that can take several minutes, rather than the few seconds in more traditional surgery, means that the ability to think ahead and communicate anticipated requirements amongst the team are particularly important. Our recent findings suggest that more experienced surgeons have considerably smoother operations, in part because they are able to minimize the number of times instruments are changed.

Task, technology and teamwork in trauma care

Finally, direct observational work in the trauma setting reveals yet another interaction between technology, team and process. We followed patients from when they arrive into the ED to their eventual arrival on a ward, ICU, post-operative care, or discharge. Trauma is defined through the requirements to orchestrating a complex set of processes, involving ED nurses, ED doctors, trauma doctors, Ambulance/Fire crews, Radiography and OR staff, and the associated varying technological tools in a variety of geographical locations, while working under substantial uncertainty and time pressure. Consequently, we found most frequent problems relate not to communication, but to co-ordination of all those components. There is a major information management component to this work, which we have been exploring through the use of technology, as well as developing and delivering trauma-specific teamwork training.

Conclusions

Direct observation of teamwork and process in surgical care is a vital and developing methodology that helps to understand work as performed rather than work as imagined. This has revealed key differences in the teamwork and technological requirements for different types of surgical care, that go well beyond the simplistic models adopted from other high-risk industries. Further study should focus not only on what goes wrong – but especially on what goes right, and extend team interventions beyond extra tasks, single checklists, or training.

References

Awad, S.S. et al., 2005. Bridging the communication gap in the operating room with medical team training. *American Journal of Surgery*, 190(5), pp. 770–774.

Catchpole, K. & Wiegmann, D., 2012. Understanding safety and performance in the cardiac operating room: from "sharp end" to "blunt end." *BMJ Quality & Safety*, 21(10), pp. 807–809.

Catchpole, K.R. et al., 2006. Identification of systems failures in successful paediatric cardiac surgery. *Ergonomics*, 49(5–6), pp. 567–588.

Catchpole, K.R. et al., 2007. Improving patient safety by identifying latent failures in successful operations. *Surgery*, 142(1), pp. 102–110.

Catchpole, Ken R., 2011. Task, team and technology integration in the paediatric cardiac operating room. *Progress in Pediatric Cardiology*, 32, pp. 85–88.

Fletcher, G.C.L. et al., 2003. Anaesthetists' Non-Technical Skills (ANTS): Evaluation of a behavioural marker system. *British Journal of Anaesthesia*, 90(5), pp. 580–588.

Grogan, E.L. et al., 2004. The impact of aviation-based teamwork training on the attitudes of health-care professionals. *J. Am. Coll. Surg.*, 199(6), pp. 843–848.

Leonard, M., Graham, S. & Bonacum, D., 2004. The human factor: the critical importance of effective teamwork and communication in providing safe care. *Quality and Safety in Health Care*, 13 Suppl 1(1475-3898 (Print)), pp. i85–i90.

de Leval, M.R. et al., 2000. Human Factors and Cardiac Surgery: A Multicenter Study. *Journal of Thoracic and Cardiovascular Surgery*, 119(4), pp. 661–672.

McCulloch, P., Rathbone, J. & Catchpole, K., 2011. Interventions to improve teamwork and communications among healthcare staff. *Br. J. Surg.*, (0007-1323 (Linking)). Available at: PM:21305537.

Young-Xu, Y. et al., 2011. Association between implementation of a medical team training program and surgical morbidity. *Archives of Surgery* (Chicago, Ill.: 1960), 146(12), pp. 1368–1373.

Yule, S. et al., 2006. Development of a rating system for surgeons' non-technical skills. *Med. Educ.*, 40(11), pp. 1098–1104.

HCI

FOUR GO MAD IN THE YORKSHIRE DALES: INVESTIGATING INTERACTIONS ON TABLETOP DISPLAYS

Chrisminder Hare[1], Sarah Sharples[1], Alex Stedmon[2] & Peter Talbot-Jones[3]

[1] *Human Factors Research Group & Horizon Doctoral Centre, University of Nottingham, UK*
[2] *Cultural, Communication & Computing Research Institute (C3RI), Sheffield Hallam University, UK*
[3] *EADS Innovation Works, University of Nottingham, UK*

Interactive tabletops are an ideal shared space for collaboration, coordination and problem solving. There is a need to understand how individuals interact with the information that these devices display. This paper presents an exploratory study that looks at the level of detail individuals understand from different types of maps displayed on an interactive tabletop. Participants were asked to collaboratively plan a route between two points according to constraints. Two types of map were used; a direct manipulation map, which used Google Earth and a passive map, which used a zoom-restricted Google Map. Results showed that users who could directly manipulate the map were able to pick out finer details in the environment but also experienced a higher mental workload.

Introduction

Developments in technology have seen an expansion in the use of interactive systems. One particular example has been the evolution of interactive tabletops (Muller-Tomfelde and Fjeld, 2012). They take the established metaphor of a traditional tabletop that users are familiar with and provide a convenient physical interface for people to meet, share information, look over maps and carry out tasks that require face-to-face collaboration (Shen, 2006). This paper explores the use of direct manipulation and passive maps displayed using a tabletop system. The motivation for this research arose from issues identified by EADS Innovation Works concerning how individuals in a group scenario appear to understand varying levels of detail from traditional paper-based maps.

Maps are a method of expressing mental concepts and images (Robinson and Petchenik, 1976) and the process of map reading is a skilled task that can involve considerable cognitive effort (Zipf and Richter, 2002). As technology has developed, tabletop interfaces allow a number of users to interact around a shared space (Ryall et al., 2004). Interactive tabletops support direct input through display, enhanced collocated collaboration for groups, positive working styles and group dynamics (Shen et al., 2006).

Furuichi et al. (2005) explored a prototype multi-user geospatial map (DTMap) presented as an interactive tabletop cartographic analysis application. DTMap can overlay different views of the same areas of a map and allow for simultaneous multiple user interaction. Ryall et al. (2006) implemented a solution that allowed multiple users to simultaneously open and use their own personalised zooming lenses on a single map interface. Another approach was developed using specialised software (eGrid) to present a multi-touch tabletop environment that incorporated finger touches and natural hand gestures to facilitate the collaboration of control centre team members for utility companies (Selim and Maurer, 2010). Using a similar interaction metaphor but with different software (GeoLens) multi-touch, multi-user GIS systems have been demonstrated in museum contexts (Zadow et al., 2011).

Current literature focuses on improving the technology driving the use of tabletop interactions, however, there seems to be limited focus on understanding how individuals use and interact with the geospatial data presented on interactive tabletops. For this reason an exploratory study was designed to understand the level of detail understood by users, their mental workload, time taken to complete the task and users' feedback.

Investigating group interactions with different map formats

In this exploratory study 16 paid participants (four women and twelve men) were allocated into four experimental groups. The groups comprised of four members who knew each other from either a work or social environment. Whilst participants were specifically recruited with no military and/or formal map reading training, two participants reported that they had knowledge of the Yorkshire Dales, four participants felt they had excellent map reading skills and five participants were familiar with using Google Earth.

A between-subjects study design was adopted. The independent variable was the type of map used on the interactive tabletop: direct manipulation or passive map. The direct manipulation map used several maps including Google Earth, Google Maps and Google Directions and the passive map was integrated from Google Maps into a single image that was displayed on an html page. The notable difference in this map was the level of interactivity; individuals were neither able to zoom into the map nor customise it. The dependent variables were the number of utterances about elements of the environment, distance of a planned route task, method of transport chosen for the route, time taken to complete task and mental workload. In addition, qualitative analyses were conducted on map interaction, use of the tabletop and participant feedback. It was anticipated that individuals would gain a higher level of use and interaction whilst using a direct manipulation map due to reduced mental workload and the ability to view at a higher level of detail.

Table 1.   Utterances from Direct Manipulation and Passive Map Users.

Utterances Referring To:	Total Number of Utterances from Direct Manipulation Map	Total Number of Utterances from Passive Map
Roads	9	4
Paths	54	4
Rivers	9	2
Laybys	0	1
Elevation	1	16
Cove	6	4
Flat Ground	3	10
Trees	1	0
Walls	7	4
Shadow	3	1

Participants completed a consent form and questionnaire to collect demographic information, assess their experience with maps and their knowledge regarding the Yorkshire Dales. A short tutorial on how to use the IntuiFace TouchTable was given and general instructions regarding the task. Participants were presented with an emergency scenario where they were asked to attend to an injured person located on the map and to calculate the fastest route between a point of departure and the destination. Each participant was given a task sheet that contained slightly different information from each other. This was done to encourage collaboration and mimic each individual having different prior knowledge. Once the task was completed, participants completed a NASA-TLX questionnaire (Hart and Staveland, 1988) and participated in a semi-structured feedback session.

Results

The results show that there were more utterances made by users of direct manipulation maps who were able to use a higher level of detail presented by the map (Table 1). Direct manipulation map users spent a considerable amount of time zooming in and exploring the finer details of their planned route. For example, they examined the paths that could be walked along and those that could not (as demonstrated by the majority of their utterances focussed on 'paths'). Although passive map users had a lower level of utterances they focussed more on aspects of elevation and flat land, by trying to get a broader understanding of the landscape and terrain. The group explored overall elevated and non-elevated areas of land to determine areas they could traverse.

The distances planned by Groups One, Two and Three were approximately the same, whilst Group Four planned a slightly longer route (Table 2). Group Two was the only group who elected to walk the entire distance, as they did not find the suitable parking area that the other groups identified. On average, direct manipulation and passive map users spent the same time exploring the maps (26.5 mins), which

Table 2.   Planned Route, Method of Transport and Time Taken to Complete Task.

Map Type	Group Number	Distance of Planned Route (Miles)	Method of Transport	Group Time Taken to Complete Study (Mins)	Average Time Taken to Complete Study (Mins)
Direct Manipulation Maps	One	1.24	Car and Walk	22	26.5
Passive Maps	Two	1.2	Walk	31	26.5
	Three	1.19	Car and Walk	24	
	Four	1.37	Car and Walk	29	

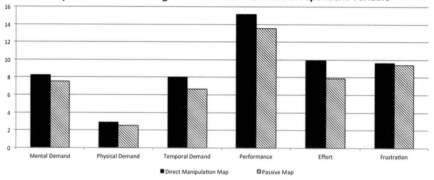

Figure 1.   A Graph to Show the Average NASA TLX Score for Each Independent Variable.

illustrated that there was no major difference for the distance of the planned route or the time taken to complete the task.

NASA-TLX multi-dimensional ratings were recorded to investigate subjective workload whilst undertaking the task. NASA-TLX uses six task demand subscales (e.g. mental, physical, temporal, performance, effort and frustration). The overall results indicated that direct manipulation maps scored highest in each of the six workload factors (Figure 1). Direct manipulation map users on average felt that the map was more demanding mentally, physically and felt both hurried and rushed when completing the task. However, they also felt that they were more success-ful in accomplishing the task but required more effort and felt more frustrated. A T-test showed that for each of the subscales the data was not statistically different ($p > 0.05$). This is most likely due to the fact that there were only eight participants testing each type of map.

Discussion

Overall, the study has provided an insight into how groups of individuals interact with direct manipulation and passive maps. All groups were able to gain an understanding of the Yorkshire Dales environment. Those users interacting with the direct manipulation map were able to get a better understanding of specific elements from the map by zooming into specific details, whilst passive map users were able to generate an overview of the environment. As part of a qualitative analysis, participant interactions with map, tabletop and each other are summarised (Table 3).

Overall, Group One had a negative view of using the direct manipulation map as they felt Ordnance Survey maps present roads and paths more clearly than the direct manipulation map. Group One felt that they could not get a good overview of the area without zooming in and out and they therefore had to make assumptions about the environment (e.g. specific road characteristics). However, contrary to Group One's negative view, Group Two felt they got a good overview of the area through the different views and components of the map such as the ability to zoom, changing the map scale, calculating distance and the searching function that they felt would make planning the route easier than on a paper map. It is important to point out that Google maps are just one type of direct manipulation map and the results for mental workload that have been observed may not apply to other forms of these maps.

Both groups had concerns regarding unreliability of the software. Group Two mentioned that although they would trust the map for the mapping of roads they were unsure about the mapping of off-road images. They felt that roads were well documented but finer details such as local knowledge shortcuts could be missing from the maps. It appeared that participants were basing this opinion on prior knowledge and use of Google Maps. One participant also explained that he had heard that some of the images on Google Maps are five years old and felt that could be different now. This suggests that it is important to consider how to communicate the age of information being presented (e.g. date stamps of when the images were taken and compiled).

Group Four found that they were unsure about shadowing effects portrayed on the passive map (e.g. cloud, rock, cliff, water, a valley, or snow). Participants also reported that as they stood on all four sides of the table, they got a different perspective of the map, some felt this to be disorientating. They wanted to be able to turn the map and show each other the map from their perspective. They liked the idea of the maps being displayed on the tabletop; however, they also wanted a zoom capability and better use of the touch sensing facilities. They also reported that the resolution was not as good as having a physical map and that the tabletop had glare and reflections that made it harder to view the displayed images. This group felt that they wanted to use the tabletop like other touch devices they had used (e.g. tablets, ipads, smart phones, etc). It needs to be acknowledged that the hardware cannot be dissociated from the software; further enhancements are needed (e.g. more fluid interactivity on surfaces) to the tabletop hardware and underlying software capabilities.

Table 3.   Shows how Participants Interacted with the Map, Tabletop and Each Other.

Group*	Participants Interaction with Software	Participants Interaction with Hardware	Group Dynamics
One	• Route focused on walls and paths. • Limited exploration of the environments elements i.e. terrain, elevation, natural markings, etc. • The group only used the default geographic view and search function.	• Direct input to the touch table was abandoned due to imprecision of gesture recognition, mouse and keyboard used as input devices.	• Two dominant individuals in the group who were the main decision makers • All participants shared the same perspective by standing on the same side of the tabletop.
Two	• An in-depth conversation about the elements of the environment. • Several routes attempted using several different maps and views, for example, Google Maps and Directions to identify paths, roads, rivers, shadows and walls. • Exact measurements of the ground were taken to establish the terrain and elevation of the ground.	• One individual controlled the keyboard and mouse. Participants instructed the mouse driver where to navigate on the map and on occasions took over mouse control.	• Participants were confident in expressing their opinions and challenging each other's suggestions, lively and fast paced discussion. • Participants stood around all four sides of the table.
Four	• Several parts of the maps were explored, including the terrain of the area, elevation and determining areas of land. • Some participants had walking and map reading skills, determining the highest point on the terrain to take the car and where the elevation was consistent to walk over. • Natural elements such as rivers, the cove itself were not explored in detail.	• One participant was the driver of the keyboard whilst other participants were instructing the participant to switch and change the views of the map.	• All participants participated in the discussion of the route and collaborated around all four sides like Group Two.

*Unfortunately due to camera failure Group Three's verbalizations and interactions were not collected.

Some individuals in the groups had previous experience of planning walking routes, however, others had limited experience. This resulted in the participants with limited experience having to spend time learning how to carry out the activity. At times, this resulted in the more experienced users taking the lead and guiding the others. Future work could look to filter participants with different levels of route planning experience.

The findings suggest that the interactivity of group dynamics played a major role; those individuals who knew each other from a social context embarked on a lively and fast paced discussion. Individuals were confident in challenging and expanding each other's suggestions. There is also a need for multiple users to interact with map technologies. In this exploratory study, the interactive tabletop and maps lacked support for individual exploration of the map, prohibiting the simultaneous exploration of multiple ideas.

This research has been carried out for explorative purposes and could be extended in numerous ways including repeating the same study with a larger and more diverse sample of participants. The experiment could also be extended by doing a within subjects experiment. This would draw direct comparisons between using the two types of map. Another extension of the study could be to get the participants to carry out the route in real life after planning the route on the tabletop. This would allow the experimenter to understand the level of detail participants understood before following the actual route. Another area of research could look at how people understand map data on horizontal displays versus vertical displays, for example, maps that are displayed on a tabletop versus maps displayed traditionally on a wall or interactive whiteboard.

Conclusion

In conclusion, this paper presents an exploratory study that provides an insight in to the varying levels of detail individuals understand from direct manipulation and passive maps displayed on tabletops. The results illustrated that groups using direct manipulation maps were able to draw more detail about the environment (e.g. identifying roads, walls, rivers, etc.) as the map allowed for exploration of the finer details. The study also showed that there was a higher level of mental workload when using direct manipulation maps. This research is a work in progress; further analysis needs to be undertaken to understand an individual's level of understanding from maps.

Acknowledgements

This work is supported by the Horizon Doctoral Training Centre at the University of Nottingham (RCUK Grant No. EP/G037574/1) and by the RCUK's Horizon Digital Economy Research Institute (RCUK Grant No. EP/G065802/1).

References

Furuichi, M., Mihori, Y., Muraoka, F., Estenther, A. and Ryall, K. 2005, DTMap demo: Interactive tabletop maps for ubiquitous computing. *International Conference on Ubiquitous Computing (UbiComp)*, September 2005

Hart, S.G., and Staveland, L.E. 1988, Development of NASA-TLX (Task Load Index): Results of empirical and theoretical research. In, P.A. Hancock and N. Meshkati (Eds.) *Human Mental Workload*. Amsterdam: North Holland Press

Muller-Tomfelde, C. and Fjeld, M. 2012, Tabletops: Interactive horizontal displays for ubiquitous computing. *IEEE Computer Society*, 78–81

Ryall, K., Forlines, C., Shen C., and Morris, M. 2004, Exploring the effects of group size and table size on interactions with tabletop shared-display groupware. *Proc. ACM Conf. Computer-Supported Cooperative Work (CSCW)*, ACM Press, 284–293

Ryall, K., Morris, M., Everitt, K., Forlines, C., and Shen, C. 2006, Experiences with and observations of direct-touch tabletops. *Proceedings of IEEE TableTop the International Workshop on Horizontal Interactive Human Computer Systems*, 89–96

Robinson, A.H., and Petchenik, B.B. 1976, *The Nature of Maps*. Chicago: Chicago University Press, 1–16

Scott, S.D., Carpendale, S., and Inkpen, K. 2004, Territoriality in collaborative tabletop workspaces. *Proceedings CSCW'04*. ACM (2004), 294–303

Selim, E. and Maurer, F. 2010, eGrid: Supporting the control room operation of a utility company with multi-touch tables. *ITS '10 ACM International Conference on Interactive Tabletops and Surfaces*

Shen, C. 2006, Multi-user interface and interactions on direct-touch horizontal surfaces: collaborative tabletop research at MERL. *IEEE TableTop (2006)*, 53–56

Shen, C., Ryall, K., Forlines, C., Esenther, A., Vernier, F., Everitt, K., Wu, M., Wigdor, D., Morris, M.R., Hancock, M., and Tse, E. 2006, Informing the design of direct-touch tabletops. *IEEE CG&A,* 26(5), 36–46

Zadow, U., Daiber, F., Schoning, J., and Krugerm A. 2011, GeoLens: Multi-user interaction with rich geographic information. *Proceedings of the Workshop on Data Exploration for Interactive Surfaces DEXIS 2011*

Zipf, A. and Richter, K. 2002, Using focus maps to ease map reading: Developing smart applications for mobile devices. *Korrekturabzug Kunstliche Intelligenz, Heft 4/02.*

OPENING INDOORS: THE ADVENT OF INDOOR POSITIONING

Michael Brown[1], James Pinchin[1] & Chris Hide[2]

[1]*Horizon Digital Economy Research, Nottingham, UK*
[2]*Nottingham Geospatial Institute, Nottingham, UK*

Indoor positioning technologies have the potential to be a revolutionary technology in the same way that mass market access to GPS revolutionised outdoor navigation. No single indoor positioning technology currently exists to provide a solution for every indoor navigation use case, however many candidate technologies have been proposed. Future indoor navigation systems are likely to need to call upon the strengths of many technologies to provide a hybrid position solution. We review a selection of state of the art indoor positioning technologies and discuss the results of a Context of Use analysis of potential user groups in this domain. Based on the result of this analysis we suggest challenges that need to be tackled by future research in this area.

Introduction

Over the past two decades Global Navigation Satellite Systems (GNSS) have had a transformational effect on the use of navigation services by non-expert users. 'SatNavs' are now almost standard equipment for the motorist and GNSS plays a large role in applications as diverse as agriculture and surveying. Largely driven by the inclusion of GNSS receivers in smartphones, global shipments are predicted to exceed 1 billion units per annum by 2020 (European Space Agency, 2012). However the growth of mass market 'Location Based Services' is hindered by the technological limitations of GNSS. Most significantly, GNSS does not provide a robust and accurate position solution to a user trying to navigate where they spend most of their time – indoors. If realised indoor positioning has the potential to provide not only everyday indoor navigation, but also enable a wide range of applications, such as guidance for the blind, positioning for emergency service command and control, and location based services for retail.

In this paper we will review the state of the art indoor positioning technology, then go on to describe a 'Context of Use Analysis' we have performed (Bevan & Macleod, 1994). Finally we will explore the extent to which current technologies fulfil the needs of these user groups.

Indoor positioning technology

"Indoor navigation is one of the last great technical problems that hasn't been solved." (Takahashi, 2012)

Accurate indoor positioning is a considerable challenge which is currently the subject of much research. Meter level accuracies are required if a person is to be reliably directed to an indoor space or point of interest. A single dominating technology analogous to GNSS does not currently exist to solve this problem. Instead many approaches have been proposed with varying degrees of accuracy, reliability and practicality. Currently and for the foreseeable future, a range of positioning technologies will need to be considered by a developer and matched to the task at hand.

Radio based approaches

The most widely researched category of indoor positioning technologies are those based on Radio Frequency (RF) signals. Common examples are high-sensitivity GNSS, methods based on a cellular phone network and WiFi based positioning. The 'time of flight' of RF signals may be used to estimate the distance to a transmitter and subsequently position through trilateration (e.g. GNSS). Alternatively the observed signal strength may be compared to pre-mapped values and used to infer position, so called 'fingerprinting' (often performed using WiFi signals). The accuracies of these techniques vary widely, from centimetre level from dedicated beacon systems to hundreds of meters for crude cell network based methods. Higher accuracy techniques usually require local dedicated infrastructure.

Often these methods provide a position solution at little or no additional cost to the user, relying on either non-local infrastructure (e.g. GNSS) or non-dedicated infrastructure (e.g. WiFi). Hardware already integrated into mobile computing devices is often re-used for navigation. The primary disadvantages of radio based methods are lack of robustness, particularly for 'fingerprinting' methods which require a known and unchanging RF environment.

Dead reckoning approaches

Miniaturised inertial sensors (accelerometers & gyros) allow indoor navigation to be achieved by measuring the accelerations and rotations of an object to be tracked. Low accuracy methods simply count the steps taken by a person and use a compass to estimate the step direction. High accuracy systems, such as foot mounted tracking systems, can achieve sub-meter level positioning indefinitely by utilising information about the gait of a person and incorporating high accuracy mapping of the indoor environment.

Dead reckoning systems rely on an accurate initial position and orientation from an external source, for example the user or another positioning technology. Over time sensor errors mean that the position solution quality will decrease. For this reason inertial based systems are rarely used without occasional aiding.

The cost of inertial systems varies from no additional cost in the case of smartphone sensors to thousands of pounds for dedicated sensor units. Often the quality of the positioning solution varies accordingly.

Table 1.   Overview of current indoor positioning solutions.

Technology	Approach	Positional accuracy	Infrastructure requirement	Notes
Global Navigation Satellite Systems (GNSS)	Radio Based	5 m with good signal to 50 m+ with poor signal, e.g. indoors	No local requirement	Indoor positioning unreliable and inaccurate even using latest high-sensitivity equipment.
Cell Tower Positioning	Radio Based	Poor, 100–5,000 m	Cell towers	More towers provide greater accuracy.
Wi-Fi Positioning	Radio Based	Room level, 5–100 m	Wi-Fi hot spots	Hot spots must be mapped for signal strength.
Dedicated local navigation beacons	Radio Based	0.1–5 m	Local beacon installation.	Either requires dedicated receiver equipment or licensing to use GNSS spectrum.
Low grade Inertial Sensor	Dead-Reckoning	Poor	None	Useful estimating distance travelled.
Mid-range Inertial Sensor	Dead-Reckoning	Degrades over time	None	Expensive sensors: £500–£5000
High-End Inertial Sensor	Dead-Reckoning	Degrades slowly over time	None	Highly expensive sensors: £5000+

Magnetic field mapping techniques utilise a 'fingerprint' magnetic field pattern for a mapped area or measure a low frequency signal from a local beacon. Magnetic methods are prone to unmapped changes in the indoor environment caused by electronic equipment, metallic objects or even changes to the building layout itself. Some such systems allow the magnetic field sensors (digital compass) in smartphones to be appropriated for positioning.

Factors including cost, infrastructure requirements and positioning performance all play a role in matching a technology or combination of technologies to a scenario of use. In this work a range of typical technologies will be introduced. In depth surveys of indoor positioning techniques can be found in Liu (2007), Zhou (2009) or Torres-Solis (2010).

Table 1 highlights the advantages and disadvantages of some of the more popular specific technologies used by each approach. Obviously the quality of the sensor used affects the cost and positional accuracy of each system, thus we include an approximate range based on current commercially available solutions.

Indoor mapping

It is worth mentioning the significant problems involved in obtaining and main-taining indoor mapping. Mapping provides context for the position solution as well as allowing route planning to take place. At present no single source of indoor mapping exists and a major effort is required to bring together and maintain the highly fragmented datasets.

Access, privacy and security issues mean that most indoor mapping will need to be maintained by the building operator or through crowdsourcing. As a result mapping will reflect the biases and priorities of the entity doing the mapping and standards of accuracy and completeness are likely to vary from building to building.

Context of use analysis

Based on a review of the literature and discussions with twelve technology and user domain experts we have identified current and potential user groups for indoor positioning solutions and explored their needs using Bevan and Macleod's (1994) context of use analysis method. Rather than reproducing the entire context of use analysis we will first highlight some key characteristics of each user group and then discuss the fit between user needs and current technologies. Table 2 contains a general overview of the analysis.

Everyday navigators

The most obvious use for indoor positioning technologies is simply extending existing navigation solutions to work indoors (Soloview & Dickman, 2011). This extension would facilitate the creation of true end-to-end navigation tools that can optimise navigation both within and through buildings, tunnels etc. Users of these solutions would be members of the general public, so specific needs are varied. Literature suggests that pedestrian navigators prioritise a simple functional inter-action (May, Ross, Bayer & Tarkiainen, 2003). Generally, these users will require a navigation aid only when traveling through large or complex unfamiliar buildings (Nurmi et al, 2011), thus they will tend to be infrequent users. For this reason, they won't be prepared to invest in or even carry specialist equipment, but the size of this user group means that it is justifiable to invest in significant infrastructure in order to support them.

Blind/Partially sighted users

A range of indoor, outdoor and ubiquitous solutions have been adopted for use by the blind and partially sighted. These range in function from crude high level navi-gation to active object detection and avoidance (Ran, Helal & Moore, 2004). Those that are blind or partially sighted need extremely accurate location information in order for any navigation solution to be effective and safe (Miao, Spindler & Weber,

Table 2.   Indoor Positioning Context of User Analysis Overview.

User Group	Tasks	Technical Environment	Physical Environment	Other Factors
Everyday Navigators	Room Level Navigation	Unlikely to use expensive or specialist Equipment	Generally only large or complex unfamiliar buildings	Convenience is the key factor for appropriation
Blind/Partially Sighted	High accuracy navigation	Will use specialist equipment if at a low-medium cost	All types of environment	Safety critical Position update must be fast and reliable
Surveyors	High accuracy positioning, usually not time critical	Will use high cost equipment if justifiable	All types of environment, often sites lacking infrastructure	
Location Based Gamers	Varied, often fast paced	High cost specialist equipment must be supported by wider business model	Potentially all environments, but locations may be limited by specific games	
Emergency Responders	Command and Control High accuracy navigation	Will use high cost equipment if justifiable	All types of environment, infrastructure may be compromised	Safety critical

2011). Blind and partially sighted people already use a range a specialist equipment from traditional white canes to talking watches and mobile screen readers (National Federation for the Blind, 2012), so these users are much more likely to invest in specialist equipment in order to complete navigation tasks. However the potential impact of technology failure for blind and partially sighted users means that a solution must be extremely robust. Local infrastructure won't necessarily be present in all environments in which these users wish to navigate, but larger scale infrastructure such as cell towers could be used if sufficiently reliable.

Surveyors

Creating internal building models requires accurate relative positioning. As this type of activity is often a precursor to infrastructure installation, it cannot be relied on when completing tasks. As professional users, expensive specialist equipment is acceptable as long it provides accurate and reliable locations solutions within domain specific tolerances (Shiu, Lu, Kang & Hsieh, 2007).

Location based gamers

While the needs of gamers vary depending on the specific location based games they play, there are some commonalities in this user group. The first priority for gamers is a good user experience; if the technology gets in the way of enjoyment it simply will not be used. With this in mind issues such as coverage and accuracy may not be important, as long as they are dealt with appropriately within the game. For example, a lack of coverage can be implemented as a game mechanic as demonstrated in Broll and Benford's (2005) work on 'Seamful Games'.

Emergency responders

In emergency situations the priorities are very different; the environment is variable and it is vital that information is of a known accuracy and delivered in a timely manner (Brunye, Gardon, Mahoney & Taylor, 2012). Emergency responders are likely to carry specialised equipment, but cannot rely on infrastructure 'on the ground' that may be compromised or simply absent.

Discussion

By comparing the needs of various user groups with the limitations of each technology we can identify where possible opportunities and pitfalls lie in implementing various indoor positioning technologies.

High end users

There are commonalities between the *Blind/Partially Sighted, Emergency Responders* and *Surveyors* user groups. All three would benefit from a robust, accurate solution that does not require infrastructure and are likely to invest in specialist equipment. Comparing these needs to the current solutions suggests a combined approach is necessary. Mid to high end inertial sensors can provide the necessary infrastructure free accuracy, but another radio based solution is needed to give initial position and correct drift over time. Even this multi-technology solution may not be sufficiently robust, as it is ultimately still dependant on infrastructure that may not be present.

Convenience users

Everyday Navigators and *Location Based Gamers* also share some characteristics. They are both primarily interested in convenience and quality of experience and don't necessarily need high-accuracy robust solutions. A reluctance to carry dedicated hardware rules out mid and high-end inertial sensors. RF methods currently lack momentum due to a lack of investment in infrastructure by building owners, technology standards by smartphone manufacturers and mapping for fingerprinting by all interested parties.

Future challenges

This analysis reveals that no single currently available indoor positioning technology can fulfil the needs of any of the user groups identified. Three priority areas for future technology development are:

- A high accuracy, completely infrastructure free solution for those willing to use specialist equipment.
- A low cost solution that integrates utilises a suite of cheap sensors to exploits any and all local infrastructure to provide a 'good enough' solution for navigation and gaming.
- Development of industry standards for the mass market implementation and use of RF signal based positioning methods.

In conclusion, by comparing a review of current internal positioning solutions with a context of use analysis exploring user needs in this area we have identified specific shortfalls that need to be tackled by future technology developments.

Future research

Following on from these finding we intend to explore user centred design and implementation of indoor positioning solutions within specific domains. Initial work is now underway exploring applications for blind/partially sighted users and emergency responders.

Acknowledgments

This work was funded by the RCUK Horizon Digital Economy Research Hub grant, EP/G065802/1.

References

Bevan, N., Macleod, M. 1994, Usability measurement in context. *Behaviour & Information Technology*, 13(1–2), 132–145.

Broll, G., Benford, S. 2005, Seamful design for location-based mobile games. *Entertainment Computing-ICEC*, 155–166.

Brunye, T., Gardony, A., Mahoney, C., Taylow, H. 2012, Going to town: Visualized perspectives and navigation through virtual environments. *Computers in Human Behaviour*, 28(1), 257–266.

European Space Agency. 2012, *GNSS market report, issue 2*. http://www.gsa.europa.eu/, May 2012.

Liu, H., Darabi, H., Banerjee, P., Liu, J. 2007, Survey of wireless indoor positioning techniques and systems. *Systems, IEEE Transactions on Man and Cybernetics, Part C: Applications and Reviews.*

May, A., Ross, T., Bayer, S., Tarkiainen. 2003, Pedestrian navigation aids: information requirements and design implications. *Personal Ubiquitous Computing*, 7, 331–338.

Miao, M., Spindler, M., Weber, G. 2011, Requirements of indoor navigation system from blind users. *Information Quality in e-Health*, 673–679.

National Federation for the Blind. 2012, *Straightforward Answers About Blindness*, available at http://www.nfb.org/straight-forward-answers-about-blindness accessed on the 28th of September, 2012.

Nurmi, P., Salovaara, A., Bhattacharya, S., Pulkkinen, T., Kahl, G. 2011, Influence of Landmark-Based Navigation Instructions on User Attention in Indoor Smart Spaces. *Proceedings of the 16th international conference on Intelligent user interfaces,* 33–42.

Ran, L., Helal, S., Moore, S. Drishti. 2004, An Integrated Indoor/Outdoor Blind Navigation System and Service. *Proceedings of the Second IEEE Annual Conference on Pervasive Computing and Communications*, 23–30.

Soloviev A., Dickman, J. 2011, Extending GPS carrier phase availability indoors with a deeply integrated receiver architecture. *Wireless Communications*, 18(2), 36–44.

Shiu R. S., Lu, C.C., Kang, S.C., Hsieh, S.H. 2007, Using a User-Centered Approach to Redesign the User Interface of a Computer-Based Surveyor Training Tool. In *Computing in Civil Engineering*, Soibelman, L., Akinci, B. (Eds), American Society of Civil Engineers.

Takahashi, D. 2012, Finnish engineers use earths magnetic field for indoor map navigation. *Venturebeat*, available at http://venturebeat.com/2012/07/15/_nnish-engineers-used-earths-magnetic-field-for-indoor-map-navigation/ accessed on the 23rd of September 2012.

Torres-Solis, J., Falk, T. H., Chau, T. A. 2010, review of indoor localization technologies: towards navigational assistance for topographical disorientation. *Ambient Intelligence*, InTechOpen.

Zhou, J., Shi, J. 2009, RFID localization algorithms and applications—a review. *Journal of Intelligent Manufacturing,* Springer.

A PREDICTIVE METHOD TO MEASURE RELATIVE EFFECTIVENESS

Nan Jiang

Software Research Centre, Bournemouth University, UK

Errors and completion are two widely used metrics for measuring the effectiveness of a system in usability testing. Since the two measures focus on different aspects of user output, a holistic view of "effectiveness" is sometimes hard to establish in a comparative study which eventually affects understanding different systems' relative effectiveness. The paper proposes a predictive method using an adapted confusion matrix to establish a correlation model to measure a system's relative effectiveness based on its own performance prediction. A case study is also provided to demonstrate how to use this method in real-world practice.

Introduction

ISO 9244-11:1998 defines effectiveness as the accuracy and completeness with which users achieve specified goals. According to ISO/IEC 25062:2006 CIF, the two factors are often reflected by errors and completion respectively in usability testing. In practice, considering the complex causes of human errors in task performance (Newell et al, 1958) (Rasmussen & Vicente, 1989), a system will be generally considered as effective if users made few errors and achieved high task completion rate on the system in a performance test. The guidance seems fine to be used for understanding a system's effectiveness independently or comparing the system before and after improvements. However, this holistic view may not be easily applied to a comparative study to understand a system's relative effectiveness for two reasons. First, a correlation between errors and completion is complex to model as they reflect different aspects of user output. Second, as Coll and Wingertsman (1990) stated, when systems show differences in magnitude and complexity, not only can user's performance on these systems be affected by such variables, their impact on effectiveness is also hard to measure.

This paper proposes a predictive method to measure a system's relative effectiveness based on the general grounds of an 'effective' system described above. This is achieved through using an adapted confusion matrix to present a practical correlation model between the two metrics and apply standard confusion matrix measurements to predict a system's relative effectiveness in a self-benchmarking process. The main benefits are obvious. First, it provides a holistic view of effectiveness with its two common measures. Second, it establishes a simplified measure for assessing a system's relative effectiveness by predicting at which error margin the system achieved its best performance. Third, it is a universal method which is more robust to the environment.

This paper proceeds as follows. Section 2 describes the overall methodology. Section 3 provides a case study where results are analysed and discussed in Section 4. Conclusion and future work are offered in Section 5.

Method

Correlation and prediction

Errors are usually classified as critical errors and non-critical errors based on their severity or seriousness in the measurement (Usability.gov, 2012). Obviously, only critical errors will result in task incompletion or incorrect outcome. In other words, a correlation can only be established between critical errors and completion. Let m be an error margin and:

$$m \in (0. \max Errors)$$

where 0 means error-free and *max* is the maximum number of errors made by a user in the test.

Suppose there is an m where a system reported relatively the fewest critical errors and most task completions, which indicates that relatively most users have completed the task on the system while each of them made no more than m errors. Then the system's relative effectiveness can be considered as the performance it has achieved at **this** m. In theory, m should be found to be as close to zero as possible. This is because it will also imply that users have made very few non-critical errors at the same time. In that case, m can be used to present the concept of both low critical and non-critical errors and high task completions in effectiveness assessment. Furthermore, when comparing different systems, it can rely on comparing at which m they have reported their 'best' relative performance. Therefore, measuring a system's relative effectiveness can be determined by predicting its m.

It should note that sometimes error rate margin can be more appropriate than error margin under certain circumstances. For example, when multiple goals can be achieved through various user paths in a semi-open task, error rate will be a more accurate measure than the number of errors. If this is the case, *max* will be the highest error rate made by a user in the usability test.

Adapted confusion matrix

Confusion matrix (Kohavi & Provost, 1998) is a predictive method used to understand the performance of different classification algorithms (classifiers) in artificial intelligence, information retrieval and data mining. As shown in Table 1, each row of the matrix represents an actual or observed class while each column represents a predicted or ideal class. Moreover, each cell counts the number of samples in the intersection for those classes noted as *tp, fn, fp* and *tn*.

Table 1.   Example of a confusion matrix used in Information Retrieval.

		Predicted	
		Positive	Negative
Classifier A			
Actual	Positive	*True positive cases (tp)*	*False negative cases (fn)*
	Negative	*False positive cases (fp)*	*True negative cases (tn)*

Table 2.   Adapted confusion matrix.

		Predicted	
		Completion	Incompletion
Error margin A			
Actual	Completion	*True positive cases (tp)*	*False negative cases (fn)*
	Incompletion	*False positive cases (fp)*	*True negative cases (tn)*

As shown in Table 2, the confusion matrix is adapted by using different error margins as classifiers and the number of completion and incompletion as classes so that it can be used to understand the performance of an error margin (m).

In terms of confusion matrix calculation, if an actual completion happened when a user made errors within the error margin, it would be considered as a true positive case (tp). Otherwise, it would be considered as a false negative case (fn) because the completion actually happened. Similarly, if an incompletion happened when a user made errors out of the margin, it would be considered as a true negative case (tn). Otherwise, it would be considered as a false positive case (fp) as it was not a completion.

Measurement

Understanding the measurement relies on interpreting the two standard measures used in the confusion matrix: *recall* and *precision*. Recall is defined as:

$$\frac{tp}{tp + fn}$$

which is the proportion of completion that were correctly identified at the specified error margin. It represents how accurate predicted completions are found in the actual observation. Recall tends to grow when error margin expands as more completion in the observation will be included into the prediction. Therefore, for a given error margin, the higher the recall is, the more accurate the prediction was and therefore more task completion have occurred.

Precision is defined as:

$$\frac{tp}{tp + fp}$$

which is the proportion of predicted completion that were correct at the specified error margin. It considers how accurate these actual completions were predicted. For a given error margin, the higher the precision is, the fewer task incompletion was found. Since incompletion is mainly driven by critical errors, it can also be used to imply the proportion of critical errors found in the margin.

In order to understand the relative effectiveness of a system, it needs to find out at which error margin the system reports relatively the highest recall and precision through analysing the differences between actual and predicted task completion results. The reason is obvious: for a given margin, if both measures are high, it will indicate most task completions have occurred (*recall*) and users made the fewest critical errors (*precision*) at the same time. Moreover, the smaller the margin is, the fewer non-critical errors were also revealed in the measurement. Therefore, different systems can be compared by understanding their 'best' error margins as a result of their individual performance prediction. In information retrieval, the concept of high recall and precision are presented by using F1 score, which is the harmonic mean of precision and recall, as defined below:

$$\frac{2 \cdot (precision \times recall)}{(precision + recall)}$$

In other words, the comparison can be simplified by studying the maximum F1 scores of different systems at a given margin.

Case study

A case study comparing Play.com and Amazon.co.uk, which are two popular online retailers in the UK, was conducted to demonstrate how to use this method to measure systems' relative effectiveness in a real-life practice.

Participants, tasks and data collection

A total of 194 first-year computing students with an age between 18 and 24 from Bournemouth University were asked to take participate in this study. A background analysis found that they all have Web and online shopping experience.

Two tasks were designed to reflect some common uses of e-commerce websites (Nielsen, 2011) and task design criteria (Usability.gov, 2011). Task A was a simple bargain hunting task for known item purchase based on built-in search engine and Task B was a complex browsing and inspiration task for unknown item purchase.

Table 3.   Samples taken on the two websites (p > 0.05).

Website	Task A	Task B	Total
Play.com	55	53	108
Amazon.co.uk	46	40	86

Table 3 shows the number of samples taken in each task on each website and there was no significant difference found in terms of sample size (Fisher's Exact, $p > 0.05$).

Test data were collected by volunteer observers who were asked to sit next to the participants during the latter's task performance. An online submission form provided by Google Docs was used to record test data into a spreadsheet for further analysis.

Metrics

Completion was measured as task completion rate, which is defined as the percentage of participants who successfully achieved the correct outcome. This is determined by using a binary measure (Nielsen, 2001) (Bevan, 2006) (Olmsted-Hawala et al, 2011). The calculation was straightforward on Task A as there was only one expected product (outcome). However, since Task B is semi-open task with multiple expected results, each participant's outcome was manually checked by volunteer observers against the success criteria given in the task description. In addition, it also considered a participant's landing category when granting a success.

An error was defined as a user attempt led by a mouse click which was not needed for performing error-free task completion with the minimal effort. The minimal effort was determined by a user's landing page – either a search engine for Task A or a category for Task B. This is because a landing page reflects the user's interest. In this study, error rate was used for both tasks as Task B has more than one expected outcome with more than one approach (i.e., user path). This was defined as the percentage of these attempts made by participants during the process of task completion.

Since Task B was a semi-open task, error rate margin was selected for the confusion matrix measurement.

Results and discussion

Overview

Figure 1 shows key descriptive data for the two metrics measured as task completion rate (CR) and error rate (ER). It is clear that Play.com outperformed Amazon.co.uk

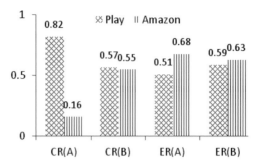

Figure 1.   Completion Rate (CR) and Error Rate (ER).

Table 4.   Correlation matrix (p < 0.01).

		Errors
Completion (Task A)	Play.com	−.709
	Amazon.co.uk	−.388
Completion (Task B)	Play.com	−.346
	Amazon.co.uk	.153

on Task A as it reported much higher task completion rate with lower error rate than the latter (82% with 51% vs. 16% and 68%), which is also in line with statistical analysis. However, the two websites delivered very similar results on Task B which cannot lead to a conclusion as the difference is only marginal (57% with 59% vs. 55% with 61%).

Relationships between metrics

A correlation matrix with the two metrics was generated by using the Pearson's product-moment correlation coefficient. As illustrated in Table 4, in top-down order, strong relationship between task completion and errors is found for Play.com on Task A (first row) while weak relationships are found in the rest of rows (i.e., Play.com on Task B, Amazon.co.uk on both tasks). Moreover, opposite relationships are reported for the two websites on Task B. This can also be seen an example that a holistic view is not always explicit in real world.

Confusion matrix measurement

For illustration purposes, the confusion matrix measurement will be focused on Task B as the two websites reported similar performance with only marginal differences. Error rate margins were set at 0%, 10%, 20%, 30%, 40%, 50%, 60%, 70%, 80%, 90% and 99% to cover all major possibilities. In fact, it is better to set these boundaries based on the normal distribution of error rates or confidence intervals

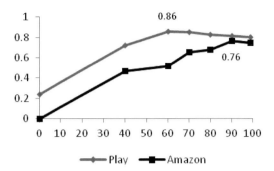

Figure 2.   F1 score trend curves on both websites on Task B.

(CI). Figure 2 shows F1 score (*y-axis*) trend over error rate margins (*x-axis*) of both websites on Task B. The highest F1 score reported on Play.com (0.86 at 60%) shows the website has a better relative effectiveness than Amazon.co.uk (0.76 at 90%) as it has a lower error margin. This suggests that participants made less critical errors on Play.com than Amazon.co.uk for achieving relatively highest task completions. In other words, users had to make more errors on Amazon.co.uk for achieving similar effectiveness performance as reported on Play.com. In addition, error-free task completion (0.24 at 0%) was found on Play.com while it was missing in Amazon.co.uk.

Conclusion and future work

When both accuracy and completeness are needed for understanding systems' relative effectiveness in a comparative usability testing, it is important to obtain a holistic view of the two measures. This should be first achieved by establishing a correlation model between their representative metrics in performance measurement. This predictive method provides a way to present this correlation based on a common understanding of 'effective' system. Moreover, the method measures a system's relative effectiveness by predicting at which error margin it has achieved its relatively highest recall and precision of task completion as they reflect most task completion and fewest critical errors respectively. Obviously, the lower the margin is, the fewer the critical errors as well as non-critical errors users made on the system thus the better relative effectiveness the system has achieved. The case study, which shows some typical comparative situations, is used to demonstrate how to apply this method and the findings are in line with in-test observations. Certainly, more user studies focusing on different environment settings (e.g., task difficulties) are needed to improve and consolidate this method in the future.

Acknowledgements

The author is grateful to Dr Haibin Liu and Dr Amanda Schierz for sharing their expertise in validating this method and to Dr John Beavis for offering continuously support on statistics. Further thanks to all first year students in the 2011–2012 cohort who were involved in the case study. Thanks should also be given to a number of usability practitioners in UXPA (User Experience Professionals' Association) networks who generously shared their industrial experience and inspired the work. Finally, the author would like to express appreciation to conference reviewers for their valuable advice on improving the paper.

References

Bevan, N. 2006. "Practical issues in usability measurement." *Interactions* 13(6): 42–43.

Coll, R. and Wingertsman, J. C. 1990. "The effect of screen complexity on user preference and performance." *International Journal of Human-Computer Interaction* 2(3): 255–265.

HowTo.gov. 2011. "Website usability testing." Retrieved November 19th, 2012, from: http://www.howto.gov/customer-service/collecting-feedback/usability-testing-fact-sheet.

International Standards Organisation. 1998, "ISO 9241-11 Ergonomic requirements for office work with visual display terminals (VDTs) – Part 11: Guidance on usability." (ed. 1), (International Standards Organisation, Geneva).

International Standards Organisation. 2006. "ISO/IEC 25062:2006 Software engineering – software product quality requirements and evaluation (SQuaRE) – Common Industry Format (CIF) for usability test reports." (ed.1.0), (International Standards Organisation, Geneva).

Kohavi, R. and Provost, F. 1998. "Glossary of terms." *Machine Learning* 30: 2–3.

Newell, A., Shaw, J. C. and Simon, H. A. 1958. "Elements of a theory of human problem solving." *Psychological Review* 65: 151–166.

Nielsen, J. 2001. "Success rate: the simplest usability metric." Retrieved November 19th, 2012, from: http://www.useit.com/alertbox/20010218.html.

Nielsen, J. 2011. "E-Commerce usability." Retrieved November 19th, 2012, from: http://www.useit.com/alertbox/ecommerce.html.

Olmsted-Hawala, E., Bergstrom, J.R., Chen J. and Murphy E.D. 2011. "Conducting iterative usability testing on a web site: challenges and benefits." *Journal of Usability Studies* 7(1): 9–30.

Rasmussen, J. and Vicente, K. 1989. "Coping with human errors through system design: implications for ecological interface design." *International Journal of Man-Machine Studies* 31(5): 517–534.

Usability.gov. 2012. "Usability test plan template." Retrieved November 19th, 2012 from: http://www.usability.gov/templates/docs/u-test_plan_template.doc.

VISUALLY INDUCED MOTION SICKNESS DURING COMPUTER GAME PLAYING

**C.T. Guo, Chi Wai Tsoi, Yiu Lun Wong,
Kwok Chun Yu & R.H.Y. So**

*Department of Industrial Engineering and Logistics
Management, Hong Kong University
of Science & Technology, Hong Kong*

The objective of this study is to examine the effects of active control verses passive watching on visually induced motion sickness (VIMS) during game playing, as well as the types of eye motions. Two experiments are reported: Experiment one compared sickness levels in active and passive conditions and Experiment two studied the effects of eye fixation during passive watching. Preliminary results show that: (i) rated nausea levels during passive watching were significantly higher than active playing ($p < 0.05$) and (ii) nausea level during passive watching was significantly suppressed by eye fixation ($p < 0.05$).

Introduction

Motion sickness can be characterized by a series of symptoms like nausea, sweating, dizziness, vomiting, etc. Study of motion sickness dated back more than one hundred years ago (Reason, 1978). Levels of motion sickness reported by individuals can be affected by susceptibility, age, gender, type of motion stimulus, frequency of exposure, etc. (e.g., Kennedy *et al.*, 2010; So and Ujike, 2010). Stimulus that can provoke motion sickness, according to the sensory organs being explicitly involved, can be categorized into three types: (i) visual motion, (ii) vestibular motion, and (iii) combined visual and vestibular motions. Motion sickness that is purely provoked by visual motion is called visually induced motion sickness (VIMS). With the development of virtual reality (VR) technique, incidences of VIMS when watching movies, VR simulation or computer games have been reported in recent years. Over eighty percent of individuals exposed to VR simulations of 20 minutes reported increases in sickness symptoms (e.g., Cobb *et al.*, 1999; Kennedy and Stanney, 1998; Wilson *et al.*, 2000). Incidents of VIMS have also been reported among users of various immersive video game systems, such as Xbox, PlayStation, and Wii (e.g., Merhi *et al.*, 2007; Stoffregen *et al.*, 2008). No doubt, the occurrence of VIMS can reduce the joy of entertainment. More importantly, efficiency and accuracy of performance can be reduced in scientific applications of VR technology. Hence, research to study factors affecting VIMS is desirable and meaningful.

Active playing and passive watching

The sensory conflict theory is widely cited as a theory to describe the needed conditions for motion sickness to occur (Reason, 1978). The theory explains that motion sickness is likely to occur when there is conflict among signals coming from different sense organs, or when there is conflict between real and expected forms of these signals. Since the "conflict" as defined in the theory is a neural psychological signal that is difficult to quantify, direct verification of this theory can be difficult. However, its ability to describe situations that can provoke motion sickness has been widely accepted. One such example is that a taxi driver should experience less motion sickness than the passengers. Rolnick and Lubow (1991) reported a "driver" and "passengers" experiment. The drivers executed a planned series of vehicle rotations and the passengers just watched and experienced the motion. Results showed that those who had controls of the rotations reported less symptoms of motion sickness than those who just experienced the motion. Xiao *et al.* (2011) extended the hypothesis to see whether the "control effect" was just limited to physical motion or could be extended to visual motion. One group of subjects drove (the driver) a computer driving simulation game and another group of subjects watched the video (the passenger). Results showed that the influence of control on motion sickness incidence is not limited to vehicle controls. It also applied to the control of virtual environment. Given the importance of eye motion in the generation of VIMS (e.g., Ebenholtz *et al.*, 1994, Ebenholtz, 2001; Hu *et al.*; Ji *et al.*, 2009; Yang *et al.*, 2011; Guo *et al.*, 2011), we hypothesized that eye motions of active and passive gaming is different and this difference could be related to the occurrence of VIMS.

Eye motions and VIMS

The presence of eye motion had been shown to increase VIMS (e.g., Ji *et al.*, 2009; Yang *et al.*, 2011). However, eye motions in these studies were provoked by watching rotating striped patterns on optokinetic drums. Studies related to eye movements and their effects on VIMS during game playing could not be found. g In other words, although eye fixation has been shown to significantly reduce VIMS when watching rotating drums, its effect on VIMS among players of computer games is not known. If, indeed, eye fixation can be verified to reduce VIMS among game players, it would provide a way to design a VIMS-free computer game.

Objectives and hypothesizes

The objectives of the first experiment were: (1) to compare VIMS levels between subjects who were actively playing computer games and subjects who were passively watching recorded videos; (2) to compare eye motions under these two modes of playing; and (3) to study effects of eye motion by using eye fixation on VIMS during active game playing. The objectives of the second experiment were: (1) to study types of eye motion during passive viewing conditions; and (2) to investigate whether eye fixation can reduce VIMS during passive viewing conditions. VIMS levels provoked by passive watching are hypothesized to be higher than those provoked by active game playing (Hypothesis I), and VIMS levels are hypothesized to be reduced by eye fixations (Hypothesis II).

Methods

3 males and 9 female, participated in the first experiment; 3 males and 4 females, participated in Experiment two. All of the subjects were undergraduate students or postgraduate students at the Hong Kong University of Science & Technology. They all signed subject consent forms and passed eye acuity tests before participating the experiment. They received money compensation of 50 HKD/hour. The experiment had been approved by the human subject committee of the Hong Kong University of Science & Technology.

Design of experiment

The video game named Mirror's Edge, developed by EA Digital Illusions CE (DICE) was used in our experiment. Mirror's Edge is a single-player first person action-adventure video game allowing for a wide range of actions. Players enjoy great freedom of movement. In this study, subjects used mouse and keyboard to control the avatar in the game from a first-person perspective. Due to rapid visual motion involved in the game, players have reported incidences of motion sickness (http://forums.steampowered.com/forums/showthread.php?t=1683987). All subjects in experiment one and two did not have experience in the game before training.

Experiment one used a within-subject design with 3 conditions: (A) active game playing with no eye fixation; (B) active game playing with eye fixation; and (C) passive viewing of recorded videos of game playing. Participants rested for at least 7 days between two successive conditions to avoid adaptation. A training session one week before the first condition was conducted to help the participants to be familiar with the game. The 12 participants were equally divided into 2 groups. The video that subjects watched in the passive viewing condition were recorded by another subjects during their active game playing conditions. Procedures were followed to match the game playing styles of these two groups of subjects. In short, the frequencies of pressing keys during the training were recorded to classify participants into "fast" moving players and "slow" moving players. In this study, a "fast" player would watch recording from another "fast" player during the passive viewing conditions. Similarly, "slow" players would watch recordings from another "slow" players during passive viewing conditions. Experiment two also used within subject design with two conditions: (A) passive watching without eye fixation; and (B) passive watching with eye fixation.

Procedure and apparatus

The procedure and apparatus of experiment one and two was the same. The game Mirror's Edge was run in a computer with Windows 7 Enterprise System and a 27-inch LCD monitor 50 cm away from subjects' eyes. During the exposure, all light was turned off except that from the stimulus. Subjects' eye motions were measured by a 16-mm eye tracker VT1 produced by Eye Tech Digital System.

During each session, participants were required to place their heads on a chin-rest stand to minimize head motion (Stoffregen and Smart, 1994). At the beginning, each subject went through the calibration of the eye tracker and fill in a pre-exposure simulator sickness questionnaire (SSQ: Kennedy *et al.*, 1993). Subjects were then exposed to the respective conditions according to a balanced presentation order and with 7-day rest in between. During the 30 minutes exposure, rated levels of nausea were measured using a 7-point nausea rating adopted from Golding and Kerguelen (1992) at every five minutes. Each player's performance in terms of mouse position and keystrokes was recorded by software named Macro Recorder, and their screenshot of playing were recorded in the form of "avi" video by software named PlayClaw. Also, their head positions were recorded by a webcam to verify that their heads did not move. After the 30-minute exposure, all participants were instructed to fill in the post-exposure SSQ.

Results

Pearson correlation test results show that there was significant correlation between the two dependent variables: post SSQ total score (SSQT) and the average 7 point nausea rating across 30 minutes, in both active (*Pearson Corr.*: 0.715, $p = 0.009$) and passive conditions (*Pearson Corr.*: 0.812, $p = 0.001$).

In Experiment one, the average 7 point nausea rating across 30 minutes in passive watching (mean: 2.2) was significantly higher than active playing (mean: 1.7) ($t = -3.592$, $p < 0.01$) and the same significant result was found in 8-people sub-group where all subjects' susceptibility rating were equal, or lower than, 3 ($t = -3.173$, $p < 0.05$, left part of Figure 1); the 7 point nausea rating at 30 minute in passive watching (mean: 3.5) was significantly higher than active playing (mean: 2.7) ($z = -2.232$, $p < 0.05$, right part of Figure 1), and the same significant result could also be found in low susceptibility sub-group ($z = -2.041$, $p < 0.05$). The average 7 point nausea data were verified to be normal and with equal variance, so the significant results were based on paired t-test, and the 7 point nausea data at 30 minute was non-normal and also failed to apply Box-cox transformation, so the significant results were based on Wilcoxon Signed Rank Tests. For sub-groups with higher sickness susceptibility ($>=4$), no significant difference between the active and passive conditions was found. By adding eye fixation, neither sub-groups nor the whole group's data showed significant reduction on 7 point nausea rating. For post SSQ scores, no significant result was found.

In Experiment two, the post SSQ sub-scores of nausea (SSQN) was significantly reduced by eye fixation (nfix: 40.4, fix: 27.2, $t = -2.8$, $p < 0.05$, Figure 2) based on paired t-test result. There was no other significant result between fixation and

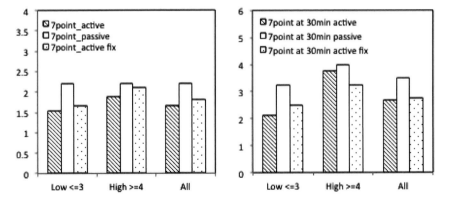

Figure 1.   **The left figure is the average 7 point nausea data across 30 minutes, and the right figure is the 7 point nausea data at 30 minute.**

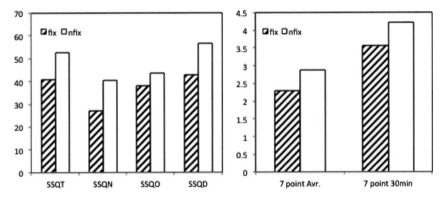

Figure 2.   **Left part the figure is the average SSQ total score (SSQT), SSQ nausea score (SSQN), SSQ oculomotor score (SSQO) and the SSQ disorientation score (SSQD) of all subjects. The right part of the figure is the average 7 point nausea rating across 30 minutes and the rating at 30 minute of all subjects.**

non-fixation conditions. However, a trend could be observed for all the dependent variables that eye fixation did reduce VIMS.

The eye tracker recorded both the left and right eye position in terms of pixels. Base upon the individual calibration data, linear regressions were conducted to obtain the gaze positions in degrees using left and right eye data. Regression results showed that linear model was valid to predict the gaze position, with left eye (x, y) and right (x, y) explained more than 95% of the variance (RSS > 95%).

For Experiment one, overall variance of eye motion during active game playing conditions and passive viewing conditions were statistically the same. We would like to remind the readers that participants did not watch their own recorded videos but videos of another game players with similar playing styles. For the active game playing condition with eye fixation, which was supposed to suppress eye motions, no significant reduction in variance of eye motion was found. This indicated that even with an eye fixation pointer placed at the centre of the screen and with instructions to fixate ones' eyes, it was difficult to fixate ones' eyes while actively playing the game.

For Experiment two, eye fixation significantly reduced eye motion in both horizontal and vertical directions. A Matlab program was run to identify duration of stop ($<=2$ dps), smooth pursuit ($<=40$ dps and >2dps) and saccadic eye motion (>40 dps). On average, 62% of the time, eye motion was categorized as "stop", 37% of the time categorized as "smooth pursuit" and no more than 1% of the time categorized to "saccadic". There was no significant correlation between these durations and sickness level.

In both experiments, there is not significant difference in VIMS score between genders.

Discussion and conclusion

Hypothesis I was supported by the nausea data from sub-group with low motion sickness susceptibility and the whole group. In other words, VIMS levels among participants with low sickness susceptibility were significantly lower when they were actively playing the game than when they were watching recorded videos of the same game. This could be explained by a ceiling effect on participants with high susceptibility, in which subjects already got very sick in the active condition, and even there is still room for them to get higher sickness in the passive condition, the difference between the two cases could be small.

Hypothesis II was supported only by SSQ nausea sub-scores. In other words, eye fixation was only associated with significant reduction in post-exposure nausea sub-scores. Compared with past studies (Ji *et al.*, 2009; Yang *et al.*, 2010; Guo *et al.*, 2011, 2012) in which OKN (optokinetic nystagmus, a type of eye motion produced by large screen moving scene) occurred nearly 100% during the whole 30-minute period when participants were watching rotating drums, OKN period in Experiment two was lower than 40% of the time. Subjects tended to relax their eyes on the screen most of the time. These observations raised a possible relationship between OKN and VIMS. Further studies are desirable.

In summary, the comparison between active and passive condition with more subjects are needs to further verify the hypothesis. Besides, further investigation on eye motion in terms of retinal slip in the passive condition, as well as the types of eye motion needs to be done.

Acknowledgement

The authors would like to thank the Research Grant Councils of the Hong Kong Government for partially supporting the study.

References

Barrett, J. (2004). Side Effects of Virtual Environments: A Review of the Literature. Australian Government, Department of Defence, Defence Science and Technology Organization, Command and Control Division Information Sciences Laboratory, DSTO-TR-1419.

Coady, E.A. (2010). The effect of drug mitigated motion sickness on physiological and psychophysical performance: Thesis of School of Graduate Studies of Master of Science in Kinesiology Department of Human Kinetics and Recreation Memorial University of Newfoundland, pp. 1–39.

Cobb, S.V.G., Nichols, S., Ramsey, A. and Wilson, J.R. (1999). Virtual reality induced symptoms and effects (VRISE). Presence, 8(2), 169. 186.

Chen, Y.C, Xiao Dong, Jens Hagstrom and Stoffregen T.A. (2011). Control of a virtual ambulation influences body movement and motionsickness. BIO Web of Conferences vol. 1, pp. 16 (2011).

Ebenholtz, S.M. (1992). Motion sickness and oculomotor systems in virtual environments, Teleoperators and Virtual Environments 1(3), 302–305.

Ebenholtz, S.M., Cohen, M.M. and Linder, B.J. (1994). The possible role of nystagmus in motion sickness: a hypothesis. Aviat Space Environ Med.: 65(11):1032–1035.

Ebenholtz, S.M. (2001). Oculomotor systems and perception. Cambridge University Press, c2001.

Guo, C.T., Ji, J.T.T. and So, R.H.Y. (2011). Could OKAN be an objective indicator of the susceptibility to visually induced motion sickness? Proceedings of IEEE Virtual Reality 2011, 19–23 March, Singapore.

Guo, C.T., Yang, J.X and So, R.H.Y. (2011). "The effects of optokinetic nystagmus on visually induced motion sickness after the confounding effects of retinal slip is controlled" i-Perception 2(4), 215.

Guo, C.T. and So, R.H.Y. (2012). "Effects of foveal retinal slip on visually induced motion sickness: a pilot study" Proceedings of 56th Annual Meeting of the Human Factors and Ergonomics Society October 22–26, 2012, Boston.

Golding, J. F. (1998). Motion sickness susceptibility questionnaire revised and its relationship to other forms of sickness. Brain Research Bulletin, 47(5), 507–516.

Golding, J.F. (2006). Motion sickness susceptibility. Auton. Neurosci: Basic and Clinical, 129, pp. 67–76.

Hu, S., Stern R.M., Vasey M.W. and Koch K.L. (1989). Motion sickness and gastric myoelectric activity as a function of speed of rotation of a circular vection drum. Aviat Space Environ Med. May, 60(5): 411–414.

Ji, J.T.T., So, R.H.Y., Lor, F., Cheung, T.F.R., Howarth, P. and Stanney, K. (2005). A Search for Possible neural pathways Leading to Visually Induced Motion Sickness. VISION, 17, 2, pp. 131–134.

Ji, J.T.T., So, R.H.Y. and Cheung, R.T.F. (2009). Isolating the Effects of Vection and Optokinetic Nystagmus on Optokinetic Rotation-Induced Motion Sickness. Human Factors, vol. 51, no. 5, 739–751.

Kennedy, R.S., and Kennedy, R.C. (2007). The Past, Present and Future of Research in Visually Induced MotionSickness. Proceedings (VIMS2007), 10–12 Dec., Hong Kong. pp. 3–8.

Kennedy, R.S., Drexler, J. and Kennedy, R.C. (2010). Research in visually induced motion sickness. Applied Ergonomics, 41(4), pp. 494–503.

Kennedy, R.S., Lane, N. E., Berbaum, K. S. and Lilienthal, M. G. (1993). Simulator sickness questionnaire: An enhanced method for quantifying simulator sickness. International Journal of Aviation Psychology, 3, 203–220.

Rolnick, A., and Lubow, R. E. (1991). Why the driver is rarely motion sick? The role of controllability in motion sickness. Ergonomics, 34, 867–879.

Scibora, L.M., Villard, S., Bardy, B., and Stoffregen, T.A. (2007). Wider Stance Reduces Body Sway and Motion Sickness. Proceedings (VIMS2007), 10–12 Dec., Hong Kong. pp. 18–23.

So, R.H., Ho, A. and Lo, W.T. (2001). A metric to quantify virtual scene movement for the study of cybersickness: definition, implementation, and verification. Presence, 10, pp. 192–215.

So, R.H.Y. and Ujike, H. (2010). Visually induced motion sickness, visual stress and photosensitive epileptic seizures: what do they have in common? – Preface to the special issue. Applied Ergonomics, 41(4), pp. 491–393.

Stanney, K.M., Mourant, R.R. and Kennedy, R.S. (1998). Human Factors Issues in Virtual Environments:A Review of the Literature. MIT Press Journals, 328–351.

Steele, J.E. (1961). Motion Sickness and Spatial perception, A Theoretical Study. Aeronautical Systems Division, Air Force Systems Command, United States Air Force, Wright-Patterson Air Force Base, Ohio.

Stoffregen, T.A., and Smart, L.J. (1998). Postural instability precedes motion sickness. Brain Research Bulletin, 47, 437–448.

Stoffregen, T.A., Faugloire, E., Yoshida, K., Flanagan, M.B. and Merhi, O. (2008). Motion Sickness and Postural Sway in Console Video Games. Human Factors: The Journal of the Human Factors and Ergonomics Society April 2008 50: 322–331.

de Vries S.C., Bos, J.E., van Emmerik, M.L. and Groen, E.L. (2007). Internal and external Field of View: computer games and cybersickness. Proceedings of the First International Symposium on Visually Induced Motion Sickness, Fatigue, and Photosensitive Epileptic Seizures (VIMS2007), 10–12 Dec., Hong Kong. pp. 89–95.

Xiao D., Yoshida, K., and Stoffregen, T.A. (2011). Control of a Virtual Vehicle Influences Postural Activity and Motion Sickness. J Exp Psychol Appl. 2011 Jun; 17(2): 128–38, pp. 1–45.

Yang, J.X., Guo, C.T., So, R.H.Y. and Cheung, R.T.F. (2011). Effects of Eye Fixation on Visually Induced Motion Sickness: Are they caused by Changes in Retinal Slip Velocity? Proceedings of the Human Factors and Ergonomics Society Annual Meeting, September, vol. 55, no. 1, 1220–1224.

HUMAN-AUTOMATION COLLABORATION IN MANUFACTURING: IDENTIFYING KEY IMPLEMENTATION FACTORS

George Charalambous, Sarah Fletcher & Philip Webb

Department of Integrated Systems, Cranfield University, UK

Human-automation collaboration has commanded significant attention in manufacturing because of the potential applications, such as the installation of large sub-assemblies. However, the key human factors relevant to human-automation collaboration have not yet been fully investigated. To maximise effective implementation and reduce development costs for future projects these factors need to be examined. In this paper, a collection of human factors likely to influence human-automation collaboration is identified from current literature and proposed future work to test validity and explore further factors associated with implementation success is presented.

Introduction

The manufacturing industry has always relied on skilled manual labour because automation has not been able to successfully replace the human dexterity needed for complex assembly tasks. Where intelligent automation is used, it is typically kept behind physical barriers or at safe separation distances from operators for health and safety reasons. However, recent advances in intelligent automation have allowed true collaborative working with human operators. In light of this, health and safety regulations and legislation are also being advanced and updated to reflect that in some circumstances it is safe and viable for humans to work more closely to industrial robots. Hence, the introduction of human-automation collaboration (HAC) in production lines has become an attractive proposition as it will enable greater flexibility in the design of large sub-assemblies and will reduce the current perturbations to manufacturing flow caused by the separation of human and robot.

Although advanced technology and legislation will allow the development of HAC systems, we do not yet fully understand how human operators will behave in these collaborative environments. It has long been realised that neglecting human factors issues can be detrimental to the successful introduction of advanced automated systems (Schonberger, 1986; Sheridan, 1990; Friscia, 1990). Parasuraman and Wickens (2008) claim that despite advances in automation, the vital link for successful operation of intelligent automation is still human presence. However, as the development of HAC systems in manufacturing is still at an emergent stage there remains a paucity of literature identifying the key human factors that are likely to influence successful adoption.

The aim of the investigation described in this paper is to address the current gap in knowledge by identifying and evaluating the human factors of most influence in relation to the introduction of HAC in manufacturing. Thus, in this research the term 'human factors' refers to the human related issues, at the individual and organisational level, that promote the successful implementation of HAC systems.

The outline of this paper is as follows: the literature review process to identify the human factors most likely to influence human-automation collaboration are presented first, followed by the future work planned for development of a suitable methodology and, finally, the paper ends with a conclusion.

Key human factors in relation to human-automation collaboration

This section describes the process employed to identify the human factors likely to influence the adoption of HAC. First, the databases used to collect a number of studies are presented. Then, the criteria for including the studies are presented. Finally, a list of key factors identified as enablers or barriers to the introduction of HAC on the shopfloor is presented.

Collection of sample studies

To identify the human factors at individual and organisational levels, which are likely to influence the use of HAC on a shopfloor, a literature search was performed using a range of databases such as: PsycINFO, ScienceDirect, IEEE and journals available at Cranfield University library and the British library. The primary search terms included: *human-automation collaboration, human-robot collaboration, successful adoption of automated systems* and *human factors in advanced manufacturing technology*. This process yielded a list of articles from a variety of domains such as: aviation, human-robot collaboration, use of personal robots and advanced manufacturing technology. On the basis of these identified articles, secondary search terms appeared such as: *situation awareness, workload, trust, organisational structure, worker preparation* and *communication*. Then, specific searches were performed using the primary and secondary search terms.

Following this initial procedure, the collected literature was examined and identified potential factors associated with the successful adoption and use of HAC on the shopfloor.

Criteria for study consideration

To include the human factors identified by an article, all studies were inspected to ensure they fulfilled the following criteria: (i) each study had to report an empirical/field examination, (ii) the study had to incorporate human participants who either viewed or participated directly in interactions with automation through physical, virtual or augmented means.

Identification of possible human factors affecting HAC

The study is aiming to capture human factors at different levels across the organisation. The factors identified were classified into two sub-categories based on whether they affect the human operator or the structure of the organisation:

(i) Factors at the individual level: influencing the human operator.
(ii) Factors at the organisational level: influencing the organisation.

Factors at the individual level

Trust in automation: The less trust an individual places in the automation, the sooner they will intervene to disuse the system if necessary (De Visser, et al., 2006; Steinfeld, et al., 2006). An example is the development of The Special Weapons Observation Reconnaissance Detection (SWORD) system. Although fully operational, it was never used because soldiers did not trust it could function safely due to unexpected movements. Therefore, trust calibration can be a catalyst for the efficacy of the cooperation between man and machine.

Mental Workload: An automated system not properly designed, can produce additional workload (Miller and Parasuraman, 2007). Endsley, et al. (1997) found increased workload and a trend for a higher number of operational errors under conditions of reduced involvement. Hence, the introduction of automation can influence mental demands placed on the operators.

Loss of Situation Awareness: Situation awareness (SA) has received extensive attention in the aviation and air traffic control domains. Loss of SA has been quoted in the literature as the root cause of various accidents such as controlled flight into terrain (Woodhouse and Woodhouse, 1995) and operational errors in air traffic control (Durso, et al., 1998). Loss of SA can pose a major risk towards the successful cooperation of human operators and automated systems.

Skill Degradation: Altering the active involvement levels of the operators has been theorised to result in skill degradation (Wiener, 1988). The deskilling debate offers contrasting evidence. Wilson and Buchanan (1988) found deskilling evidence on a number of operators. On the other hand, Mazney, et al. (2009) reported that introduction of an automated system only affected novice users. As skill degradation appears to have a range of effects on operators further investigation is required in relation to HAC.

Automation-Induced Complacency: The introduction of automation introduces additional cognitive demands due to monitoring activities. These have been found to result in AIC, especially in relation to the performance of concurrent tasks (Parasuraman, et al., 1993). In addition, the potential for an individual to exhibit AIC was investigated in an attempt to capture antecedents of complacency (Singh, et al., 1993). Hence, AIC appears to be a major negative impact when shifting from manual to an automated process.

Stress, Anxiety and Safety due to HAC: Collaboration in real time between humans and robotic assistants is bound to have the two agents in close proximity with the evident the possibility of collision. Also, the size and the sudden moves of an industrial robot can generate fear and surprise (Aria, et al., 2010). Hence, similar stress, anxiety and safety effects can be present during cooperation with a manufacturing automated system.

Effects of Adaptive Automation: It has been considered that by keeping the operator in-the-loop using variable levels of automation, enhanced performance can be achieved (Endsley and Kiris, 1995). Endsley and Kaber, 1999 indicated that using intermediate levels of automation during task implementation can keep operators in the loop and augment their performance.

Effects of Imperfect Automation: Investigations of the effects of imperfect automation have shown mixed results. Some studies indicate that operators are affected by the level of automation reliability (Parasuraman et al., 1993; Singh, et al., 2005). On the other hand, Dzindolet and colleagues (2001) did not find a significant effect of imperfect automation on operators' performance.

Perceived Attentional Control (PAC): It has been theorised that PAC level can influence reliance on automation and effectiveness of multitasking. More recently, Chen (2011) indicated that individuals' PAC appears to influence their reliance on automation particularly when workload level is high.

Attitudes towards Robots/Automation: Investigating people's attitudes towards automation can be an important topic for developing effective HAC. If operators are negatively biased against automation, this will later endanger implementation and effective use.

Factors at the organisational level

Communication to the Workforce: Communication of quality information during an organisational change can alleviate the associated uncertainty which has been linked with negative effect on psychological well-being (DiFonzo and Bordia, 1998; Rafferty, 2002). Hence, lack of communication can induce a psychological state of doubt and endanger acceptance of the system.

Training and Development of the Workforce: A key theme appearing on the literature is the training and development of the workforce (Waldeck, 2007). It has been declared that, workforce development is an important feature during AMT implementation that can promote organisational performance (Boyer, et al., 1997).

Formation of a Multidisciplinary Team: Investigation of the factors that enabled the successful implementation of AMT on the shopfloor indicated that the establishment of multidisciplinary teams from various departments can have a major positive impact (Bidanda, et al., 2005).

Worker Involvement in the Implementation: Worker involvement has been linked with successful implementation of a technological system (Sohal, 1999). Therefore, the degree of employee participation can be a potential enabler for successful adoption of HAC.

Identification of a Process Champion: The presence of a technology champion during the implementation phase can be a significant factor. This can be a knowledgeable person, who understands the technology and its benefits so that it can motivate people around them. Also, it has to appear credible to the workers in order to reduce resistance (Chung, 1996).

Organisational Flexibility and Top Management Commitment: According to a number of studies, top management commitment can be a key factor. The management team needs to be involved throughout the process and not simply with technical aspects of the project. Also, Zammuto and O'Connor, 1992 support that an organic culture is the key to exploiting the benefits of advanced automated technologies.

Trade Unions Involvement: The degree of a firm's unionisation level has been reported to influence the introduction of automated systems. Certain studies have indicated that unions do not always resist the introduction of technology (Small and Yasin, 2000). Some others have found that union involvement from the beginning can be a way to handle such a sensitive issue.

Future work for the development of the methodology

The ultimate objective of the current research is to develop a human factors tool suitable for implementing HAC on a manufacturing shopfloor. To do this the identified key factors outlined above need to be explored within a range of appropriate production processes in various manufacturing organisations to test their validity across industrial settings. Two types of processes in terms of stage of maturity will be explored via case studies: (i) a relatively mature, post-automation process and (ii) an example of in-progress automation implementation.

Post-automation case study

Processes that have already been automated will be studied. Semi-structured interviews with shopfloor operators will be conducted to collect retrospective self-report data regarding their experiences of the transition where their manual work was replaced by automation. Manufacturing engineers, system designers and management personnel will also be interviewed to gather their experiences and recollections regarding the introduction of the automated process. In addition, historical organisational data will be gathered where possible regarding the impacts of automation changes on key performance indicators and human resources.

Pre-automation case study

The transformation of fully manual processes to automated processes will be observed so that development and implementation can be studied at key stages. A full task analysis of the wholly manual process which first be performed in order to identify changes to the human role that result from the new automation. This will enable identification of specific manual task elements that have been replaced and analysis of the effects on operators. The study will also involve an on-going series of semi-structured interviews with shopfloor operators, engineers, system designers and management personnel regarding the implementation of the automation and gathering of available organisational data so that impacts through the change can be monitored.

The collected data from each case study will be analysed both separately and together in order to evaluate key factors related to automation implementation. To analyse the qualitative data, template analysis will be used to establish key themes, emergent themes and the relationship between them. This data collection and analysis strategy will not only establish the validity of the factors that have been identified in current literature as most likely to influence the implementation of human-robot automation but will also reveal any factors that have not yet been acknowledged.

Conclusion

This paper has presented research work in progress that is being conducted to advance the development of HAC in manufacturing processes. It has presented the key human factors that were identified through a literature review as most relevant to successful implementation as well as the future work being planned to develop a suitable methodology for validating these key factors. It is clear that the limited range of factors that have been identified thus far need to be validated across organisations and situations in order to test their relevance in different contexts, and that attempts should be made to explore any new factors.

References

Aria, T., Kato, R. and Fujita, M. (2010). Assessment of operator stress induced by robot collaboration in assembly. *CIRP Annals– Manufacturing Technology, 59*, 5–8.

Bidanda, B., Ariyawongrat, P., Needy, K. L., Norman, B. and Tharmmaphornphilas, W. (2005). Human related issues in manufacturing cell design, implementation, and operation: A review and survey. *Computation and Industrial Engineering, 48*, 507–523.

Boyer, K. K., Leong, G. K., Ward, P. T. and Krajewski, L. J. (1997). Unlocking the potential of advanced manufacturing technologies. *Journal of Operations Management, 15*(4), 331–347.

Chen, J. (2011). Individual Differences in Human-Robot Interaction in a Military Multitasking Environment. *Journal of Cognitive Engineering and Decision Making*, 83–105.

Chung, C. (1996). Human issues influencing the successful implementation of advanced manufacturing technology. *Journal of Engineering and Technology Management, 13*, 283–299.

De Visser, E., Parasuraman, R., Freedy, A., Freedy, E. and Weltman, G. (2006). A comprehensive methodology for assessing human-robot team performance for use in training and simulation. Santa Monica, CA: Human Factors and Ergonomics Society.

DiFonzo, N. and Bordia, P. (1998). A tale of two corporations: managing uncertainty during organisational change. *Human Resource Management, 37*(3), 295–303.

Durso, F. T., Truitt, T. R., Hackworth, C., Crutchfield, J. and Manning, C. A. (1998). En route operational errors and situation awareness. *International Journal of Aviaton Psychology, 8*, 177–194.

Dzindolet, M., Peterson, S., Pomranky, R., Pierce, L. and Beck, H. (2003). The role of trust in automation reliance. *International Journal of Human-Computer Studies, 58*(6), 697–718.

Endsley, M. R. and Kaber, D. B. (1999). Level of automation effects on performance, situation awareness and workload in a dynamic control task. *Ergonomics, 42*(3), 462–492.

Endsley, M. R. and Kiris, E. O. (1995). The out-of-the-loop performance problem and level of control in automation. *Human Factors, 37*(2), 381–394.

Endsley, M., English, T. and Sundararajan, M. (1997). The modelling of expertise: The use of situation models for knowledge engineering. *International Journal of Cognitive Ergonomics, 1*(2), 119–136.

Friscia, A. (1990). Systems integration: What it is. What it isn't. *Automation*, 32–38.

Mazney, D., Rottger, S., Bahner-Heyne, E. J., Schulze-Kissing, D., Dietz, A., Meixensberger, J. and Strauss, G. (2009). Image-guided navigation: the surgeon's perspective on performance consequences and human factors issues. *International Journal of Medical Robotics and Computer Assisted Surgery, 5*, 297–308.

Miller, C. and Parasuraman, R. (2007). Designing for flexible interaction between humans and automation: Delegation interfaces for supervisory control. *Human Factors, 49*, 57–75.

Parasuraman, R. and Wickens, C. (2008). Humans: Still vital after all these years of automation. *Human Factors*, 511–520.

Parasuraman, R., Molloy, R. and Singh, L. I. (1993). Performance Consequences of Automation-Induced "Complacency". *The International Journal of Aviation Psychology, 3*(1), 1–23.

Parasuraman, R., Mouloua, M., Molloy, R. and Hilburn, B. (1992). *Training and adaptiveautomation. Vol. 2: Adaptive manual training.* The Catholic University of America.

Rafferty, A. (2002). Relating different types of organisational change to measures of employee well-being. Singapore: Paper presented at the XXV international congress of applied psychology.

Schonberger, R. (1986). *World class manufacturing.* New York: Free Press.

Sheridan, J. (1990). The new Luddites? *Industry Week*, 62–63.

Singh, I. L., Sharman, H. O. and Singh, A. L. (2005). Effect of training on workload in flight simulation task performance. *Journal of the Indian Academy of Applied Psychology, 31*, 81–90.

Singh, L. I., Molloy, R. and Parasuraman, R. (1993). Automation-Induced "Complacency": Development of the Complacency-Potential Rating Scale. *The International Journal of Aviation Psychology, 3*(2), 111–112.

Small, M. and Yasin, M. (2000). Human factors in the adoption and performance of advanced manufacturing technology in unionised firms. *Industrial Management and Data Systems*, 389–402.

Sohal, A. (1999). Introducing new technology in to a small business: a case study. *Technovation*, 187–193.

Steinfeld, A., Fong, T., Kaber, D., Lewis, M., Scholtz, J., Schultz, A. and Goodrich, M. (2006). Common metrcs for human-robot interaction. Salt Lake, UT: ACM.

Waldeck, N. E. (2007). Worker assessment and the provision of developmental activities with advanced technology: An empirical study. *International Journal of Production Economics, 107*, 540–554.

Wiener, E. L. (1988). Cockpit automation. In E. L. Nagel (Ed.), *Human factors in aviation* (pp. 433–459). San Diego, CA: Academic Press.

Wilson, F. and Buchanan, D. (1988). The effect of new technology in the engineering industry: Cases of control and constraint. *Work, Employment and Society, 2*, 366–380.

Woodhouse, R. and Woodhouse, R. A. (1995). Navigation errors in relation to controlled flight into terrain (CFIT) accidents. (pp. 1403–1406). Columbus, OH: The 8th International Symposium on Aviation Psychology.

Zammuto, R. F. and O'Connor, E. J. (1992). Gaining Advanced Manufacturing Technologies' benefits: The Role of Organization Design Culture. *Academy of Management Review, 17*, 701–728.

WORK AND WELLBEING

BUS DRIVING – CAN IT BE A GOOD JOB?

Wendy Jones[1], Roger A. Haslam[1] & Cheryl Haslam[2]

[1]*Loughborough Design School, Loughborough University, UK*
[2]*School of Sport, Exercise and Health Sciences,*
Loughborough University, UK

Bus driving is recognised as an occupation where jobs are typically of poor quality and can have adverse effects on health. The current study explored how job quality differed for bus and coach drivers from three companies, identifying the most realistic areas for improvement, based on the similarities and differences between the companies. It also confirmed the usefulness of this approach for ergonomics in general. In areas of stress management and low control there was found to be limited potential for change. Scope for improvement was found in planning of working hours, health and safety, and vehicle/maintenance quality in some companies. However, it was acknowledged that change was unlikely to occur unless employers could be persuaded that it would be beneficial to their organisation.

Introduction

Assessing and improving job design and job quality have long been of interest to ergonomists, with the aim of reducing the adverse effects of work on health. More recently, consideration has been given to how work can be positively beneficial for health (Smith et al., 2011). There is much discussion about which features are most important in making a job 'good': Rose (2003) identifies pay and security as key, Lowe (2001) finds relationships to be a critical factor, and Clark (2005) highlights the importance of the actual work done. There are also concerns about the impact of sedentary work, and the need for ergonomics to move away from its 'less is better' paradigm (Straker & Mathiassen, 2009).

Bus drivers are an interesting group to study in this respect, given the inactive nature of the role and the long association with poor health. Morris et al. (1953) found the risk of heart disease for bus drivers to be twice that of their conductor colleagues. They also suffer from gastrointestinal disorders, musculoskeletal problems and poor mental health (Tse et al, 2006). Bus drivers have low job satisfaction (Rose, 2003); report stress and fatigue which they associate with passengers, traffic, and timetables (Biggs et al., 2009; Tse et al., 2007); and suffer from obesity which persists despite provision of exercise facilities, healthy food and education (French et al., 2010). The European Working Conditions Survey, considering employment sectors in Europe, identifies Land Transport (which includes bus driving) as one of the worst. Working conditions are poor, including long and non-standard working hours, low levels of job control, and risks of physical violence. Poor health outcomes include work related stress and musculoskeletal problems (Jettinghoff & Houtman, 2009).

This study aimed to assess the potential for improving job quality in bus driving. The study design compared different companies and the experiences of their drivers to identify instances of good working practices which may serve as examples for others in the industry. Sharing best practice is an approach widely taken by organisations such as the Health and Safety Executive, Business in the Community, and Eurofound as a means of improving job quality.

Method

Organisations

Three UK bus companies were recruited for this study. The companies are identified in this paper by pseudonyms as some of the data presented is commercially sensitive.

- BigBus is a large independent company which employs around 800 drivers across three depots and provides timetabled bus services.
- LittleBus is a small, family run company which employs around 100 drivers, and provides timetabled bus services as well as private hire coaches.
- LittleCoach is a small family run company which employs around 60 drivers and runs private hire and holiday excursions in the UK and overseas.

Formal interviews

Semi structured interviews lasting around 25 minutes each were carried out with 50 drivers (9–11 from each of the smaller companies, and from each depot of the larger company) to explore their experiences of working in their current and previous jobs. 43 of the interviewees were male, and 7 female. The average age was 45 years, (range 23 to 64), with a similar age profile across each company. Maximum length of service for interviewees was 22 years, although the average varied from 8 years in BigBus to less than 2 years in LittleBus.

At LittleBus and LittleCoach, interviewees were recruited by the researcher based on availability. At BigBus, interviews were scheduled by depot managers, based on availability. Male, female, recently recruited and long serving drivers were represented in the sample at each company/depot.

Interviews were also carried out with ten managers (3 at LittleBus, 2 at LittleCoach, 5 at BigBus), about their particular roles within the organisation. Copies of company policies and procedures were obtained where possible.

Observation, informal interviews and other data

Visits to each depot were used as opportunities for unstructured observation and informal discussion with drivers, and to view artifacts such as rotas. A total of 33 visits were carried out, facilitating conversations with 62 drivers and supervisors in addition to those formally interviewed. Observation was also undertaken on 12 bus journeys with BigBus and LittleBus during the period of study. It had been intended to gather and compare outcome data such as sickness absence, staff turnover and accident rates, but the two smaller companies did not keep records which would permit this.

Analysis

Recorded interviews were transcribed. For informal discussions and observations, brief notes were made as soon as possible, and more complete notes written up within 24 hours. Observation notes and interview transcripts were imported into NVivo and coded. Initial coding was done against a framework of key work aspects identified from the literature, the template was revised as additional themes were identified (King, 2004) and the coding was reviewed by a second researcher. Other measures taken to improve the trustworthiness of the findings include the large sample size of interviewees (formal and informal), the inclusion of observational data as well as interviews, and reporting the key findings back to the organisations to invite comments.

Results and discussion

Hours

Working patterns varied widely. At BigBus, most drivers worked a 3 shift pattern with full weeks on earlies, lates or 'middles'. Within this structure, there could be wide variation in start times such as 6 am one day followed by 11 am and 7 am on the second and third days respectively. At LittleBus there were more fixed shifts and slow rotation, e.g. starting each week slightly later than the previous week. At LittleCoach there was less variation in the working days, most shifts started between 6.30 am and 9 am. Drivers at all three companies were unhappy about quick changes, e.g. nine hours or less overnight between shifts.

Typical working days varied in length from approximately 8 hours at BigBus to 10–12 hours at LittleCoach and LittleBus. Drivers at BigBus also benefitted from knowing their working patterns at least 3 months in advance. At the other extreme, work at LittleCoach was rarely scheduled more than 2 days ahead, and working hours could change at less than 24 hours' notice.

Some drivers at BigBus, found positive value in their work patterns, e.g. having early shifts that finished by lunchtime, so they could spend time with their families.

Other drivers were indifferent to the working hours, accepting them as an inevitable feature of the job they had chosen. However, irregular working patterns and last minute shift changes gave many drivers cause for concern about the impact on their health, their safety and their personal life.

Inactivity

The sedentary nature of the job and its impact on health was an issue in all three organisations. Many drivers talked about having gained weight since starting the job. BigBus provided an on-site gym and arguably, their employees had fewer barriers to undertaking exercise as their shifts were shorter: but there was no clear evidence that they were actually more active as a result.

The sedentary nature of many jobs is a major public health issue, with evidence suggesting that prolonged sitting increases mortality (Patel et al., 2010). This is likely to remain a significant factor affecting good job quality for bus drivers.

Stress

Timetabling caused stress for drivers in all companies on occasion, they disliked running late because it made passengers unhappy and angry. Drivers at LittleBus and BigBus complained that some routes did not allow sufficient time to get to destinations. However, both companies claimed to plan schedules with great care, and had recently installed bus tracking equipment so they could monitor late running and revise timetables if necessary.

Passengers were identified as a source of stress. They were perceived as being more challenging in urban areas than rural ones and on timetabled services in comparison with private hires or scheduled coaches. BigBus was seen by some of its drivers as judging its employees unduly harshly if, for example, they reacted to bad passenger behaviour.

There were clear individual differences in the way drivers handled these issues in their work. Over half saw passengers as being the aspect of their job that made it good, and made little or no comment about the potential challenges. They talked about how they built relationships with regular customers. Others saw passengers as the biggest difficulty they faced, emphasising the fact that they could be rude, aggressive or elderly and slow, which made it difficult to keep to the timetable. There were differences too in the way the stress of running late affected drivers, with some finding it a significant problem and others accepting that it was beyond their control.

The key causes of stress in bus driving seem unlikely to change but there is scope to minimise their impact on drivers through well planned and regularly reviewed timetables and good support for staff when difficulties arise. The findings in this study regarding the differences between individuals echo those in the literature about the impact of different coping styles (Machin & Hoare, 2008). This suggests there are benefits to be achieved from training staff to deal constructively with such challenges, and also to consider the ideal personality characteristics for successful drivers when recruiting.

Physical ergonomics and comfort

Bus quality and maintenance were generally good at BigBus and LittleCoach, but appeared less satisfactory at LittleBus, where drivers complained about old buses, cold buses and poor maintenance. This was confirmed by observation. There was a wide range of vehicle models and drivers reported that some were particularly uncomfortable or difficult to drive. At BigBus there were standard bus models and driver representatives had been involved in their selection in the past, although some believed that bus quality was deteriorating. Prolonged sitting aggravated the discomfort, for example some shifts at LittleBus and BigBus involved unbroken periods of driving in excess of 5 hours.

Although the job will always involve prolonged sitting and whole body vibration, this study suggests that some companies could do more to improve physical comfort. Careful selection of new vehicles, with driver involvement, and account taken of ergonomics factors in the cab could reduce risk. High standards of maintenance, including to the driver's seat, are also critical.

Pay

The hourly rate at BigBus was 20–40% higher than at the two smaller companies, although there was some potential for tips at LittleCoach and bonus payments at LittleBus. There was evidence that drivers at the smaller companies worked longer hours, which compensated for their lower hourly wage.

Many drivers at BigBus considered themselves to be well paid (particularly in relation to their level of education) and considered their pay to be a feature which made their job good. The drivers at LittleBus and LittleCoach raised concerns about the job being poorly paid, particularly because of the responsibility they had for other people's lives and the consequences of making a mistake.

Pay level is an important feature in determining job quality (Rose, 2003) and in this study there is a clear difference between the large and small organisations. At the smaller companies pay rates are below those which may be considered adequate to support an acceptable lifestyle (Davis et al., 2012).

Job security

Job security was not raised as an issue of concern at LittleBus; drivers were generally quite itinerant and were confident of getting employment elsewhere if necessary. At LittleCoach, there were some concerns about the reliability of working hours as much of the work was seasonal and there would be occasions when less work was available (and therefore pay would be lower). Drivers at BigBus had some anxieties about job security given the company's relatively low threshold for taking disciplinary action including dismissal.

Health and safety

General health and safety seemed to be taken very seriously at BigBus, with provision of formal safety training, operation of safety committees, and enforcement of rules e.g. wearing high visibility jackets, keeping to marked walkways. Driver training, provided in-house, included dealing with customers, driving skills, and periodic on-bus observations. At LittleBus and LittleCoach there appeared to be a lower emphasis on health and safety. Problems included uneven or diesel coated road surfaces in the depot, a lack of safe walking routes, and poor compliance with welfare legislation.

Fixed safety screens were built into all buses at BigBus to protect against the risk of passenger violence. Although some drivers had concerns about risk from passengers, there were also many who disliked the screens, feeling they interfered with communication with their passengers. Observation confirmed that passengers on such buses were less likely to interact with the driver than those on LittleBus, although the differing use of automatic cards and cash could also have contributed to this. At LittleBus there were no protective screens, but drivers did not raise concerns about the risk of violence during their interviews. BigBus and LittleBus had closed circuit television installed which drivers generally felt positive about, as it could be used as evidence if claims were made against them (by passengers or other drivers).

The wide variation in management of health and safety in these three companies suggests there is scope for improvement in some. This could include training to cope better with the uncontrollable risks from passengers and other traffic.

Control

In all companies, personal decision making was limited by factors such as fixed timetables, scheduled working patterns and the need to comply with rules and regulations. However drivers in all companies had some opportunities to select or apply for the working patterns and bus routes that best suited them. At LittleCoach and LittleBus there was a greater degree of autonomy whilst driving. Drivers at BigBus had radios, and were expected to contact Radio Control for advice before making any decisions.

The job of a bus driver is unavoidably inflexible, and requires a high degree of compliance from drivers. Yet freedom from close supervision is a specific reason why many chose to go into the job and there are opportunities for driver input into decision making. There is also scope for companies to intentionally recruit drivers who are most comfortable with this low level of autonomy.

Likelihood of change

Drivers at all three companies could benefit from improvements to their working hours. However, commercial pressures make it unlikely that driver working patterns will change substantially unless regulatory change forces them to: the current rules for drivers on local bus services permit 7½ hours driving without a formal break (although a maximum of 5½ hours is more usually applied), and an overnight rest period as short as 8½ hours (VOSA, 2011). The smaller companies in particular have less flexibility, as they have fewer drivers and routes to work with, and little slack in the system.

At both LittleBus and LittleCoach, other key factors to improve job quality would be improved pay and management of health and safety: also better vehicle quality (at LittleBus) and work planning (at LittleCoach).

Pay rates are unlikely to change given the competitive nature of the industry and the lack of negotiating power which the drivers have. This also makes it unlikely that changes will occur to working time, particularly as the drivers may favour long working hours to compensate for low pay rates.

The biggest barrier to change in the smaller companies is the culture and the personal style of the owner/director in each case. At LittleCoach, a priority of maximising the usage rate of vehicles leads to acceptance of last minute bookings and late changes to drivers' work schedules. At LittleBus there is a focus on running a low cost business. Both companies rely on informal discussion rather than policy to resolve problems or personal difficulties, but with differing approaches amongst managers, the results are unpredictable.

Neither organisation responded to an invitation to meet and discuss the management report provided by the researcher following data gathering. Change is unlikely at either company as long as they are able to recruit staff and to operate successfully.

Summary and conclusions

The most intractable challenges to high job quality for bus drivers are similar to those in many industries: prolonged sedentary work and the potential conflict between the needs of the employee and those of the customer or wider society. In bus driving, this results in unsociable working hours, low levels of autonomy, relatively low pay for some, and a risk of hostility or violence from passengers.

This study has highlighted other areas where there may be scope for change. It has shown that comparison between companies is a good method for exploring this, but that changes to job quality will not occur unless employers can be persuaded that there are benefits in doing this.

It is unlikely that bus driving will ever be a 'good' job, but this study has illustrated that it may be 'good enough' in some organisations. This is particularly the case for workers who are temperamentally suited to it. An ergonomics approach can usefully highlight the most promising areas for improvement.

References

Biggs, H., Dingsdag, D., & Stenson, N. (2009). Fatigue factors affecting metropolitan bus drivers: A qualitative investigation. *Work*, *32*(1), 5–10.

Davis, A., Hirsch, D., Smith, N., Beckhelling, J., & Padley, M. (2012). *A Minimum income standard for the UK in 2012. Keeping up in hard times* (Vol. 2012). York: Joseph Rowntree Foundation. Retrieved from http://www.minimumincomestandard.org/downloads/2012_launch/mis_report_2012.pdf.

French, S. A., Harnack, L. J., Hannan, P. J., Mitchell, N. R., Gerlach, A. F., & Toomey, T. L. (2010). Worksite environment intervention to prevent obesity among metropolitan transit workers. *Preventive medicine*, *50*(4), 180–185.

Jettinghoff, K., & Houtman, I. (2009). *A sector perspective on working conditions*. Eurofound. Retrieved August 25, 2011, from http://www.eurofound.europa.eu/pubdocs/2008/14/en/1/ef0814en.pdf.

King, N. (2004). Using templates in the thematic analysis of text. In C. Cassell & G. Symon (Eds.), *Essential Guide to Qualitative Methods in Organizational Research* (pp. 256–270). London: Sage Publications.

Machin, M. A., & Hoare, P. N. (2008). The role of workload and driver coping styles in predicting bus drivers' need for recovery, positive and negative affect, and physical symptoms. *Anxiety, Stress & Coping*, *21*(4), 359–375.

Morris, J. N., Heady, J. A., Raffle, P. A. B., Roberts, C. G., & Parks, J. W. (1953). Coronary heart disease and physical activity of work. *The Lancet*, *262*(6796), 1111–1120.

Patel, A. V., Bernstein, L., Deka, A., Feigelson, H. S., Campbell, P. T., Gapstur, S. M., Colditz, G. A., et al. (2010). Leisure Time Spent Sitting in Relation to Total Mortality in a Prospective Cohort of US Adults. *American Journal of Epidemiology*, *172*(4), 419–429.

Rose, M. (2003). Good Deal, Bad Deal? Job Satisfaction in Occupations. *Work, Employment & Society*, *17*(3), 503–530.

Smith, A., Wadsworth, E., Chaplin, K., Allen, P., & Mark, G. (2011). *What is a good job? The relationship between work/working and improved health and wellb* (Vol. 11.1). England: IOSH.

Straker, L., & Mathiassen, S. E. (2009). Increased physical work loads in modern work – a necessity for better health and performance? *Ergonomics*, *52*(10), 1215.

Tse, J. L. M., Flin, R., & Mearns, K. (2007). Facets of job effort in bus driver health: Deconstructing "effort" in the effort-reward imbalance model. *Journal of occupational health psychology*, *12*(1), 48–62.

Tse, J. L. M., Flin, R., & Mearns, K. (2006). Bus driver well-being review: 50 years of research. Transportation Research Part F: Traffic Psychology and Behaviour, 9(2), 89–114.

Vehicle and Operator services Agency, (2011). *Rules on Drivers' Hours and Tachographs Passenger-carrying vehicles in GB and Europe* PSV 375: VOSA/CIS/2175/FEB11. VOSA.

"IT'S JUST PART OF THE JOB!" RAFT GUIDES WORKING WITH BACK PAIN

Iain Wilson, Hilary McDermott & Fehmidah Munir

School of Sports, Exercise and Health Sciences,
Loughborough University, UK

Work-related injury and ill-health is an under-researched area in the UK outdoor industry. This study explored work-related health among white water (WW) raft guides. Semi-structured interviews were undertaken with thirteen raft guides. Back pain was reported by participants. Older workers, however, demonstrated higher levels of awareness in relation to their personal health and safety. The results suggest that early intervention is required to promote the health and well-being of white water raft guides.

Introduction

In the UK, the Outdoor Industry is a growing sector (SkillsActive, 2010). Whilst there is a focus on client health and safety, there has been little research into the health and well-being of employees. Previous research involving Mountain Leaders identified a culture where injuries are perceived as 'part of the job' (McDermott & Munir, 2012). It is not known whether such a culture exists amongst those leading water activities. This study therefore explored work-related injury among white water (WW) raft guides in the UK in order to build a foundation for a large scale study which will inform the improvement of current health and safety guidelines.

Method

Semi-structured interviews were conducted with thirteen raft guides (92.3% Male), recruited from across the UK. Participants were aged between 19 to 43 years. The interview schedule, covered topics such as previous experience, working conditions and practices, injury and attitudes. The interviews were transcribed verbatim and subsequently analysed using Thematic Analysis (Braun & Clarke, 2006).

Results

Six themes were identified, of which four were related to the onset and management of back pain. These were protective behaviours, management of pain, attitudes toward pain and personal awareness of health and safety. The other themes were related to causes of injury and the psychological benefits of work.

All interviewees felt back pain was a common problem among WW raft guides. The majority reported experiencing back pain which they felt was a direct result of their work. These participants reported that back pain was an expected and accepted aspect of their work culture. For example, a 19 year-old male stated: *"It's just part of the job you have to work with, really."*

The interviews identified the role of age and experience in the use of protective behaviours in preventing work-related injury. Awareness of personal health and safety was more evident among the older raft guides (35 years and over) with more experience. One 37 year-old male participant explained: *"We [older raft guides] are really conscious of health and safety, but it took us 20 years to get here!"*

Discussion

The findings reported here were similar to those reported in mountaineers (McDermott & Munir, 2012) where work-related injuries are perceived as 'part of the job'. This suggests that those who work in the outdoor industry are more likely to accept injury and continue working whilst managing their pain. Older WW raft guides are more aware of protective behaviours such as warming up, thus reducing the risk of work-related injury. Younger WW raft guides would benefit from the transfer of knowledge from more experienced colleagues. Although a small sample, the findings highlight a potential area for further research.

Relevance

The study results will be used to inform a large study assessing work-related injury in the outdoor WW sports industry in order to improve health and safety guidelines.

References

Braun, V., & Clarke, V. 2006, Using thematic analysis in psychology. *Qualitative Research in Psychology*, *3*(2), 77–101.

McDermott, H., & Munir, F. 2012, Work-related injury and ill-health among mountain instructors in the UK. *Safety Science*, *50*(4), 1104–1111.

SkillsActive. 2010, "The outdoors survey 2009" Retrieved 29th November, 2011, from: http://www.skillsactive.com/assets/0000/6348/SkillsActiveTheOutdoors Survey2009_FINALReport_Jun2010.pdf.

DEVELOPING SOFTWARE TO HELP SMALL BUSINESSES MANAGE OCCUPATIONAL SAFETY AND HEALTH

Sally Shalloe, Glyn Lawson, Mirabelle D'Cruz & Richard Eastgate

Human Factors Research Group, University of Nottingham, UK

European SMEs experience higher accident rates than their larger organisations which directly affects profit and competitiveness. Research has shown that SMEs may not see the benefits of implementing OSH management systems and often adopt a reactive rather than proactive approach when accidents occur. This paper describes the user centred development approach being adopted by the EU funded IMOSHION project to develop an OSH knowledge and learning management system to address issues of access to OSH knowledge, enhance OSH training and improve safety culture within SMEs.

Introduction

It is estimated that accidents at work resulted in at least 83 million calendar days of sick leave in the European Union (EU) Member States in 2007 (Council of European Union, 2011). The figure for work-related health problems is even higher, resulting in an estimated 367 million calendar days of sick leave in the EU (EUROSTAT, 2011). Within the EU, 8.1% of those aged 15 to 64 that worked or who had previously worked have reported a work-related health problem in the previous 12 months which equates to approximately 23 million people.

According to another EUROSTAT report, in the EU "the rate of accidents is higher in small companies". The report further explains that "… the incidence rate of accidents at work is higher in small and medium size local units (SMEs) as compared to local units employing more than 250 employees. This trend is particularly clear in the sectors of manufacturing, electricity, gas and water supply, and construction" (EUROSTAT, 2004). Levels of Occupational Safety and Health (OSH) management fall with decreasing company size, especially in companies with less than 100 employees (European Agency for Safety and Health at Work, 2011).

It has been shown that these accidents cost SMEs time and the availability of specific human expertise which directly translates into money and competitiveness (Beevis & Slader, 2003). Generic factors cited by Arocena & Nunez (2010) for such statistics (based on research conducted in Spanish SMEs) included:

- restricted finances and lack of OHS management skills in SME managers;
- lack of commitment to OSH from managers;

- absence of workers' representation in safety management;
- non-permanent workforce;
- reliance on outsourcing and larger firms;
- infrequent OSH inspections;
- a general preference for a non-formalised approach to preventative activities.

Research has shown that many SMEs often do not see immediate cost benefits to implementing OSH management systems and companies who have experienced OSH related incidents often tend to focus on problem solving rather than adopting a systematic approach to prevention. However, it is claimed that methods developed specifically for large firms cannot be transferred to smaller firms and a customised approach is required to promote awareness and management of OSH in SMEs (Champoux & Brun, 2003).

The IMOSHION (IMproving Occupational Safety & Health in European SMEs with help of simulatION and Virtual Reality – SME-2-243481) project was launched in early 2011, with partners including SME Associations and SMEs from Spain, Germany and Bulgaria, and Research and Technology developers (RTD) from the Netherlands, France, Germany and the UK. This three year EU Seventh Framework Programme funded project was conceived from discussions between a group of European SME associations who identified a common concern about safety amongst their member organisations. The SME associations conducted a survey with their members to identify their specific needs and requirements. The survey resulted in a list of the most recurrent problems which led to the definition of six main needs:

1. improve access to information related to OSH regulation in a didactic and comprehensive way, for efficient implementation of the different regulations
2. provide scenarios with references to regulations and standards, with a focus on correct, efficient implementation of the regulations
3. organise meetings on safety issues around accident or near-misses to learn from experience
4. provide efficient training for machine operators (either on production or maintenance operations)
5. design OSH-sound work environments
6. create and foster a safety culture in the workplace, embedded in every level of an organisation.

These needs were then employed by the RTD performers, together with the SME associations and the SMEs, to define the tools to be developed during the project. For each need, a corresponding system requirement was detailed:

- an OSH documentation management system
- a visual simulation of OSH practices for a given workplace
- a simulation of significant incidents at work, to learn from experience
- a user-friendly training tool for production/maintenance operations

- an easy-to-use planning tool for workstations and workplace, taking into account OSH practices
- a didactic and enjoyable tool, such as a serious game, for nurturing safety culture.

These requirements were used to outline four different tools to be developed by the RTD performers during the project:

Tool A: an OSH knowledge and learning management system
Tool B: a workplace simulation for experimenting and training on OSH issues
Tool C: an immersive training and learning tool for OSH procedures
Tool D: a planning tool for OSH prevention at workstation and workplace levels.

This paper focuses on the user-centred development of Tool A.

Development methodology

The overall strategy of the IMOSHION work plan is to follow an agile software development methodology (Beck et al. 2001). This approach allows the development process to progress with minimal impact from possible changes in requirements that often occur in such exploratory projects. It is characterised by short research and development cycles and constant validation procedures. Similar to participatory approaches (Haines et al. 2002; Vink et al. 2005), agile development enables the active involvement of end-users in the development process to ensure that their needs are addressed, and to increase user acceptance. In IMOSHION, the agile methodology was based on the following core elements: establishment of working groups which focus on particular end-user groups during regular meetings; iterative usability studies to allow a continuous validation and better acceptance of the results; end-user stories providing usage scenarios that are representative of the use of the IMOSHION solutions in specific situations; and dynamic road mapping to determine the priorities of each story. The work plan of IMOSHION follows a six-month iterative process. At the end of each cycle, both user and technical evaluation studies are performed. These are followed by regional roundtables to assess the results of these demonstration cases.

Identification of OSH issues

Five workshops and questionnaire studies were conducted in Spain, Bulgaria and Germany to establish the OSH related issues faced by SMEs. The aim was also to gather baseline data for later evaluation and to explore user and technical requirements for the tools. In total there were 128 participants from a range of SME sectors. The main OSH issues that SMEs reported included:

- Finding legislation
- Coordinating OSH procedures
- Employee attitudes towards OSH

- Obtaining information on current potential dangers with new equipment or means of production
- Personal Protective Equipment (PPE) carelessness by employees
- A lack of time for training
- Difficulties encouraging employees to attend OSH courses
- Encouraging employees to stop neglecting OSH issues and comply with OSH related directives
- Lack of general OSH knowledge amongst employees
- Lack of sector specific OSH knowledge amongst employees

SMEs highlighted requirements for the IMOSHION toolset to provide up-to-date country and sector specific information; be easy to use and accessible to all employees; provide cost effective training, planning and risk assessment process and help management comply with legislation.

Tool A OSH knowledge and learning management system development

Developed by the University of Nottingham, Tool A is a knowledge and learning management system to address the need for access to OSH knowledge, enhance OSH training and promote a safety culture.

To support the development of Tool A, more detailed user requirements were captured by SME Associations using questionnaires, interviews, site visits and round-table discussions during a second cycle of workshops. Additional requirements were collected by interview and questionnaires conducted by the SME Associations themselves. A summary of the initial requirements is shown in Table 1.

The software and hardware requirements were used to develop control flows and user dialogs which were used to support the design specification. A review of existing software knowledge and learning management systems revealed that Moodle best matched the user requirements. Moodle is open source and therefore an affordable learning management system which can be adapted to meet the knowledge and learning management needs of SME Associations and SMEs.

Different user profiles such as employee, system administrator and OSH expert were identified and specific use cases and scenarios defined. The use cases are illustrated in Figure 1.

An example screen shot from the demonstration version of Tool A (developed to support evaluation activities) highlighting the web feed function is shown in Figure 2. In this instance, the web feed function is highlighted.

This demonstration version was evaluated by the SME Associations and SMEs in a third round of workshops. User testing was carried out using both hands-on evaluation activities and video based demonstrations of the functions available in the first versions of the tool; the latter being used when large participant numbers meant hands-on evaluation was not practical. Demonstration OSH topics were

Table 1.   Initial user requirements for Tool A.

Requirement category	User requirement
System access	Provide different levels of user access including SME visitor access Off-site access
System administration	Updated information should be highlighted to users at each new log in The system should be regularly updated with new EU and OSH information Integrated email system needed Search function needed Discussion forum should be available
Accessibility	Information should be available in each users' language Accessible using Internet Express and Usable with Firefox web browsers Usable by non OSH specialists Easy to use by inexperienced computer users
Training management	Use of quizzes/test to assess knowledge and retention Training tools to include practice opportunities and illustrations of cause/effect
OSH Administration	Record of preventative actions taken Record of corrective actions taken

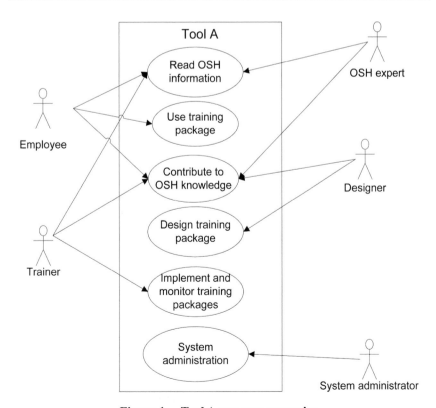

Figure 1.   Tool A use case scenarios.

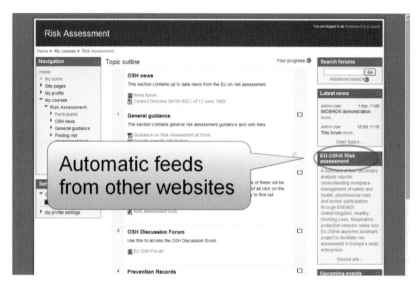

Figure 2.   Example screen shot from Tool A.

Table 2.   Usefulness evaluation summary from SMEs.

Function	Selected comments by SMEs
Latest news	Useful function Needs to be more prominent on the screen
Automatic feeds from other web sites	Good for providing updated information
Upcoming events	Helpful function
Recent activity	Useful function for monitoring system use Reporting systems need to be easy to use
Discussion forum	May not be as useful for smaller companies Can be helpful for knowledge sharing
Setting up a quiz	Good for evaluating training Would like a repository for sharing quizzes and training materials
Accessing web pages	Useful for directing employees to OSH resources
Different levels of access with different functions available	Important to have this Good that it can be customised
See who has read documents or web pages	Good for tracking training Need a simple-to-use reporting system
File store	Would like to have version control system Useful for managing all OSH documentation
See who is on-line	Useful when using messaging function Reports of user activity are helpful

set up in Tool A and screen capture software used to collect video footage of each function being used. Questionnaires were used to gather user ratings of the previously identified user requirements, as shown in Table 1.

SME Association members assessed how well each requirement had been met; they also provided comments and feedback about each function, as did a total of 56 SMEs. A summary is shown in Table 2.

Discussion

Overall the feedback was very positive. These early results suggest that SMEs would find Tool A useful and they could see how it would support their OSH activities. Some suggested that it would save time, allowing flexibility for employees to learn at a convenient time and place and would allow fast and easy access to information. Positive comments were received as to how the tool might enable easier management of OSH information and support greater knowledge sharing within both individual organizations and industrial sectors.

Specific feedback focused on improving the look and feel of the system, by including more icons and less text, and the ability to add company branding. Users found the system easy to use and liked the fact it was web based and so easy to access both on and off site. The flexibility of the system was cited as a positive attribute as SMEs differ widely in their training and OSH management practices. However, a number of specific new functions were requested, such as a detailed reporting facility for knowledge and training management, version control, compatibility with handheld devices, and country/sector-specific information. These will be addressed in future releases of Tool A.

Conclusions

The user-centred iterative development approach employed within IMOSHION for these early versions of Tool A has enabled the capture and design of relevant functionalities with good usability. This has contributed to the positive feedback received from the SME Associations and SMEs. Additional requirements suggested by users will be incorporated in the development of the second version of Tool A, which will be evaluated at the next series of workshops. Development will continue with the iterative approach, with the final system available from early 2014.

References

Arocena, P. & Núñez, I. 2010. An empirical analysis of the effectiveness of occupational health and safety management systems in *SMEs. International Small Business Journal*, Vol. 28, No. 4, 398–419.

Beck, K. et al. (2001) *Manifesto for Agile Software Development*, available online on http://www.agilemanifesto.org/.

Beevis D and Slader IM, 2003, Ergonomics–costs and benefits. *Applied Ergonomics*, 34, 413–418.

Champoux, D., & Brun, J.P. (2003). Occupational health and safety management in small size enterprises: an overview of the situation and avenues for intervention and research. *Safety Science* (41), pp 301–318.

Council of the European Union, (2011). COMMISSION STAFF WORKING PAPER. *Mid-term review of the European strategy 2007–2012 on health and safety at work*, 9518/11, Brussels, 27.4.2011.

European Agency for Safety and Health at Work, (2011). *Understanding workplace management of safety and health, psychosocial risks and worker participation through ESENER,* TE-AM-11-001-EN-C, Spain, Bilbao

EUROSTAT, *Work and health in the EU, A statistical portrait, Data 1994–2002* (2004), Luxembourg: Office for Official Publications of the European Union.

EUROSTAT (2011). *Europe in figures. Eurostat yearbook 2011.* Luxembourg: Publications Office of the European Union.

Haines, H., Wilson, J., Vink, P., & Konigsveld, E.A.P. (2002). Validating a framework for participatory ergonomics. *Ergonomics*, 45:309–327.

Vink, P., Nichols S., & Davies, R.C. (2005). Participatory Ergonomics and Comfort. In P. Vink (Eds.), *Comfort and Design* (pp. 41–54). Florida: CRC Press.

MEASURING WELLBEING IN THE WORKPLACE: SINGLE-ITEM SCALES OF DEPRESSION AND ANXIETY

G.M. Williams & A.P. Smith

School of Psychology, Cardiff University, UK

This paper examines the use of single-items to assess the degree of depression and anxiety in a workforce. One hundred and twenty university staff were assessed for depression and anxiety using the Hospital Anxiety and Depression Scale and a single-item measure for each factor. Results show that scores from 1–10 on a single-item can be used to identify likelihood of depression or anxiety with a sensitivity between 71% and 87% and a specificity between 73% and 88%. The findings provide initial support for the use of a single item to measure these common mental disorders in a way that is quick and easy, and therefore reduces barriers to assessing mental well-being in the workplace.

Introduction

Common Mental Disorders (CMDs) including depression and anxiety are the most widespread of mental health conditions, with a household survey in 2007 showing that 16.2% of adults in England met the diagnostic criteria for at least one CMD in the week prior to interview, more than half of which presented with mixed anxiety and depressive disorder (HSCIC, 2009). In Wales 9% of adults aged 16 and above reported depression and 6% reported anxiety (Welsh Government, 2011). The prevalence of mental health problems in the workforce is not much different from that in the population at large, with the total cost of mental health problems to employers estimated at almost £26 billion per year in terms of sickness absence, reduced productivity, and staff turnover (SCMH, 2007). Despite this, many mental health problems go undiagnosed and untreated, and a survey of senior managers suggested that nearly half thought that none of their workers would ever suffer from a mental health problem during their working life (SCMH, 2007). Being able to identify CMDs such as depression or anxiety in employees would therefore be of significant benefit to employers wishing to reduce the prevalence and impact of poor mental health in the workplace. Furthermore, being able to do this quickly and easily would allow organizations to measure the impact of factors that may be affecting employee well-being, such as job demands or control (Häusser, Mojzisch, Niesel, & Schulz-Hardt, 2010) without lengthy questionnaire and analysis processes.

There are a number of existing measures of depression and anxiety, such as the Beck Depression Inventory (Beck, Ward, Mendelson, Mock, & Erbaugh, 1961), and the

Hospital Anxiety and Depression Scale (HADS) (Zigmond & Snaith, 1983), which provide cut-off scores to indicate depression and, in the case of the HADS, group respondents into mild, moderate, or severe depression. While these and other similar measures have been in use for some time, in general they take longer to administer and are more cumbersome to score than a two-item depression questionnaire that has also been found to perform equally well at identifying depression (Whooley, Avins, Miranda, & Browner, 1997). Using structured interview by a professional as a benchmark, Whooley et al. (1997) demonstrated that a two-item questionnaire was not significantly different in its identification of the presence or absence of depression to 6 other measures ranging from 5 to 21 items, including the BDI and BDI short form. Using a similar procedure Kroenke, Spitzer, and Williams (2003) also demonstrated comparable performance in a two item measure when compared to a 9-item and 20-item scale in 6000 primary care clinic patients, supporting the idea that a very brief measure can provide a screening tool for depression.

The current study is intended to build on the research into very brief measures of mental health by examining the performance of single-item measures of depression as well as anxiety in an organizational context. The new measures are designed to provide respondents with examples of thoughts or feelings that are related to a diagnosis of depression or anxiety, allowing them to consider these examples and respond with an overall score for depression or anxiety from 1 to 10. This approach combines the two questions considered to be an essential feature of a major depressive episode used in Whooley et al. (1997) and Kroenke et al. (2003) into examples as part of the single question, while the use of a 10-point scale also provides the opportunity to create 3 groups based on the position of responses on the scale, complimenting conclusions by Skoogh et al. (2010) that a 3-group 'yes/no/maybe' approach would be better than a 'depressed/not depressed' approach. The same method is used for the anxiety item with wording in the examples taken from particular items in the HADS. The purpose is therefore to determine whether a single-item approach would provide similar performance to the HADS in identifying depression or anxiety, and whether the design of the scale also facilitates the use of a low-medium-high risk format.

Study design

Ethical approval was provided by the Cardiff University School of Psychology ethics committee. A total of 120 university staff completed the HADS depression and anxiety scales along with single-item measures of depression and anxiety, as part of a larger questionnaire on well-being predictors and outcomes. The HADS scales contain 14 items, 7 for depression and 7 for anxiety, with responses on a 4 point scale providing a score of 0–3 for each item. Cut-off points for the HADS are as follows: 0–7 normal; 8–10 mild depression/anxiety; 11–14 moderate depression/anxiety; 15–21 severe depression/anxiety. The single-item versions took the form of a question with examples which respondents were instructed to use as guidance (for example: "On a scale of one to ten, how depressed would you say you are in general? (e.g. feeling 'down', no longer looking forward to things or enjoying things that you used to)"). Responses ranged from 1 (Not at all depressed/anxious) to 10 (Extremely depressed/anxious).

Results

The age of the respondents ranged from 21 to 64, with the majority being married or living with a partner (62.5%), earning £20,000 to £29,999 (35%) and educated to degree or higher degree level (73.3%). In total, 116 respondents completed the depression section of the HADS, while 114 completed the anxiety section. Of these, 21 (18.1%) had mild to severe depression, while 51 (44.7%) had mild to severe anxiety. Table 1 shows the number of those who also completed the single-item versions of the scales and where they scored on those scales.

Table 1 shows that using a cut-off of 1–5 for 'not depressed' and 6–10 as 'depressed' on the single-item measure gave a sensitivity of 71.4% and specificity of 87.2%. This means that 71.4% were correctly classified as depressed while 87.2% were correctly classified as not depressed. For anxiety, the single-item anxiety scale, using a cut-off of 1–5, had a sensitivity of 86.3% and a specificity of 73%.

In order to examine whether the single-item scales could be used to identify the degree of depression or anxiety rather than just their presence or absence, a regression was performed to determine the relative cut-off points. For this comparison the 'normal' and 'mild depression' groups from the HADS data were combined in order to compare 3 groups from each measurement method representing low, medium, and high risk of a significant level of depression or anxiety. Figure 1 and figure 2 show the scatter plots from the depression and anxiety regressions respectively.

Using the results from the regression plots in figure 1 and figure 2, a score of 1–5 on the single-item scale was considered best fit for the normal to mild depression range on the HADS, 6–8 for moderate, and 9–10 for severe depression. This approach correctly accounted for 81.1% of the normal to mild group and 83.3%

Table 1.   Comparison of measures using 2-group scales.

	Single-item Score 1–5	Single-item Score 6–10	Totals
Normal Range			
Depression	82 (87.2%)	12 (12.8%)	94 (81.7%)
Anxiety	46 (73%)	17 (27%)	63 (55.3%)
Mild-Severe			
Depression	6 (28.6%)	15 (71.4%)	21 (18.3%)
Anxiety	7 (13.7%)	44 (86.3%)	51 (44.7%)
Totals			
Depression	88 (76.5%)	27 (23.5%)	115 (100%)
Anxiety	53 (46.5%)	61 (53.5%)	114 (100%)

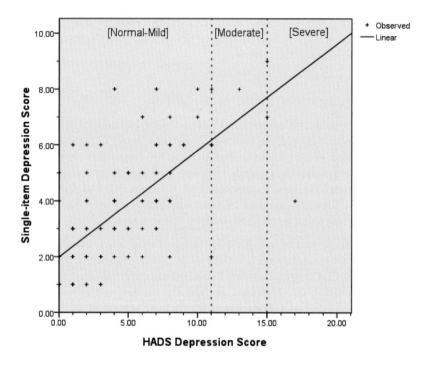

**Figure 1.   Shows linear regression for HADS depression score on single-item
depression score. Dashed lines represent cut-off points for HADS groups.**

of the moderate group for depression, however those scoring in the severe range
on the HADS scored across the scale on the single-item measure (table 2). For the
anxiety measure, the same cut-off scores correctly accounted for 63.4% of those
in the normal to mild range, 80% of the moderate range, but only 8.3% of the
severe range, with the majority of those scoring severe falling in the 6–8 group on
the single-item scale (table 2). The regression also shows that a score of 5 on the
single-item scale acts as the pivotal point where classification on the HADS moves
from 'normal' to mild depression/anxiety.

Discussion

The results demonstrate that at least 71.4% of those completing the HADS would
be similarly classified in terms of their depression or anxiety by using a single-
item scale scored from 1–10 and grouped as 1–5 and 6–10. The sensitivity of
71.4% and specificity of 87.2% in the depression item is almost identical to a
similar larger scale study in which 971 patients completed a single-item 'are you
depressed' question and in which a sensitivity of 73% and specificity of 87% was

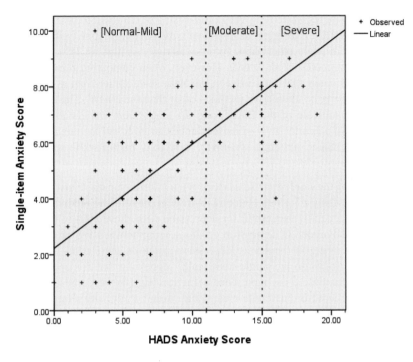

Figure 2. **Shows linear regression for HADS anxiety score on single-item anxiety score. Dashed lines represent cut-off points for HADS groups.**

Table 2.   Comparison of measures using 3-group scales.

	Single-item Score 1–5	Single-item Score 6–8	Single-item Score 9–10	Totals
Normal-Mild				
Depression	86 (81.1%)	20 (18.9%)	0 (0%)	106 (92.2%)
Anxiety	52 (63.4%)	28 (34.1%)	2 (2.4%)	82 (71.9%)
Moderate				
Depression	1 (16.7%)	5 (83.3%)	0 (0%)	6 (5.2%)
Anxiety	0 (0%)	16 (80%)	4 (20%)	20 (17.5%)
Severe				
Depression	1 (33%)	1 (33%)	1 (33%)	3 (2.6%)
Anxiety	1 (8.3%)	10 (83.3%)	1 (8.3%)	12 (10.5%)
Total				
Depression	88 (76.5%)	26 (22.6%)	1 (.9%)	115 (100%)
Anxiety	53 (46.5%)	54 (47.4%)	7 (6.1%)	114 (100%)

found compared to the HADS (Skoogh et al., 2010). Skoogh et al. (2010) did however find that this level of performance required both 'yes' and 'I don't know' responses to be combined, while the 1–10 response format used in the current study, along with examples of depressed thoughts or feelings, seems to provide similar results. Other studies examining very brief scales of depression have also found very similar results, for example 82.1% sensitivity and 80.4% specificity (Kroenke et al., 2003), and while others have found sensitivity as high as 96%, specificity falls to 57% (Whooley et al., 1997). However, when comparing these studies it should also be noted that they used a two-item scale rather than a single-item scale and compared to alternative methods other than the HADS. The comparison of the anxiety subscale of the HADS and the single-item anxiety measure also demonstrate that a similar level of performance can be found for this approach to anxiety measurement, although sensitivity is slightly higher while specificity is slightly lower at 86.3% and 73% respectively.

In terms of using the single-item scales to provide information on severity of depression or anxiety, using a 3 group structure in order to provide low, medium, and high risk groups is generally successful for depression with above 80% of those in the normal-mild and moderate depression groups scoring between 1–5 or 6–8 respectively. Those scoring in the severe depression range on the HADS were spread evenly across groups on the single-item scale, suggesting that this approach may not be reliable for those at the highest end of the spectrum, however the fact that only 3 participants scored in this range means that further study with a larger sample, or in a group with a higher depression prevalence, is needed. Similar performance was found for the anxiety item in the moderate anxiety group, although a larger proportion of those in all anxiety groups based on the HADS scored within the 6–8 range on the single-item measure, meaning that compared to the depression scale, a larger proportion of those with normal-mild or severe anxiety would be identified as medium risk.

Overall, the results from this study suggest that a single-item depression or anxiety scale would provide a valid tool for investigating depression or anxiety in an organisation, particularly when concerned with identifying depressed/non-depressed or anxious/non-anxious groups using a score of 6 out of 10 as a cut-off. The results also suggest that a 1–10 scale may provide a suitable approach to identifying risk of significant depression or anxiety, with scores between 1–5 and 6–8 generally serve as low and medium risk respectively, however further research is needed before relying on a score of 9–10 to identify those likely to have severe depression or anxiety. Finally, the relationship between the single-item and HADS measures suggests that these measures may also be suitable for detecting the impact of factors related to well-being in the workplace, and when these very brief outcome measures are combined with similarly short measures of occupational factors (e.g. Williams (2012); Williams and Smith (2012)) an indication of the well-being process in an organisation is possible for those with limited resources.

While this study provides evidence for the use of these single-item measures in the identification of depression and anxiety when resources are limited, further research is necessary in order to test the measures on a larger scale and in different work groups, in order to provide more confidence in the replicability of the findings and the applicability to different populations. Furthermore, comparison to other measures of depression and anxiety is also warranted, as although the HADS is a valid comparison (Bjelland, Dahl, Haug, & Neckelmann, 2002), comparison with other depression and anxiety measures will further support their validity.

Conclusion

An emphasis on the importance of well-being in the workplace needs to be supported by practical tools that can be used in environments where time and cost restrict the choice of methods available. Expanding research evidence for the use of such tools provides confidence in organisations to assess employees' well-being without being deterred by potentially long and complex data collection and analysis. The results here show that in terms of depression and anxiety – two of the most common issues in mental health – a single item can help to identify the presence of these issues in practice, and while such items should be used cautiously, their brevity may enable those who are restricted in their resources to begin reducing the personal and economic impact of poor mental health in the workplace.

References

Beck, A. T., Ward, C. H., Mendelson, M., Mock, J., & Erbaugh, J. 1961. "An inventory for measuring depression". *Archives of general psychiatry*, 4, 561–571.

Bjelland, I., Dahl, A. A., Haug, T. T., & Neckelmann, D. 2002. "The validity of the Hospital Anxiety and Depression Scale: An updated literature review". *Journal of Psychosomatic Research*, 52(2), 69–77.

Häusser, J. A., Mojzisch, A., Niesel, M., & Schulz-Hardt, S. 2010. "Ten years on: A review of recent research on the Job Demand-Control (-Support) model and psychological well-being". *Work and Stress*, 24(1), 1–35.

HSCIC. 2009. "Health and Social Care Information Centre: Adult Psychiatric Morbidity Survey 2007".

Kroenke, K., Spitzer, R. L., & Williams, J. B. W. 2003. "The patient health questionnaire-2: Validity of a two-item depression screener". *Medical Care*, 41(11), 1284–1292.

SCMH, 2007. *Policy Paper 8: Mental Health at Work: Developing the business case.* (The Sainsbury Centre for Mental Health, London).

Skoogh, J., Ylitalo, N., Larsson omeróv, P., Hauksdóttir, A., Nyberg, U., Wilderäng, U., et al. 2010. "'A no means no'-measuring depression using a single-item question versus Hospital Anxiety and Depression Scale (HADS-D)". *Annals of Oncology*, 21(9), 1905–1909.

Welsh Government. (2011) "Welsh Health Survey 2010". Retrieved 1st September, 2012, from: http://wales.gov.uk/topics/statistics/headlines/health2011/110913/?lang=en.

Whooley, M. A., Avins, A. L., Miranda, J., & Browner, W. S. 1997. "Case-finding instruments for depression: Two questions are as good as many". *Journal of General Internal Medicine*, 12(7), 439–445.

Williams, G. M. 2012, Developing short, practical measures of well-being. In M Anderson (Ed.), *Contemporary Ergonomics and Human Factors 2012,* (Taylor & Francis, London), 203–210.

Williams, G. M., & Smith, A. P. 2012. "A holistic approach to stress and well-being. Part 6: The wellbeing process questionnaire (WPQ Short Form)." *Occupational Health (At Work)*, 9(1), 29–31.

Zigmond, A. S., & Snaith, R. P. 1983. "The hospital anxiety and depression scale". *Acta Psychiatrica Scandinavica*, 67(6), 361–370.

STRESS, JOB SATISFACTION AND MENTAL HEALTH OF NHS NURSES

Jo Williams & Andrew P. Smith

School of Psychology, Cardiff University, UK

This study is a further analysis of workplace stress comparing NHS nurses with different occupational groups. The study had three main aims: (1) to compare the effects of demand-control-support conditions, effort-reward imbalance and job satisfaction on work stress in nurses and other occupational groups; (2) to compare the effect of stress on mental health outcomes in nurses and other groups; and (3) to examine the effects of confounding variables on stress levels. The results supported previous studies showing that stress levels were higher for nurses than the general population. Nurses reported higher demands and higher levels of extrinsic effort. However, nurses reported higher job satisfaction than other workers. Other groups of workers experienced higher levels of anxiety and depression than nurses.

Introduction

The Health and Safety Executive (HSE, 2011) state that the industries reporting the highest rates of work-related stress in the last three years were health, social work, education and public administration (HSE, 2011). Current sickness absence rates in the NHS remain at about 4% (NHS Information Centre, 2011). The Boorman Review estimated that the direct cost of NHS staff sickness was £1.7 billion. (Department of Health, 2009). The National Health Service is one of the UKs largest employers, so the issue of work stress is crucially important, both in terms of employees' health and well-being, as well as costs to the NHS due to employee absence. Nurses have been identified as one of the groups with the highest sickness absence rates at around 5%. (Audit commission, 2011). In particular, it has long been acknowledged that NHS nurses have been experiencing a higher incidence of stress-related illnesses than the general population (Smith, 2001). Much of previous and current research into stress levels in the NHS has been carried out using nurses as the occupational group of choice. Nurses were among 20% of employees who experienced high levels of negative stress in the workplace as far back as 2000 (Smith et al, 2000), with 31.8% feeling extremely stressed. Despite organizational interventions, stress levels in nurses remain persistently high. Research has shown that health professionals are a group at significant risk from the negative effects of stressful workplaces. Of health workers, nurses are particularly at risk from stress-related problems, with high rates of turnover, absenteeism and burnout. Kirkcaldy and Martin (2000) state that nurses have higher than normal rates of physical illness, mortality, and psychiatric admissions. Clegg (2001) cites figures from 1979–83 showing that suicide rates for nurses were significantly higher than the national average, and life expectancy for nurses was approximately 72, only one year more than miners. The CBI (1995, cited in Clegg, 2001) reports that in the private sector in the UK in 1994, 3.4% (8 days) of working time was lost through absence per employee, compared to 6% (14 days) in the health service. Kunkler and Whittick (cited in Clegg, 2001) state that some UK health service trusts can lose in excess of £1 Billion per year due to sickness absence in nurses.

Nurses can be exposed on a daily basis to a large number of potent stressors, including conflict with physicians, discrimination, high workload, and dealing with death and patients and their relatives. McVicar (2003) states that many such situations encountered by nurses at work have a high cost in "emotional labour". Shift working and bullying are also common stressors in nursing, and both are thought to be related to numerous health problems (Boggild and Knuttson, 1999; Ball, Pike, Cuff, Mellor-Clark, and Connell, 2002).

It is clear that some of the working conditions that characterise nursing may be implicated in stress-related issues. Two of the most influential theories commonly used in studying work-related stress are the Demands-Control-Support model (Karasek & Theorell, 1990) and the Effort-Reward imbalance model (Siegrist, 1996). Both models have been found to predict many physical and psychological health outcomes, including heart disease and mortality, depression (Van Der Doef & Maes, 1999) and have also been used in nursing populations (Weyers, Peter, Boggild, Jeppesen, Jeppe and Siegrist, 2006, and Rijk, Le Blanc, Schaufeli, & de Jonge, 1998).

The DCS model predicts that those exposed to high levels of psychological demand, and low levels of job control and social support, are likely to suffer negative health outcomes. Karasek (1979) proposed an interaction effect between demands and control, so that when demands are high and control is low, a high-strain situation develops, exposure to which is particularly likely to lead to negative health outcomes. High control (from the sub-factors of skill discretion and decision authority) is also proposed to buffer the effect of high demands on health outcomes. This model may be well applied to nursing samples, because a lack of social support as well as excessive demands are common in nursing (Muncer, Taylor, Green, & McManus, 2001) and control may well vary by occupational grade.

The Effort-Reward Imbalance model (Siegrist, 1996) is also popular and influential in work-stress research. Based on the concept of reciprocity, the ERI model proposes that high levels of work-related effort should be matched by high levels of reward (economic, recognition, promotion prospects, job security, etc). It is proposed that if efforts (external demands or internal motivations) are high, but rewards are low, then strain and negative health outcomes are likely to ensue. Like the DCS model, the relationship between efforts and rewards is proposed to be interactive, so that high levels of reward buffers high levels of effort (Peter & Siegrist, 1999).

The ERI model may be suited to studying work-related stress in nurses, as there is much evidence that nursing is a demanding occupation and thus requires effort, and levels of pay in newly qualified nurses may be lower than other high-stress occupational groups, such as teachers and police officers (Demerouti, Bakker, Nachreiner, & Schaufeli, 2000).

Despite the popularity of the above two models, they are largely focused on job characteristics or environmental factors (Cox, Griffiths, & Rial-Gonzalez, 2000) and generally fail to take account of individual factors (intrinsic effort from ERI being an exception). These models cannot readily explain for example, how different individuals exposed to the same levels of stressors, may suffer different health outcomes (Perrewe & Zellars, 1999).

Aims and objectives

1. To compare the associations between job characteristics, job satisfaction and work stress in nurses with other occupational groups.
2. To compare the effect of stress on mental health outcomes in nurses and other groups
3. To compare the effects of confounding variables such as demographics on stress levels in nurses with other groups.

These aims and objectives were investigated by conducting secondary analyses of existing surveys of occupational health.

Secondary analyses of the Bristol and Cardiff stress at work studies

In total 17,000 people were selected at random from the Electoral Register for the Bristol study in 1998 (details of the samples are given in Smith et al., 2000). For the Cardiff study, thirty thousand people were selected at random from the Electoral Registers (22,500 Cardiff and 7,500 from Merthyr Tydfil in 2001; Smith et al., 2004). There were 8007 in all other occupational groups, and 277 in the nursing group. In a further analysis, nurses were compared to all other matched occupational groups (based on CASOC, which provides a socio-economic classification based on the nature of a person's job), with 277 nurses and 930 in the matched occupational group.

Nurses were firstly compared to all other groups and then to matched occupational groups with regard to stress levels, job characteristics, job satisfaction, clinical anxiety and clinical depression levels.

Measures

The Job Content Questionnaire (JCQ) is comprised of three subscales that measure job demands (e.g. workload); job control (from sub-factors of decision authority and skill discretion); and levels of social support. A 21-item version of the Effort-Reward Imbalance (ERI) Questionnaire was used. The three subscales measure intrinsic effort (internal motivation or "over-commitment" to work) extrinsic effort (from external demands) and internal reward (perceptions that rewards are adequate).

Stress at work was measured with the following single item questions (from Smith et al., 2000, see below). High stress was categorized as being in the very or extremely stressful category.

In general, how stressful do you find your job?

Not at all stressful	Mildly stressful	Moderately stressful	Very stressful	Extremely stressful
ϖ_0	ϖ_1	ϖ_2	ϖ_3	ϖ_4

Anxiety and depression were measured using the Hospital Anxiety and Depression Scale (HADS: Zigmond & Snaith, 1983).

Results

Nurses versus all the other workers

Stress

A comparison was made between NHS nurses and all the other workers (details of these samples are given in Smith et al., 2000). The findings supported many other studies and revealed that 34% of Nurses experience high stress levels (very or extremely stressed) compared to 19% of all other occupational groups (Chi-Square $= 40.7$, df $= 1$, p $< 0\ .001$).

Job characteristics

Job characteristic scores were split into quartiles. 43% of nurses experienced high demands at quartile 4 (Q4) whereas only 28% of other workers had scores in Q4 (Chi-Square $= 50.8$, df $= 3$, p < 0.001). However, nurses reported more social support, with only 12% in the lowest quartile compared to 24% of the other workers (Chi-Square $= 26.7$, df $= 3$, p < 0.001). Similarly, only 4% of Nurses had a low level of control in quartile 1 compared to 26% of other workers in quartile 1 (Chi-Square $= 94.0$, df $= 3$, p $< .0001$). 32% of nurses reported high levels of extrinsic effort (Q4) compared to 20 % of other workers (Chi-Square $= 49.9$, df $= 3$, p < 0.001). However, there were no significant differences between nurses and other groups for intrinsic effort, or in the rewards they received.

Job satisfaction

Different aspects of job satisfaction were measured. 13% of all other workers felt dissatisfied with people compared to only 8% of nurses (Chi-Square $= 6.9$, df $= 1$, p < 0.01). 19% of all groups were dissatisfied with skill and interest compared to just 7% of nurses (Chi-Square = 26.3, df = 1, p < 0.001). 32% of nurses were dissatisfied with management compared to 42% of all other workers (Chi-Square = 11.6, df = 1, p < 0.001). However, there were no significant differences between the groups in satisfaction with pay and prospects and working conditions.

Mental health

Nurses had lower levels of clinical anxiety (19%) than all other workers (25%). A similar pattern was observed for clinical depression, with 4% of nurses being in this category compared to 6% of other workers.

Nurses versus groups from the same CASOC category

The above findings were replicated in comparisons of nurses with those from the same CASOC category. In summary, nurses reported greater job demands and intrinsic effort but more support and control. They also reported more job satisfaction and the higher stress levels did not translate into more mental health problems. Indeed, levels of clinical anxiety and depression were actually lower in nurses which may reflect the buffering effects of support, control and job satisfaction. The next section examines whether the same psychosocial models apply to both nurses and other workers.

Associations between job characteristics and stress

Further analyses were carried out, using backward stepwise logistic regressions, to discover whether there were differences between groups regarding job characteristics and links with stress. When job characteristics such as job demand, control, support, effort and reward were examined controlling for demographics, nurses were found to be 1.8 times as likely to become highly stressed compared to other groups. Females were 1.3 times as likely, those on the highest incomes 2.5 times as likely. Part-time workers were 0.7 times as likely to be stressed as full-timers, and manual workers were .8 times as likely to become highly stressed. Job demands, control, support, extrinsic and intrinsic effort and reward were found to still have an effect on stress when controlling for demographics. Those with the highest job demands were 2.7 times as likely to be stressed, and those with the highest extrinsic effort were 8.9 times as likely to be stressed, with stress levels doubling at each quartile, from 2.19 to 4.5 to 8.9. Those with the highest intrinsic effort were 3.3 times as likely, those with the highest rewards were still 1.3 times as likely to be highly stressed. When job satisfaction components of job characteristics were included, satisfaction with working conditions and pay and prospects were found to have the most resilient effect, with those dissatisfied with working conditions and pay and prospects 1.4 times as likely to become highly stressed. Bullying at work was then included. Nurses were still 1.6 times as likely to be stressed and those bullied were 1.7 times as likely. None of the variables were removed this time, demonstrating their strong effect.

Separate analyses of the nurses and other workers showed that job and personal characteristics were related to stress in the same way in each group. In other words, nurses report higher stress levels because of greater exposure to stressful working conditions rather than showing different associations between working conditions and reported stress.

Stress and mental health

Further logistic regressions were carried out to examine whether the association between stress and clinical anxiety/depression differed in nurses and other workers. Those experiencing high stress were 3.85 times as likely to suffer from clinical anxiety. However, nurses were half as likely to suffer from clinical anxiety when stressed as the other groups. For clinical depression, those experiencing work stress were 3.22 times as likely to suffer from clinical depression. Nurses were again half as likely to experience clinical depression as other groups.

Discussion

Nurses reported higher stress levels (34%) compared to all workers (19%) and matched occupational groups (19%). It has been widely acknowledged that high work stress levels are associated with high demands, low support, medium levels of control, high extrinsic and intrinsic effort, dissatisfaction with working conditions and pay and prospects, and workplace bullying. High work stress has also been associated with mental health outcomes such as clinical anxiety and clinical depression. As expected, Nurses reported higher job demands than the general population and matched occupational groups. Similarly, they reported higher levels of extrinsic effort. This difference in exposure to potentially demanding working conditions plausibly account for the higher levels of perceived stress reported by the nurses. It is widely acknowledged that nursing has high emotional job content and it has also been established that emotional labour is associated with high stress levels. This analysis, however, has not been able to determine whether the high demands placed on Nurses are emotional or just reflect a high workload.

Unlike perceived stress, nurses report high job satisfaction and greater control and support than other workers. Levels of clinical anxiety and depression were also lower in the nurses and this may reflect the positive aspects of the job. Other research (Mark and Smith, 2012) has shown that positive coping skills reduce the likelihood of developing mental health problems. The effects of coping, control and support appeared to be independent which suggests that one must consider the additive effects of these psychological resources. Further research is needed to establish other potential causes of persistently high work stress in nurses. For example, we still do not know why some nurses become stressed and other nurses remain relatively unstressed. We also do not know whether disposition (characteristics such as personality and attributional style) has a greater impact on the wellbeing of nurses than for other groups. Further investigations are, therefore, required into the emotional demands placed on nurses and how they cope with these.

This analysis has demonstrated that higher job demands and higher levels of extrinsic effort explain to some extent why nurses experience higher stress. The analysis has also shown how other groups experienced more job dissatisfaction than nurses. Clinical anxiety and depression were found to be associated with work stress but other groups of workers experienced higher levels of anxiety and depression than nurses. The analysis drew attention to two important moderators of the stress/mental health pathway in nurses. Social support has already been acknowledged as buffering stress effects, so this result supported much previous research. However, the high job satisfaction of nurses, as providers of care, was shown in this investigation to be an important moderator of the effects of negative stress, despite stress levels remaining high for nurses. The way forward, therefore, is to attempt to establish further potential causes of persistently high work stress in nurses. In addition, one needs to identify the positive characteristics that prevent perceptions of stress leading to clinically significant mental health problems.

References

Audit Commission 2011, "Managing sickness absence in the NHS". Health Briefing, February 2011.

Ball, J., Pike G., Cuff, C., Mellor-Clark, J. and Connell, J. 2002, *RCN Working Well Survey*. http://www.rcn.org.uk/pub-lications/pdf/working_well_survey_inside1/pdf.

Boggild, H. and Knutsson, A. 1999, Shift work, risk factors and cardiovascular disease. *Scand J Work Environ Health* 25(2): 85–99.

Boorman, S. 2009, NHS Health and Wellbeing. *Department of Health.*

Clegg, A. 2001, Occupational stress in Nursing: A review of the literature. *Journal of Nursing Management* 9: 101–106.

Cox, T., Griffiths, A. and Rial-Gonzalez, E. 2000, *Research on Work Related Stress*. European Agency for Health and Safety at Work.

Demerouti, E., Bakker, A. B., Nachreiner, F. and Schaufeli, W. B. 2000, A model of burnout and life satisfaction among nurses. *Journal of Advanced Nursing* 32: 454–464.

Karasek, R. 1979, Job demands, job decision latitude and mental strain: Implications for job redesign. *Administrative Science Quarterly* 24: 285–306.

Karasek, R. and Theorell, T. 1990, *Healthy work: Stress, productivity and the reconstruction of working life.* New York: Basic Books.

Kirkcaldy, B. D. and Martin, T. 2000, Job stress and satisfaction among nurses: individual differences. *Stress Medicine* 16: 77–89.

Mark, G. and Smith, A. P. 2012, Occupational stress, job characteristics, coping and mental health of nurses. *British Journal of Health Psychology* 17: 505–521.

McVicar, A. 2003. Workplace Stress in Nursing: A literature Review. *Journal of Advanced Nursing* 44(6): 633–642.

Muncer, S., Taylor, S., Green, D. W. and McManus, I. C. 2001, Nurses' representations of the perceived causes of work-related stress: a network drawing approach. *Work & Stress* 15(1): 40–52.

Perrewe, P. L. and Zellars, K. L. 1999, An examination of attributions and emotions in the transactional approach to the organizational stress process. *Journal of Organizational Behavior* 20: 739–752.

Peter, R. and Siegrist, J. 1999, Chronic Psychosocial stress at work and cardio-vascular disease: the role of effort-reward imbalance. *International Journal of Law & Psychiatry* 22: 441–449.

Rijk, A. E. de, Blanc, P. M. le, Schaufeli, W. B. and Jonge, J. de 1998, Active coping and need for control as moderators of the job demand-control model: Effects on burnout. *Journal of Occupational and Organizational Psychology* 71: 1–18.

Siegrist. J. 1996, Adverse health effects of high-effort/low-reward conditions. *Journal of Occupational Health Psychology* 1: 27–41.

Smith, A. P. 2001, "Perceptions of Stress at Work." *Human Resource Management Journal* 11(4): 74–78 (13).

Smith, A., Johal, S. S., Wadsworth, E., Davey Smith, G. and Peters, T. 2000, The Scale of Occupational Stress: the Bristol Stress and Health at Work Study. *HSE Books. Report 265/2000.*

Smith, A., Wadsworth, E., Moss, S. and Simpson, S. 2004, The scale and impact of medication use by workers. *HSE Research Report 282. HSE Books. ISBN 07176 29163*

Van Der Doef, M. and Maes, S. 1999, The Job-Demand (-Support) Model and psychological well-being: a review of 20 years of empirical research. *Work & Stress* 13(2): 87–114.

Weyers, S., Peter, R. M. A., Boggild, H., Jeppe, H. and Jeppesen, Siegrist, J. 2006, Psychosocial work stress is associated with poor self-rated health in Danish nurses: a test of the effort–reward imbalance model. *Scandinavian Journal of Caring Sciences* 20: 26–34.

Zigmond, A. S. and Snaith, R. P. 1983, The Hospital Anxiety and Depression Scale. *Acta Psychiatrica Scandinavia* 67: 361–370.

DESIGN APPROACHES

INCLUSIVE DESIGN WITHIN A LARGE ORGANISATION

Anna Mieczakowski[1], Sue Hessey[2] & P. John Clarkson[1]

[1]*Engineering Design Centre, University of Cambridge, UK*
[2]*BT Innovate & Design, UK*

Ensuring that inclusivity is practised in consumer product design requires a fine balance between the needs of the business and those of the users. This paper explores ten best-practice principles derived from auditing the practice of inclusive design within BT, supported by over a decade of research on the uptake of user-centred design in industry. The principles described herein are likely to benefit other organisations operating under similar pressures associated with cost reduction and time to market.

Introduction

Inclusive design has many well-known benefits: it can lead to greater customer satisfaction, cheaper running costs long-term and enhanced corporate social responsibility. Previous studies have examined the uptake of the ethos and practice of inclusive design in industry, describing both existing success stories and barriers to adoption (Dong et al., 2004; Goodman-Deane et al., 2009). They have mainly focused on companies which do not yet practise the principles of inclusive design, however. We therefore conducted an audit of the current state of inclusive design at BT, which was an early adopter having concentrated on inclusion over the past eight years, to understand more about how it is practised in everyday business. Fourteen employees, who are inclusion practitioners and have good understanding of the internal processes at the company, took part in the study. The interactions with them focused on successful inclusivity-led processes, products and services developed to date and existing challenges to employing inclusive design. The investigation was carried out through semi-structured interviews, which were then coded using the general inductive approach (Thomas, 2006). This resulted in the identification of three overarching themes – People, Process and Practice – with users at the heart of them, and ten specific principles for (further) development of inclusive design practice.

Study and findings

Overall, the results show that the company studied has made good progress in its mission to be inclusive. This is despite some of the widely known organisational challenges, such as: (1) having managers juggle many, often conflicting, priorities; (2) the need to deliver products and services to market in restricted timescales and within finite budgets, both of which do not always allow for as thorough user requirements capture and testing as is optimal; and (3) despite existence of

Figure 1.   Ten Principles for the Development of Inclusive Design Practice.

numerous support resources for design (e.g. templates, standards, personas and user scenarios), they are not centralised, may be difficult to access and are not updated in line with the speed at which the market is changing.

Inclusion is considered to be a long-term activity at BT, the key objective of which is to become an inclusive company in a holistic sense by addressing the needs of a diverse customer base in both product design and service support, alongside (and not at the expense of) demonstrable business benefits. The following ten principles for change are key to any organisation wishing to achieve these goals.

Industrial application of the ten principles for change

The findings from this study were translated into ten principles for change shown in Figure 1, which are likely to be applicable to other large and small-to-medium-sized enterprises with consumer product or service offerings.

In line with the dominant themes identified in the interviews, these ten principles are grouped into three overarching areas – People, Process and Practice – placed within the wider context of *user experience*. These principles are provided to enhance a company-wide design of products and services for intuitive, efficient and satisfying use by all users. They are further explained in the next sections.

1. Think Users! – Think about All Users Throughout the Design Process

There is a need to ensure that the company's commercial goals and consideration of users' needs are tightly linked. It is critical that companies accommodate users' needs and balance them with product and service requirements. A greater focus on users is the most important principle; the other nine principles, which sit around this key element indicate how this can be pragmatically achieved.

2. People – Identify a Senior Level Inclusion Executive

Especially in larger companies, there is a need to consider appointing a senior level inclusion executive who will drive change in how inclusive design is practised throughout the business. This person will need to be knowledgeable and enthusiastic about inclusive design, push it towards acceptance, bring together all existing inclusion practitioners, enhance communication between them and support existing inclusion initiatives.

3. People – Ensure the Authority of Internal Experts

Where inclusion expertise exists within a company, there is a need to ensure the authority of internal inclusion experts and allow them to influence product and service-related decisions. An effective way of doing this could be by providing an independent reporting line to the senior level inclusion executive in order to get support and backup from the local management. Internal resources can offer a better insight into the business and into where application of inclusive design is most important than an external agency, for example.

4. People – Employ Accessibility and Usability Experts

There is a need to employ staff with good knowledge and experience of inclusive design from day one. Where possible, inclusion experts should be recruited so that they are available 'in house', saving the time and cost implications of commissioning an agency for the same work. These people should have very specific and expert skills, which can be shared by means of mentoring others in the company's user-centred design community.

5. Process – Promote the Cost Benefits of Inclusive Design

There is a need to make the cost benefits and implications of inclusive design clear to commercial managers in terms of their bottom line. Changing a product late in the design process, or correcting issues once the product is in-life is more expensive than making informed, business-led design decisions early on (Mynott et al., 1994). Therefore, it is key to justify the cost by articulating the benefits and also understand the cost implications of not considering such an approach.

6. Process – Ensure the Focus is on Achieving User Requirement Targets

There is a need to focus on achieving user requirement targets and define a process that will sustain such focus at every stage. A process that allows development teams to focus on designing with users in mind, maintain this focus and refine it when necessary will have a significant impact, leading to better and more inclusive products and services.

7. Process – Evaluate Inclusivity Throughout the Design Process

There is a need to ensure that designers, developers, managers and external suppliers evaluate inclusivity throughout the design process. It is important that inclusion is considered and measured from the proposition creation stage to ensure that it is adopted consistently from the outset. The successful, measurable delivery of inclusion requires collaboration between all stakeholders, application and adherence to specific acceptance criteria and access to inclusivity resources such as the Inclusive Design Toolkit (http://www.inclusivedesigntoolkit.com/).

8. Practice – Nurture and Develop Communities of Practice

There is a need to develop communities of practice made up of designers, developers, internal and external experts and real users. Collaboration helps people and companies grow and achieve greater things. An inclusion expert may offer greater insight into how a 75-year-old, widowed, computer illiterate person with an onset of dementia would access and use a computer than, for example, a software engineer, but by collaborating they can complement each other's skills.

9. Practice – Maintain and Develop Competence

There is a need to continually enhance knowledge and skills relating to inclusive design, including the provision of training and time spent to keep up with new ideas, in order to foster innovation. It is critical to ensure that company staff have the time to learn about the latest findings, tools and resources in inclusive design to grow their expertise. For example, BT is raising awareness of inclusive design through online courses and face-to-face trainings.

10. Practice – Provide a Repository of Inclusivity Data

Provision of an accessible, well researched and regularly updated store of information on how to design inclusively can enhance the quality of the design output within a company. These resources may include user personas, journeys, observations and interaction models. Designers and developers could make best use of such data by having it centralised and having improved access to it, which in turn enables re-use of existing knowledge and contribution of new material.

Depending on the strategic intent of the organisation in question, these principles can be broken down into more specific steps, with more effort required in some areas than others to reduce the gap between current and desired practice.

References

Dong, H., Clarkson, P. J., Ahmed, S. and Keates, S. 2004. "Investigating perceptions of manufacturers and retailers to inclusive design." The Design Journal 7 (3): 3–15.

Goodman-Deane, J., Langdon, P. and Clarkson, P. J. 2009. "Key influences on the user-centred design process." Journal of Engineering Design 21 (2/3): 345–373.

Mynott C., Smith J., Benson J., Allen D. and Farish M. 1994. "Successful Product Development: Management Case Studies." The Design Council, London, UK.

Thomas, D. R. 2006. "A general inductive approach for analyzing qualitative evaluation data." American Journal of Evaluation 27 (2): 237–246.

COMBINING HUMAN INFORMATION PROCESSING METHODS FOR PRODUCT DEVELOPMENT

Hsiu-Lin Liu, Wolfgang Friesdorf & Duojin Wang

Department for Human Factors Engineering and Product Ergonomics
Berlin Technical University, Germany

User-centred design (UCD) and usability are two design concepts that are essential to the product development process, but there are few methods that directly carry out these principles. The aim of this study is to determine corresponding methods to implement these concepts' substantial potential, especially in terms of human performance. These methods will be combined into a new approach, and will then be applied to the product design of a household product.

Introduction

The concept of user-centred design (UCD) has been widely used in product design over past decades. It offers an organised principle for the creation of product designs, and tries to ensure that the given goals can be accomplished effectively, efficiently and satisfactorily for a specified context of use, namely the International Organization for Standardization, 1999. The UCD argues that the context of use should be clearly understood, thus enabling the whole process of product design to be based around the human factor. The design process should involve the user by considering the cognitive and physical characteristics of users, with the aim that they can easily and intuitively use the product.

Several UCD methods have been proved helpful in reducing the costs and time involved in product development, as well as improving usability (Wixon et al. 2002). These methods can be categorised according to their purpose, including for instance methods for understanding the context of use (task analysis, user interviews, user requirements analysis, etc.) or methods for evaluation (usability evaluation, prototype without user testing, etc.). However, although UCD states that the user should be at the center of product design, the most widely used methods do not take the cognitive ability of the user into consideration.

Among the numerous UCD principles and rules that should be taken into account during the product development process, there are few concrete methods involving rules that refer to human cognitive abilities. This paper argues that if a method were to combine the principles and rules of UCD while paying special attention to human cognitive ability, the resultant product development process would be far more efficient and user-oriented. This article therefore aims to discover means to combine existing methods into a new approach.

In the next section, we outline the main concepts of UCD and the stages of information processing, and further describe the relationship between them. Building on this foundation, the second part of this paper goes on to characterise appropriate methods that correspond to the essential principles of UCD. Finally, the aforementioned methods are combined to create a new approach.

User-centred design

The fundamental concept behind user-centred design is to place the user at the heart of the development process (Haklay 2010). Therefore, the particular skills and abilities of a given user must be taken into account during proper product development There is a correlation between the workload involved in the performance of the task, regardless of what physical and psychological burden it entails, with a lighter workload making it easier for the user to perform the task.

Norman (1988) has outlined several rules that express the design concept of UCD, these include simplifying the structure of tasks; making things visible; and using knowledge from memory. Shneiderman (2005) also suggested some golden rules for designing user interfaces, such as offering informative feedback, permitting easy reversal of actions, and reducing short-term memory load.

These two lists of principles share certain key features, notably the importance in product development of considering the human ability to memorize while performing a task. Related to this are two other key points of a product's development, which are as follows: the importance of using common sense and the importance of reducing as much as possible the burden of information processing during product use. With the help of these factors the user is able to efficiently perform a task and reach the specified goal without having to expend too much mental energy.

Human information processing stages

The performance of a given task does not only depend on the product used, but also varies depending on factors such as system design and environmental differences. Before a user is able to correctly carry out a task, a series of stages must occur, between the reception of the signal from the environment and the motor process that performs information processing in the brain. The performance of information processing might also be affected by the product's design (Wickens, Hollands 2000).

Research into information processing (Lindsay, Norman 1977) argues that there are several stages of information processing, which stand between a raw signal originating in an environment to the occurrence of a corresponding active reaction. The raw signal is initially received by a receptor and analyzed to determine its meaning. This process is also called situation awareness (SA) and can be influenced in a variety of ways, by factors such as time, pressure and mental workload (Endsley 2002). Sometimes multiple sources of information are available in an environment, which are processed simultaneously in order for a single task to be performed.

After the completion of SA, analyzed information is compared with pattern memory that was laid down as memory in the past. Through comparison an individual is able decide upon the most appropriate solution for reacting to the present situation. In order to avoid an inconvenient product and any consequent associated risks, the two key factors, namely the requirement for human memory and the information processing workload during use of the product, should be regarded as invaluable criteria during the evaluating process. Hence, a suitable method for the product development process should possess not only a function for creative design, but also take human information processing into consideration.

In the following step, morphological analysis is combined with a flowchart, where the morphological analysis will be used as a means of product development, and the flowchart will be used as a means of describing a logical sequence, for example a cognitive process. The aim of this approach is to assist in choosing a small number of solutions from a myriad of alternatives during product development, as well as to aid the evaluation of the solutions' usability.

Methods

Morphological analysis

Morphological analysis (Zwicky 1989) is a method that can be used to structure and investigate complex relationships in a system that contains a multi-dimensional problem. It was applied in the astrophysical field in the 1970s, and was later also applied in various fields, ranging from problem structuring (Ritchey 2006) to developing strategy alternatives (Pousttchi et al. 2002) and developing innovative product designs (Dowlatshahi 1994; Eversheim 2003). Using this method, a researcher can easily suggest various initial conditions, define possible solutions or generate and decide on a final solution.

The morphological box is a system of cross consistency assessment based on a grid formation, which can produce various combinations of existing concepts. A morphological box comprises rows and columns, where every row presents a parameter in the leftmost column, followed by the possible attributes of this parameter. The box thus aids in the presentation of all the parameters and attributes, since one can connect every attribute from each parameter with a line, so that several combinations can be created. Figure 1 presents a system with five parameters and their respective attributes.

Flowchart

Flowcharts have been a part of computer programming since the early stages of computer use. They involve using symbols and arrows to represent a high-level

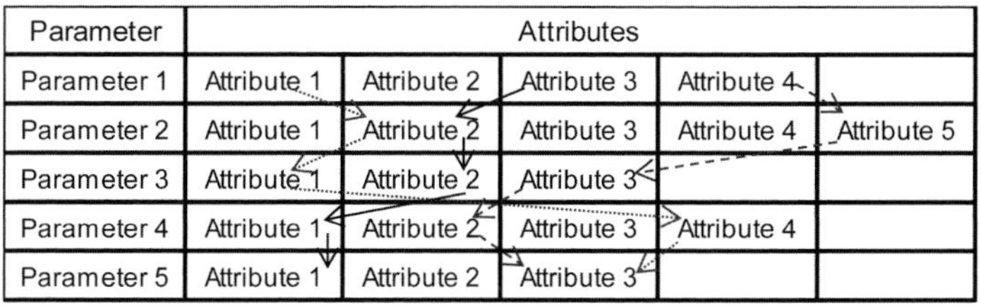

Parameter	Attributes				
Parameter 1	Attribute 1	Attribute 2	Attribute 3	Attribute 4	
Parameter 2	Attribute 1	Attribute 2	Attribute 3	Attribute 4	Attribute 5
Parameter 3	Attribute 1	Attribute 2	Attribute 3		
Parameter 4	Attribute 1	Attribute 2	Attribute 3	Attribute 4	
Parameter 5	Attribute 1	Attribute 2	Attribute 3		

Figure 1.   Structure of a morphological box.

definition of the solution that is to be implemented. It has been suggested both that the flowchart is a tool that developed before programming and that it deserves to be used more (Farina 1970). The flowchart is generally acknowledged to facilitate the comprehension of arithmetic and data movement instructions, and it is also often advantageous to use a flowchart to portray a complex condition in the next stage of programming (Scanlan 1989).

Flowcharts use different symbols to denote different types of instructions. Arrows then connect the symbols to show the flow of operations, so that errors in the logic of performance can be easily detected. Flowchart symbols, standardised by the American National Standards Institute (ANSI), can be divided into terminal, input/output, processing, flow lines, and the decision, among other things. The pictorial representation of a problem in a flowchart offers a visual depiction of the logical sequence of a system, thus allowing for effective analysis. It can also be used as a tool for systemic testing, to confirm whether results conform to expectations. Considering these features of the flowchart, it can also be seen as a method to analyze the utility of products during usage.

Result

In order to show the effects of both methods on the product development process, an example product design will be used here. The example is a household product: a faucet in the bathroom. The approach can be divided into three phases, which are shown in detail below.

Phase 1

The faucet has two major functions: adjusting the temperature and the volume of water. Let us take the case of a task in the bathroom: getting warm water from the faucet. A morphological box can be established based on the following four steps:

1. Define the parameters. Parameters include: inherent characteristics (material); method of manipulation (grasp, one-lever); form (shape and size); labeling (instruction); and method of installation (built-in, mounted).
2. Define the attributes. The corresponding attributes of each parameter will be listed in this step. Some attributes can be subdivided into further parameters. For instance, the handhold of the grasp faucet can be further subdivided in terms of shape and size, which pertains exclusively to the handhold of the grasp faucet (grasp diameter and grasp surface).
3. Construct a grid divided into rows and columns. The number of rows depends on the quantity of parameters and possible parameters, while the number of columns depends on the quantity of attributes for the respective parameter.
4. Fill the morphological box by entering the parameters/attributes into the rows and columns.

In order to allow better interpretation of the details of the attributes, there is no limitation on the style used to present them. As shown in Figure 2, attributes can be presented by letter, numeral, photo and pictogram.

Parameter	Attributes					
Manipulating way	grasp faucet	single	sensible	touch panel		
grasp diameter		4 cm	5 cm	6 cm	7 cm	
grasp surface						
lever		1 cm		1.5 cm	2 cm	
				Labeling for the degree of temperature	1 2 3 4 5 6 ▮▮▮►	▮▮▮▮► 1 2 3 4 5 6
Material	copper	stainless	ceramic	brass , chrome-plated		
Temperature color	● ●	○ ○	● ●	● ●	● ●	
Pictogram						
Assembly						
⋮						

Figure 2.   Example of constructing a morphological box with the product: faucet.

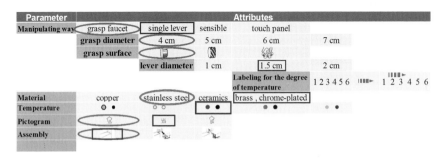

Figure 3.   Two different kinds of the form.

Phase 2

After constructing the morphological box with all of the parameters and attributes, the next task is to ensure the compatibility of attributes between different parameters. One innovative attribute of a parameter might not co-exist with the attribute of another parameter. In a discussion of the compatibility of attributes, professionals will be able to define more than one form of faucet, in terms of design principles and customer requirements. The combinations of different attributes, namely alternatives of product forms, are shown in Figure 3. One combination of various attributes of parameters is presented with pertinent information circled to refer to the first alternative of product design. In the case of a second combination of attributes, one can mark them with a rectangle to present this alternative. The manipulation of both product designs will be transferred to the next phase to simulate the information processing undergone by users during product use.

Phase 3

The alternatives will then be evaluated against the criteria of human information processing. The suggested task is to obtain warm water. The complete depiction of the manipulation process of two alternatives in the form of a flowchart is shown in Figure 4 and Figure 5. The rhombus symbol in the flowchart relates to the cognitive ability of decision-making. The size of the rhombus in Figure 4 and Figure 5

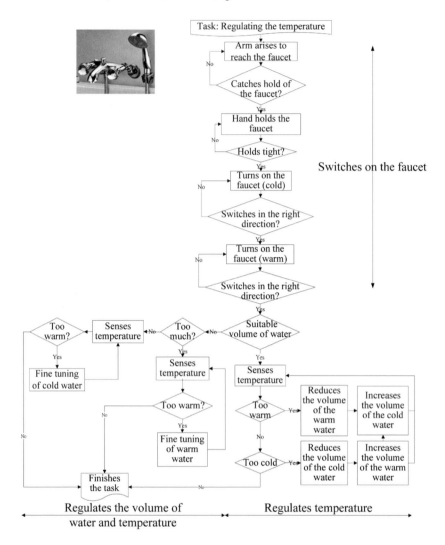

Figure 4.   Manipulation process for the task of regulation of temperature using the grasp faucet in form of a flowchart.

depicting the manipulation of the grasp faucet is obviously much larger than that of the single lever faucet, indicating that the single lever facet is easier to use.

Conclusion

Morphological analysis has been used many times in various fields for different purposes. Through the use of various techniques, including presenting attributes through words, numerals and pictograms, interdisciplinary professionals may find it quite simple to depict an attribute that might be incomprehensible for them. The flowchart was designed to interpret a logical process and is therefore especially

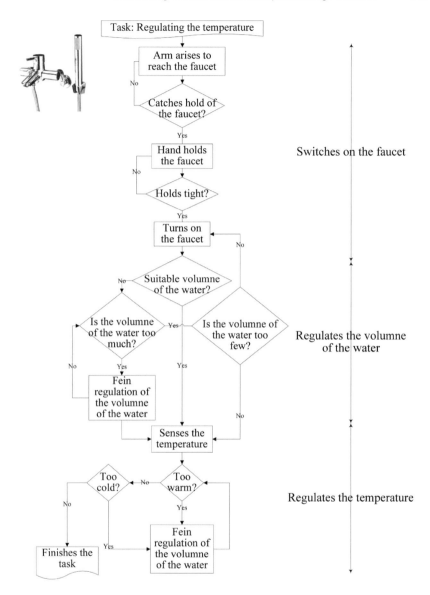

Figure 5. **Manipulation process for the task of regulation of temperature using the single lever faucet in form of a flowchart.**

appropriate for interpreting an information process. Those two methods each have their own special functions and purposes, and furthermore the combination of the methods offers an extra evaluation tool for a new product in the product development process.

This combination of two methods considers users' information processing, and can help the product designer consider whether the user has too much of a psychological burden while using a product. This is especially relevant for elderly people, who may suffer from memory loss. If a designer is able to consider human information processing factors before a product is launched onto the market, it will prevent an unused product from coming onto the market and causing unnecessary losses.

This article has sought to find a method by which the content of the design principle can be interpreted. However, the two methods combined here are insufficient for the purpose of expressing and carrying out all of the UCD principles and the full extent of its functionality. Thus, the combination of the methods in this article can be regarded as a start point from which further development of a systemic methodology can begin to integrate other methods with the UCD principles, so that product designers can effectively create UCD products.

References

Dowlatshahi, S. 1994. A morphological approach to product design in a concurrent engineering environment. *Int J Adv Manuf Technol* 9(5), 324–332.

Endsley, M. R. 2002, *Designing for situation awareness*, (London: Taylor & Francis). 31–42.

Eversheim, W. 2003, *Innovationsmanagement für technische Produkte. Mit Fall-beispielen*, (Springer, Berlin, Heidelberg, New York, Hongkong, London, Mailand, Paris, Tokio), 32–36.

Farina, M. V. 1970, *Flowcharting*, (Prentice-Hall Englewood Cliffs, N.J.), 1–15.

Haklay, M. 2010, *Interacting with geospatial technologies*, (John Wiley, Chichester, West Sussex, UK, Hoboken, NJ), 97–100.

IBM 1970, *Flowcharting Techniques*, (International Business Machines Corporation), 8.

International Organization for Standardization (Ed.) 1999, *Human-centred design processes for interactive systems (ISO 13407)*, (International Organization for Standardization), 3–10.

Lindsay, P. H.; Norman, D. A. 1977, *Human information processing. An introduction to psychology 2nd ed.*, (Academic Press, New York), 369.

Norman, D. A. 2002, *The design of everyday things. 1st ed*, (Basic Books New York), 188.

Pousttchi, K.; Selk, B.; Turowski, K. 2002, *Acceptance criterias for mobile payment procedures*, (Mobile and Collaboratvie Business), 51–67.

Scanlan, D.A. 1989. Structured flowcharts outperform pseudocode: an experimental comparison. *In IEEE Softw* 6(5): 28–36.

Shneiderman, B. 2005, *Designing the User interface*, (Verl. Moderne Industrie, Bonn), 74–75.

Wickens, C. D.; Hollands, J. G. 2000, *Engineering psychology and human performance. 3rd ed*, (Prentice Hall, Upper Saddle River, NJ), 10–12.

Wixon, D.; Vredenburg, K.; Mao, J.; Smith, P. W.; Carey, T. 2002, *A survey of user-centered design practice.* Proceedings of the SIGCHI conference on Human factors in computing systems Changing our world, changing ourselves – CHI'02: ACM Press, 471.

Zwicky, F. 1989, *Entdecken, Erfinden, Forschen im morphologischen Weltbild*, (Baeschlin), 88–105.

HOW INDUSTRIAL DESIGNERS USE DATA DURING A POWER TOOL DESIGN PROCESS

Jiangjun Zhang[1], Hua Dong[1,2] & Long Liu[1]

[1] *College of Design and Innovation, Tongji University, Shanghai, China*
[2] *School of Engineering and Design, Brunel University, UK*

When designing, industrial designers process a variety of data in order to achieve an appropriate design. This paper describes a study focusing on how industrial designers use data during a power tool design process. A survey was conducted with paper-and-pencil and online questionnaire, followed by structured interviews with professional industrial designers with power tool design experiences. A design flow analysis was carried out to identify the key design phases of a typical power tool design process. The findings indicated that most industrial designers considered the phase of translating 2D design data to 3D design data as a key step to realize the design concept, and data sharing and discussion between industrial designers and engineers plays an indispensable role in the power tool design process.

Introduction

Hand tool design benefits greatly from the breakthroughs in technology like plastic technology, computer aided design and manufacturing methods (Haapalainen *et al.* 1999). As one branch of hand tools, power tools have reached maturity and the associated design process is highly systematic and organized.

In a general product design process, given the differences in their knowledge frames and experiential backgrounds (Restrepo and Christiaans 2004), designers have individualised approaches to design (Goodman *et al.* 2006). Industrial designers' diverse and idiosyncratic behaviours towards data (e.g. data collection, use and management) have a great influence on the design development process. In this paper, data which needs to be processed refers to the any raw input (e.g. facts, measurements, statistics) received in the whole power tool design process. In different phases of a power tool design process, industrial designers process numerous data which is acquired by themselves or provided by a third party. Although designers get data from different sources, the first point of reference is often the client, in particular through the design brief (Goodman *et al.* 2007). Designers rely heavily on the Internet and other Web-based resources (Hirsh 2004, Mason and Robinson 2011, Zhang 2011), and their data acquisition is generally from documentary sources (Aurisicchio *et al.* 2009).

While existing literature has discussed designers' diverse data-related behaviours in a general product design process (e.g. Baya, 1996; Song, 2004; Nickpour, 2012), this study focuses on industrial designers' data-related behaviour during a power tool design process, aiming to develop better understanding of the power tool design process, and hence to develop more appropriate tools to support the design process.

Methods

Questionnaire and structured interview survey were conducted to understand data-related issues in a power tool design process.

Questionnaire survey

A questionnaire survey, in both paper-and-pencil and online formats, was conducted to gain the general information of the role industrial designer played in a specific power tool design project. The questionnaire was distributed to a number of professional designers by paper-based questionnaire and online questionnaire platform Survey Gizmo (www.surveygizmo.com). Typical questions included:

"Do you arrange the schedule before you begin a power tool design project?
To what extent the arrangement of schedule will affect the whole design
*   project?*
How do you allocate your time during your power tool design process?
Which of the following design and research methods did you use in your
*   design process? (multiple choices provided)*
What kind of data do you have during your design process?
Which of the following items (a selection of items) has the most significant
*   influence in your power tool design process?*
What will you do if you encounter problems in the design process?"

Structured interview

Structured interviews were carried out with eight industrial designers, both novice and experienced ones, in order to gain a better understanding of how they collect, use and manage data in a power tool design process. Typical questions included:

"Can you briefly introduce the process of a power tool design project?
How do you collect data in every phase in your design process?
Which phase is the most important in the whole design process, and why?
What do you think of the product style guideline book's influence on your
*   design? What kind of data can you get from it?*
Do you share the data with other people? If yes, how do you share the data?
If not, why don't you share the data?
What problems do you have when you are using the data in a power tool
*   design project?*
How do you manage the data in a power tool design project? (e.g. documents,
*   data files, solid models, etc.)"*

A comprehensive flow analysis was made to identify the main phases of a power tool design process based on the data collected from the questionnaire and the interview.

Results

In total 14 participants with power tool design backgrounds answered the questionnaire survey in full. The respondents were young designers, with the age range from 23 to 32. Most of the respondents had short experiences in power tool design, with five having less than one-year experience, and only two having more than five years' experience. The relatively small numbers of the participants in the study was because of the difficulty in accessing industrial designers with the experiences of power tool design.

- Three quarters of the designers arranged the schedule before the start of a new power tool project.
- The three quarters of the designers all agreed that the schedule arrangement had a great influence on the whole design project.
- All the designers used methods like sketching, three dimensional model building, physical prototyping, and ergonomic testing. Most of them also mentioned methods such as online image searching, usability testing, photographing and brainstorming. Emails were the most frequently used method to communicate with client among all the designers, however, experienced designers chose regular meetings as the main method to communicate. In the early sketching stages, conventional sketch tools were selected like paper and pencil. Most of the designers chose software such as Adobe Photoshop and Sketchbook (graphic software), Rhinoceros and Keyshot (3D digital modeling and rendering software) to better support their design.
- The most time-consuming design steps during a power tool design process were three-dimensional modeling and sketching (on tablet and on paper). Table 1 shows the average percentage of time designers spent on each design step.

Table 1.   Average percentage of time spent on each design step.

	Specific design step	Average percentage of time
1	Three-dimensional modelling	28.2%
2	Sketch on tablet	15.8%
3	Sketch on paper	10.9%
4	Physical model making	10.4%
5	Ergonomic and internal structure analysis	6.9%
6	Inspiration collection, exploration and management	6.1%
7	Communication with client	4.9%
8	Communication with engineers	4.8%
9	Communication with team and project manager	4.5%
10	Final deliverables	4.4%
11	Project kick-off meeting	2.9%

- In terms of the sources of data, all the designers regarded the following as the most important ones: design briefing from the client, discussion with team and project manager, and visual brand language book. Most of them also referred to relevant data from competitors' products and other industrial design, like car design, heavy machinery design for inspiration.
- When asked about the most significant influences on the power tool design process, designers mentioned: 1) client's requirements, 2) project manager's suggestion, 3) time limitation, and 4) cost. No one mentioned design tools.
- Nearly all the designers tended to enquire power tool design professionals when they had problems during their design process.

Eight designers participated in the structured interview. The interviewees were composed of half Chinese industrial designers and half European industrial designers. Both novice and experienced industrial designers with power tool design experience were included. Prevailing issues included the following:

1. A power tool design process could be divided into four main design phases, conventional sketch phase, digital sketch phase, 3D digital modeling phase, and CAD engineering phase.
2. In the "conventional sketch phase", data mainly came from three sources: 1) client briefing 2) team discussion 3) product style guidelines. The importance of design briefing provided by the client had been recognized by all the designers. In the "digital sketch phase" and the "3D digital modeling phase", data was collected mainly from building and testing mock-ups, and in the "CAD engineering phase", only technical data was referenced.

Figure 1 presents the general workflow of an industrial designer during a power tool design process. It starts from the project kick-off meeting and ends with final deliverables. The core development includes four main phases.

3. The second and the third phases are crucial. The transition from two-dimensional design data to digital three-dimensional data was a key step to reserve the original concepts and styling elements during a power tool design process.
4. The style guideline book was regarded as very important, as it extracted the essence of power tool design aesthetic philosophy which guided almost every style design detail and made the design fit into a family look.
5. Data share was typical for the power tool design process, especially between industrial designers and engineers.
6. The problems/challenges mainly came from the transition between the 2nd and the 3rd phases, and between industrial designers and engineers, e.g. "*I think especially in the rough stages, 2D sketches are not refined yet, or 3D sketches are started very roughly. In both cases, there will be a lot of changes, every person looks at it and feels a little different, like you paint a shadow, you think it is a ditch inward, but a different person looks at it and thinks it digs outward. So it can happen, and we never just send something out and done, we always send out and discuss about it, in person, or on the phone, emails are not enough ...*"

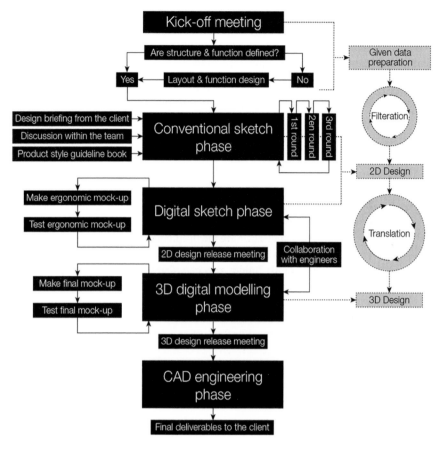

Figure 1.   Workflow of the power tool design process.

7. The data was stored both in personal computers and shared platforms, for example servers and online storage. The data was often organized by design phases and iterations (e.g. first round, second round, third round).

Discussion

A large number of mixing types of data are processed by industrial designers during the power tool design process. For example, at the beginning of a power tool design process, industrial designers are informed by the client and also the project manager. They receive the design briefing, requirements and sources from colleagues, style guideline references etc. These types of data can be defined as **given data**. Meanwhile the perception of data of requirements from clients and design directions given by project managers varies among different designers. Goodman suggests that designers can often find themselves faced with too much information which they need to filter (Goodman *et al.* 2007). In parallel with the processing of all these objective design data, designers generate relevant data for their own use as reference of the objective design data; these data can be identified as **derived data**. Data sources like web-based resources, designer's experiences, inspiration and imagination are included in this type of data.

The use of the style guideline book was prevailing during a power tool design process. One designer regarded the style guideline book as a source of aesthetic philosophy for industrial designers; another designer thought it was a genetic blueprint for keeping power tool design into a family look. As to achieve a unified visual language, the style guideline book provides almost all the detailed design elements which will determine the general design direction.

The typical power tool design process features tools in each of its four phases (i.e. conventional sketching, digital sketching, 3D digital modelling, and computer-aided design tools). Goodman brings forward six main design phases of a typical product design process: briefing/defining the problem or opportunity; analysis/data collection; creativity/synthesis; development/prototyping; evaluation/testing; and manufacturing (Goodman et al. 2006). This can also be applied to the power tool design process. Industrial designers were responsible for the first four phases and then engineers took over the last two phases. In the interviews, all the industrial designers emphasized the importance of the collaboration with engineers. Pei notes that engineers focused on technical properties and cost while industrial designers emphasized more on form and expression (Pei et al. 2009). The industrial designers interviewed indicated that "industrial designers have to negotiate with the engineers", but designers should try their best to persuade engineers to follow their design concept on the premise that all the technical issues, engineer structure and function were fulfilled.

The data conversion from 2D design to 3D design is a crucial step which decides if the design concepts industrial designers generated earlier in both conventional and digital sketch phases could be passed down and realized. Many efforts have been spent on creating new modelling software, like Naya's work which focuses on freehand sketches and drawings as a way to obtain 3D geometric models (Naya et al. 2002), and new software CATIA Version 6 which will be released in 2013 allows industrial designers to easily sketch on 3D models. Experienced industrial designer will think about 3D details while sketching on a 2D platform, while novice industrial designers are more easily get stuck here: they tend to generate less 'derived data' than the experienced industrial designers, and generally cannot accurately estimate 3D details based on 2D sketches. One of the industrial designers expressed that the hallmark of a power tool design's success was designers' ability to complete the 3D modelling alone (i.e. without the help of engineers). In the current power tool design process, collaboration between industrial designers and engineers is throughout, and data sharing between the two groups is also common practice.

One major limitation of the study is the small numbers of participants, and the fact that many of them are novice designers. This was because of the difficulty in recruiting participants from industry. The findings might not be generalizable but they provide useful insights. The participants' responses might be more strongly influenced by their education on how to design rather than their accumulated experience on how best to design.

Conclusions and future work

Our study analysed data-related issues during a power tool design process; common patterns and challenges were identified. A general workflow of the power tool design process is outlined. Most industrial designers considered the phase of translating two-dimensional design data to three-dimensional design data as a key step to realize the design concept.

In a power tool design process, the majority of the data in initial phases is provided by the client or other third parties. Data management (e.g. storage and backup) is periodically carried out by industrial designers, and data sharing and discussion between industrial designers and engineers plays an indispensable role.

The findings suggest that tools can be developed to optimize the transition of 2D design phase to 3D design phase during a power tool design process. This will be the focus of the next stage of the research.

Acknowledgements

We thank all the industrial designers who participated in our research. Hua Dong is sponsored by the Program for Professor of Special Appointment (Eastern Scholar) at Shanghai Institutions of Higher Learning.

References

Aurisicchio, M., Bracewell, R.H., and Wallace, K.M. 2010, Understanding how the information requests of aerospace engineering designers influence information seeking behaviour. *Journal of Engineering Design*, 21(6), 707–730

Baya, V. 1996, *PhD thesis: Information handling behaviour of designers during conceptual design: three experiments.* Mechanical Engineering, Stanford University

Goodman, J., Langdon, P. M. and Clarkson, P. J. 2006, Providing strategic user information for designers: methods and initial findings. In Clarkson, et al. (eds). *Designing Accessible Technology*, pp. 41–51

Goodman, J., Langdon, P. M. and Clarkson, P. J. 2007, Formats for user data in inclusive design. In Stephanidis, C. (eds). *Human Computer Interaction International 2007*, (Springer)

Hirsh, S. and Dinkelacker, J. 2004, Seeking information in order to produce information: an empirical study at Hewlett Packard Labs. *Journal of the American Society for Information Science and Technology*, 55 (9), 807–817

Mason, H. and Robinson, L. 2011, The information-related behavior of emerging artists and designers: inspiration and guidance for new practitioners. *Journal of Documentation*, 67(1), pp. 159–180

Naya, F., Jorge, J. A. and Conesa, J. 2002, *Direct Modeling: from Sketches to 3D Models.* Proceedings of the 1st Ibero-American Symposium in Computer Graphics, pp. 109–117

Nickpour, F. 2012. Data Bahaviour of Designers. Unpublished PhD dissertation, School of Engineering Design, Brunel University.

Pei, E., Campbell, I. and Evans, M. A. 2009, *Building a common ground – The use of design representation cards for enhancing collaboration between industrial designers and engineering designers.* In: Undisciplined! Design Research Society Conference 2008, Sheffield Hallam University, Sheffield, UK, 16–19 July 2008

Restrepo, J. and Christiaans, H. 2004, Problem structuring and information access in design. *Journal of Design Research*, 4(2), 1551–1569

Song, S. 2004, *PhD thesis: shared understanding, skecthing, and information seeking and sharing behaviour in the new product design process.* University of California, Berkeley

Zhang, J., Xu, F., Dong, H., Hong, Z., Wu, C., Wang, J. and Li, B. 2011, *The management of user data in the design process.* Proceedings of 2011 Tsinghua-DMI International Design Management Symposium, pp. 273–283

STANDARDS

PRODUCT DEVELOPMENT – SAFETY AND USABILITY OF MEDICAL DEVICES

C.J. Vincent

UCLIC, University College London, UK

In the field of medical devices, innovation and safety are key drivers for product suppliers. Standards provide an essential route for suppliers to demonstrate quality and consistency and provide vital specifications for comparability and testing. There are multiple approaches to standardisation, for example, specification of a product, implementation of management systems (process) or establishment of common values or principles to maximise business potential. The paper will outline the use of product standards (technical specifications) during medical device development, and examine the inherent benefit.

Introduction

Standards provide for the growth of markets. Across a market as a whole, standardisation offers efficiencies in terms of maintenance, compatibility and elimination of wasteful duplication or unproductive labour (DTI 2005). Standards also underpin the regulation of products. For example, in the European Union, the "placing onto market" of medical devices is governed by a number of European Council directives that are implemented though national law. The directives specify essential regulatory requirements, corresponding to the quality, safety and performance of medical devices. Modular, open, voluntary and harmonised standards support compliance with the regulatory requirements. Within the EU, this exemplifies a class of "new approach" directive. Contrasting "old approach" directives contain a large amount of technical detail, which adds to the challenge associated with approval and revision. For new approach directives, bodies such as CEN and CENELEC prepare consensus standards to support compliance. National Standards Bodies (NSB) are involved in the generation of consensus standards and private, independent, certification authorities or "Notified Bodies" assess conformity. This means that new approach directives need only contain essential requirements. For medical devices, the adoption of harmonised consensus standards provides benefit as a single European standard replaces numerous national standards. Harmonised standards therefore cut the cost of compliance, provide a single point of access to the market and support free trade. As the adoption of standards is voluntary, organisations are free to innovate, although in many cases, incorporation of tried and tested solutions is appropriate. In these cases, product standards provide a basis for quality, consistency, comparability and testing. For consumers, standards communicate an attribution of quality and safety that would otherwise remain hidden (e.g. use of the CE mark).

Product standards and medical devices

There are approximately 300 standards that are current, UK-specific and applicable to general medical devices. Examples include the use of standard scalpels (BS EN 27740:1992), surgical gloves (the BS EN 455 series) and standardised connector types such as the Luer conical fitting (BS EN 20594-1:1994). The final case is interesting, because the Luer fitting was originally developed in a proprietary setting by Karl Schneider, for Wülfing Luer, in 1896. It was then made available to the wider industry to promote interoperability. Without standardisation, health services fail to work together in an effective way. For example, in 1988, following the Ramstein air-show disaster, incompatibilities between the connector types used on IV catheters impeded the emergency response (Brown 2012). The Luer fitting has since become a global standard. Conversely, although the Luer connector has proved successful in allowing interconnection between multiple equipment types, it has also been implicated in several wrong-route administration errors. These are where mistaken connection of the wrong device or substance results in delivery to an unintended part of the body. A study commissioned by CEN showed that when the potential for misconnection was considered across multiple medical connector types (including the Luer), 27% could be fatal (PD CR 13825:2000).

There is therefore an inherent complexity in product standardisation, with a balance to be achieved between flexibility and control. It can be challenging to future-proof a solution and consider system-wide factors. But by incorporating sufficient flexibility, product standards can provide a force for good. For example, within England, over 15m people have Long Term Conditions (LTCs), accounting for 70% of the health and social care budget. Standards enable assisted living technologies to be deployed, to address this societal challenge. Standards can be used to incorporate the needs of the mildly to moderately impaired, design products for home use (e.g. BS EN 60601-1-11:2010) and, in combination with ergonomic data, take into account cultural, social and individual differences (e.g. ANSI/AAMI HE75:2009). Product standards can be applied to optimise user interaction and facilitate commonality in the principles of operation or the properties of the user interface (e.g. PD IEC TR 60878:2003; BS EN 60601-1-8:2007). They also allow developers to incorporate tried-and-tested electro-mechanical solutions therefore saving on development resource.

Product standards therefore generate a level playing field across the market as a whole. This is because organisations avoid the need to develop their own solutions. In this way, standards benefit Small and Medium sized Enterprises (SMEs), through a reduction in development cost. Within the EU, 80% of medical device companies are SMEs (Eucomed 2012). If a standard is unnecessarily complicated, difficult to implement or inaccessible, benefits may be outweighed by difficulties experienced during implementation. For medical devices, this may act as a barrier to safety and usability. Safety has become a recent concern within the EU and USA, due to a number of high profile device recalls (FDA 2011, EU 2012).

Mechanisms of standardisation and the safety and usability of medical devices

Research has examined the extent to which those involved in the design, development and deployment of medical devices find harmonised standards of benefit. An aspect of safety relates to the interaction between the user and the device interface. Here, the design of equipment can reduce the likelihood of errors occurring and ease of error recovery. A qualitative interview study, focussing upon the interactive properties of devices, examined the challenges associated with the adoption of standards in this area. Practitioners expressed a need for clear, concise, graphically illustrated material. They suggested that interpretation is currently constrained by lengthy annexes and complicated interdependencies (Vincent and Blandford 2011a).

In another case, relating to medical device connector types, there were delays incorporating a proposed solution into a network of products. These were partly due to concerns regarding competing or superseding standards. There was a fear that standards could be usurped at an international level (Vincent and Blandford 2011b). For example, if a standard is revised shortly after an organisation has adopted it, or a different standard comes into widespread use, there may be a significant cost associated with change. In many cases, the willingness of a manufacturer to adopt a given standard was dependent upon the extent to which customers perceived it to be beneficial.

For global markets, studies have shown that in some cases, standards fail to realise their purpose. For example, graphical symbols designed for medical contexts are sometimes supplemented with regional specific textual labels (IEC 60878:2003). This is because comprehensibility varies across symbol type and country. For example, the "bell cancel" symbol (IEC number 5576) was reported comprehendible by 100% of German users, but only 65.4% of Chinese users. The "do not reuse" symbol (IEC number 1051) was reported comprehendible by 32.5% of German users and 46.2% of Chinese users (Liu and Hoelscher 2005). In this case, manufacturers may question the benefit provided by standardisation.

Conclusions

Understanding when and why organisations do, or do not employ standards is important. This is because insight can be used to help inform future standards, confirm the suitability of current standards and guide the approach to standardisation. From the perspective of those implementing standards, there are many questions that need to be considered when deciding on an approach. For example: To what extent is the information contained within a standard accessible and how easy will it be to apply it efficiently? Is the standard adequately specified? How current is the standard and for how long is it likely to remain current? How widely adopted and recognised is a standard? Is there a need for recognition beyond a given market? In the future EU medical device regulations are likely to continue to recognise the role of harmonised standards (EU 2012). Optimising their utility will be of benefit.

The work was funded by EPSRC grant EP/G059063/1.

References

Brown, J., 2012. Air show disaster elevates Luer connector [online]. http://tinyurl.com/8pvj7lq [Accessed 24/11/12].

DTI, 2005. The empirical economics of standards London: Dept. of Trade and Industry.

EU, 2012. Proposal for a regulation of the European Parliament and of the Council on Medical Devices, and amending directive 2001/83/EC, regulation (EC) no 178/2002 and regulation (EC) no 1223/2009 [online]. http://tinyurl.com/8s45bfe [Accessed 24/11/2012].

Eucomed, 2012. Contributing to a healthy and sustainable Europe [online]. http://tinyurl.com/dytsyot [Accessed 24/11/12].

FDA, 2011. Understanding barriers to medical device quality [online]. http://tinyurl.com/c2zkwva [Accessed 24/11/12].

Liu, L. & Hoelscher, U., 2005. Evaluation of graphical symbols used in ICU [online]. http://tinyurl.com/d4snngz [Accessed 24/11/2012].

Vincent, C.J. & Blandford, A., 2011a. Designing for safety and usability: User-centered techniques in medical device design practice. Proceedings of the Human Factors and Ergonomics Society Annual Meeting. Las Vegas: HFES, 793–797.

Vincent, C.J. & Blandford, A., 2011b. Maintaining the standard: Challenges in adopting best practice when designing medical devices and systems. Workshop on Interactive Systems in Healthcare (WISH). Washington, DC: American Medical Informatics Association (AMIA).

PUTTING THE CUSTOMER FIRST – USER-CENTRED DESIGN USING ISO 9241

Tero Väänänen

HSBC Bank Plc.

User-centred design is an approach which aims to enhance the user experience of products by focusing on the users, their needs and requirements. ISO 9241-210 'Human-centred design for interactive systems' provides a framework based on six principles that a user-centred design process should follow. These principles form the foundation of the user-centred design process at HSBC. Although it may not always be possible to utilise the perfect user-centred design for every project, we have seen significant increases in customer satisfaction as a result of applying the methods to HSBC's online banking.

User-centred design

According to the ISO 9241-210 (2010, p. vi) user-centred design is:

"An approach to interactive systems development that aims to make systems usable and useful by focusing on the users, their needs and requirements."

User-centred design can enhance not just the usability and usefulness of the system, but all the elements of user experience as described by Peter Morville (2004) in his User Experience Honeycomb: *usefulness, desirability, accessibility, credibility, findability* and *usability.*

User-centred design ensures that the system can and will be used by the intended target audience and enhances effectiveness, efficiency and user satisfaction (ISO, 2010, p. vi). This provides the necessary focus to design and develop functionality that will make the biggest positive difference to the user and increases the likelihood of completing the project successfully on time and within budget (ISO, 2010, p. 4).

Part 210 of the ISO 9241 standard (2010, p. 5) provides a framework based on six key principles that a user-centred design process should follow.

The principles do not dictate which specific methods or tools should be used, but they are open enough to allow the resulting user-centred design process to complement the existing practices with an organization.

Being able to refer to a set of proven principles can provide the required authority to ensure the key aspects of user-centred design are adhered to throughout the project lifecycle. In addition, by ensuring the user-centred design process is iterative and that users are involved at every stage of the process, the standard can provide the opportunity to use creative design methods like sketching and rapid prototyping on paper or electronically.

User-centred design in practice

At HSBC our user-centred design process is based on the ISO 9241-210 principles, but it also complements the existing procedures and regulations within the bank. It is flexible enough to be adapted to fit non-ideal projects, yet enables us to benefit from user-centred design as much as possible.

In addition, we use Part 110 of the ISO 9241 standard to provide the principles we can use for heuristic evaluations in which user experience experts systematically inspect a user interface and identify issues based on their own expertise, best practices and well-known design principles (Nielsen, 2005).

In a recent project we were able to improve customer satisfaction measures of HSBC's online bank by 7%, by carrying out heuristic evaluations to identify user experience issues and applying our user-centred design process to work through them. Although it was not a perfect user-centred design process, the methods applied improved the user experience of our online bank.

References

ISO. 2006, *BS EN ISO 9241-110:2006 Ergonomics of human-system interaction – Part 110: Dialogue principles*, (BSI).
ISO. 2010, *BS EN ISO 9241-210:2010 Ergonomics of human-system interaction – Part 210: Human-centred design for interactive systems*, (BSI).
Morville, P. 2004, "User Experience Design." Retrieved September 30th, 2012, from: http://semanticstudios.com/publications/semantics/000029.php.
Nielsen, J. 2005, "Heuristic Evaluation." Retrieved September 28th, 2012, from: http://www.useit.com/papers/heuristic/.

THE CASE FOR HUMAN AND ORGANISATIONAL FACTORS STANDARDS

Robert W. Miles

Offshore Safety Division – HSE

This paper discusses the case for the offshore energy industry moving towards more performance based standards for human and organisational factors such as safety management systems, permit to work and fatigue risk management.

Introduction – the importance of standardisation

Standards influence many things that we interact with in the course of our normal working lives. We take them for granted to such an extent that we are quite confounded when they are in conflict. Standards evolved to provide an assurance of a predictable and repeatable level of performance; the drivers were interoperability and mass production. In the offshore oil and gas industry, while we see ourselves as individuals the reality is that in a globalised industry we need to transfer staff from one jurisdiction to another, from one installation or contract to another.

Some important lessons

This drive for "human interoperability" has led to a number of attempts to introduce a standard Permit to Work (PTW) system across the North Sea. Many staff transferring from one installation to another or changing clients will be faced with a different PTW and require additional training. Why attempts to standardise PTW have failed provides an important lesson.

Each multi-national wanted their PTW system to be the new standard, no one wanted to have to change their system and face all the on-costs.

PTW systems tend to evolve to become fit for purpose. A drill rig with a crew of (say) 100 can function very effectively with a paper PTW system whereas a 250 POB production platform will require a fully integrated computerised PTW and work control system.

Does this mean that there is no future in pursuing standardisation in PTWs? Far from it, the lesson that emerges is that standardisation of PTW *systems* is not the right objective. We should instead look to standardise two key elements:

1. The user interface/experience – so that each system can be navigated based upon common training and users will "recognise" where they are.

2. The system performance – so that regardless of the design of the PTW there are assured standards for key PTW objectives such as error prevention.

A performance standard; yes, a standard system; no. The same lessons can be transferred to safety management systems. As with PTW there is already comprehensive guidance that *describes* these systems but as with PTW, SMS are personal to individual organisations. No current SMS guidance sets standards for performance or user interface.

A regulatory perspective

Enforcers of regulation like standards. The assurance of performance is their goal and inspecting against standards is simple once there is consensus that the standard is effective. However, we need goal setting to raise performance and address the unpredictable with one caveat: goal setting works when founded on standards that serve to benchmark goals, goal setting without standards becomes ungrounded.

The regulation of working time is a case in point. The EU took the prescriptive route in the Working Time Directive. By and large this has failed to regulate working hours in relation to safety. The American Petroleum Institute recently introduced API Standard 755 for human fatigue risk management. This sets performance standards for risk management and as such provides a model for how to take human and organisational standards forward.

The way forward

My conclusion is that the time is right to revisit a range of human and organizational factors currently covered by industry guidance, and with the lessons from past failures and successes, draft a range of standards based on assured performance.

Reference

RP 755 – Fatigue Prevention Guidelines for the Refining and Petrochemical Industries – American Petroleum Institute 2010

USING STANDARDS TO SUPPORT HUMAN FACTORS ENGINEERING

Roland A. Barge

Rolls-Royce PLC, UK

In a safety critical context, standards and guidelines are key drivers to enable effective engineering practice resulting in safe and innovative design solutions. Multiple sources of information feed into our engineering process. The paper will outline how these sources of information are utilised within our team and examine how we develop information sources turning promising practices into engineering best practice through a process of continuous improvement.

Introduction

Rolls-Royce is a world-class organisation providing innovative power solutions to our customers across the world. Our team forms part of the plant safety department within our nuclear division, which is currently undertaking a range of high-profile projects. A range of standards, guidelines and other sources of information form a resource from which the team can draw, for the assessment of human interactions with highly complex systems, examples of which are provided in Table 1 below.

Working closely with other specialist areas, we define detailed people related requirements based on standards (e.g. MoD. 2008). Other guidelines (e.g. EPRI. 2005) provide good practice on how to undertake qualitative and quantitative assessments of human performance, from which we develop best practice and make clear recommendations within a safety critical context.

Table 1.   Example standards used.

Example standards used	Brief descriptions of use
Human Factors for the Designers of Systems (MoD. 2008)	Provides guidelines on a variety of topics, provides references to other standards and is one of our primary source materials for our design manuals.
Alarm Systems: A Guide to Design, Management and Procurement (EEMUA. 2007)	Provides useful reference for the management of alarms and the development of requirements and provides source material for our design manuals.
Ergonomic requirements for office work with visual display terminals (BSI. 1998)	Provides source material for our design manuals e.g. the presentation of information and colour display. Also used in our design and development process.

Engineering process

During our engineering process we apply knowledge of ergonomics, gained from experience and various information sources to optimise the design of equipment, environments and systems within a safety critical context. Methods are applied to minimise human error and enhance human performance, achieved through a robust engineering process, early human factors analysis and continuous assessment. Early involvement is critical, maximising potential benefits and minimising potential costs. Initially by collating requirements raised by the project, or within a contract (including mandated standards), or as a result of relevant standards (e.g. EEMUA. 2007) or regulatory requirements (e.g. Health & Safety Executive Regulations).

Any solution is dependent on the quality of information available. A range of sources can be utilised e.g. drawings, procedures, standards (e.g. BSI. 2004, BSI. 2000a), databases (MoD. 2007a) and publications. Other assessment tools e.g. hazard identification, design input, task analysis and provision of human reliability and human error data are employed in conjunction with data gathering techniques such as interviews and observation to gather relevant information from subject matter experts to aid design assurance. From the application of appropriate assessment techniques, for example; allocation of function, workload analysis (e.g. BSI. 2000b), training needs analysis (e.g. MoD. 2007b), environmental assessment (e.g. BSI. 2002), workspace assessment (e.g. MoD. 2008), human reliability analysis (e.g. Kirwan, B. 1994), human computer interaction design (e.g. BSI. 1998). From which final outcomes can be drawn. Following any project, a review of the lessons learned enables development of our product, processes and knowledge, as required.

Information sources are central to supporting engineering activities, generally forming a database of gathered information. Potentially promising practices can then be selected (e.g. through benchmarking) from this database against a range of criteria e.g. context, quality, time, cost and experience. Once identified, the selected practice can then be challenged and validated through research or use. The selected practice may require some modification, which is then marked as current best practice. Best practice has been defined as '*a set of interrelated work activities repeatedly utilised by individuals or groups that a body of knowledge demonstrates will yield an optimal result*' (Tucker *et al.*, 2007). Best practices should be current, developed, proven (by research or by use), demonstrable and effective. However current best practices do have limitations e.g. limited life and validity, potential to hinder innovation (Quayzin, X. 2011).

Once validated as current best practice, the practice can then form a benchmark for evaluation of future practices. Information sources and current best practices should be continuously reviewed, as required.

Conclusion

In a safety critical context, standards and guidelines are key drivers to enable effective engineering practice resulting in safe and innovative design solutions. Multiple sources of information feed into our engineering process and are used to select and develop promising practices into engineering best practices through a continuous improvement process. There are still some key questions open for discussion, such as how can we identify what practices are best practices and how do we know when a current best practice is no longer the best practice?

References

BSI. 2004, *Nuclear power plants – Main control room – Alarm functions and presentation* (BS IEC 62241: 2004), British Standards Institute.

BSI. 2000a, *Ergonomic Design of Control Centres* (EN ISO 11064), British Standards Institute.

BSI. 2000b, *Ergonomic principles related to mental workload. Design principles* (BS EN ISO 10075-2: 2000), British Standards Institute.

BSI. 2002, *Light and Lighting. Lighting of Work Places. Indoor Work Places* (BS EN 12464-1:2002), British Standards Institute.

BSI. 1998, *Ergonomic requirements for office work with visual display terminals* (EN ISO 9241:1998), British Standards Institute.

EEMUA. 2007, *Alarm Systems: A Guide to Design, Management and Procurement.* Engineering Equipment and Materials User Association, Publication No. 191, Second Edition, 2007, Engineering Equipment and Materials User Association.

EPRI. 2005, *Guidance for the design and use of automation in nuclear power plants*, Report 1011851, October, 2005, Electric Power Research Institute.

Kirwan, B. 1994, *A Guide to Practical Human Reliability Assessment.* CRC Press.

MoD. 2007a, *Database of Anthropometry Survey of UK Military Personnel 2006-7* (June 2007), Haldane-Spearman Consortium, version 1.1, QinetiQ.

MoD. 2007b, *Joint Service Publication (JSP) 822, Part 5, Chapter 3, Defence Training Support Manual 3 – Training Needs Analysis*, Ministry of Defence, United Kingdom.

MoD. 2008, *Defence Standard 00-250, Human Factors for the Designers of Systems*, May 2008, Ministry of Defence, United Kingdom.

Tucker, A. L. *et al.*, 2007, *Implementing new practices: An empirical study of organisational learning in hospital intensive care units*, Management Science, 53(6), pp. 894–907.

Quayzin, X. 2011, *Are best practices really best practice?* 6th IET International Conference on System Safety, September 2011, pp. 1–4.

DEFINING AND CAPTURING HUMAN FACTORS IN SUSTAINABLE DEVELOPMENT

Miles Watkins

Aggregate Industries Europe, UK

Understanding the influence of human factors on the culture of an organisation provides a strong impetus for improving its sustainability. The methodology proposed by BS 8900, the primary standard for sustainable development in the UK and the first sustainable development (SD) standard in the world, asks users to list the principles with which they identify, such as inclusivity, integrity, stewardship and transparency. A case study illustrates how the use of BS 8900 can lead to tangible improvements in the sustainability of an organisation. Adopting BS 8900 should improve awareness of the strategy, context and effects of the organization. But most ambitiously, it should also change the culture within the organisation.

Introduction

Standards have evolved beyond *product* (specifications and test methods) and *process* (codes of practice and management systems) to increase the business *potential* for organisations that use them. Recommendations and requirements in recent sustainable development standards have pushed the user to consider the context in which they operate and the principles by which they operate, allowing them to assess not only their progress against sustainability issues, but also their progress against the principles they espouse.

In the context of BS 8900, sustainable development is taken to mean 'an enduring, balanced approach to economic activity, environmental responsibility and social progress'. A successful approach to managing sustainable development helps to ensure that an organization makes high quality decisions that promote continuing and lasting success. This approach is partly a result of the view among sustainability experts that:

- SD has to be embedded within an organisation for it to be in itself sustainable;
- SD has to take account of interconnectedness—particularly of its three elements: people, planet and profit; and
- there is a moral aspect to SD that is attractive to the individuals asked to implement it.

BS 8900, currently under revision, takes these views into account by asking its users to first list the SD principles with which they identify. Those suggested by the standard as a minimum are inclusivity, integrity, stewardship and transparency. Inclusivity in particular can be taken to encompass many of the areas traditionally within the realms of human factors. Inclusivity is about ensuring that all those affected or who could be affected (stakeholders) are identified, considered and not disadvantaged.

The implementation of the SD strategy and policy with regard for these principles, should be within a process of stakeholder engagement.

Among the results of implementing the standard, in the long term at least, is that: employees, customers and suppliers understand and buy into the organisations strategy, etc.; and better relationships are fostered with all stakeholders including neighbours and local authorities, making communication and co-operation easier.

Closer examination of behavioural change

Standards have always had an effect on the behaviour of their users and the users of standardised products and services. Standardisation should bring dependability, and claims of compliance with a standard should reassure.

In the area of sustainability, influencing behaviour and organisational culture is essential to deliver real progress. Here is one example of how that has happened through the application of BS 8900.

The Chartered Institute of Environmental Health has used the guidance and recommendations in BS 8900, within the context of the third-party certifier NQA's Sustainability Assessed scheme, in order to develop sustainable principles, assess its performance against those principles, and create a sustainability framework for the organisation. In doing this it sought to underpin the whole process with one of the fundamental requirements of BS 8900, i.e. stakeholder engagement. This has meant a growing understanding of SD within the organisation—beyond simply environmental issues to touch core organisational areas such as HR and finance. Applying the standard has also meant progressive involvement of key stakeholders in corporate planning and objective setting. As such, employee behaviour has become better aligned with the SD goals of the organisation, making the behaviours that contribute to SD (such as avoiding waste and supporting new recruits) more intuitive.

Reference

British Standards Institution 2006, *BS 8900, Guidance for managing sustainable development*. http://www.bsigroup.com (currently under revision).

DEBATE: FROM KNOBS AND DIALS
TO HEARTS AND MINDS?

Scott Steedman

Director of Standards, BSI Group, UK

Human factors shape behaviours in the workplace from the board-room to the office or factory floor. Culture and motivation can have a huge impact on organisational performance. As public debate grows over the quality of leadership in the workplace and attitudes to integrity, safety and risk, there is demand for 'higher standards of behaviour' within business and industry. Best practice codes and guidance are common practice in many aspects of business process and product quality. A panel of experts, chaired by BSI's Director of Standards, will debate how voluntary consensus best practice standards might be developed to integrate human factors more effectively into business and industry and stimulate a step change in productivity and performance.

Introduction to BSI plenary lecture and symposium

BSI is the UK National Standards Body, responsible for balancing the needs of industry, consumers and end-users and government policy in the generation of consensus best practice codes, guidance and standards. Voluntary standards have been used by business and industry for over one hundred years to improve the effectiveness of markets, raise product quality and provide confidence for customers and suppliers.

Ergonomists have long been part of the standards community capturing human factors knowledge, testing it through peer review and a public comments process and disseminating it in published guidelines and codes of best practice. Over the years, several hundred standards have been developed through national and international Ergonomics Technical Committees such as the British Standards, PH/9 and the ISO TC 159. However human factors are also a valuable part of a wider range of codes and guidance published by BSI, for example on inclusive design, risk assessment and workplace stress.

In recognition of this and the development of new kinds of standards of potential interest to the community of ergonomists and human factors specialists, BSI has helped to bring together a symposium of papers on standards-related topics.

BSI's Director of Standards, Dr Scott Steedman, will present an overview plenary paper on the subject of standards and industry performance by way of introduction to the Symposium. He will discuss the opportunity for the UK to lead the development of a new generation of standards that can address business potential, pushing beyond the traditional fields of best practice in management processes and product safety into areas of behaviour and business principles. The integration of human factors will be essential in the delivery of this vision.

The subsequent Symposium will be chaired by Tom Stewart, Fellow IEHF and Past President (to 2010); Founder Systems Concepts; Chair of standards committees BS PH/9: Ergonomics and ISO TC 159 SC4 Ergonomics of human-system interaction. The Symposium will comprise presentations by a number of experts considering how human factors are already incorporated into best practice standards and the challenges and opportunities that this presents.

Papers will be from:

- **Dr Chris Vincent**, Research Assoc. UCL Interaction Centre, CHI-MED
- **Tero Väänänen**, User Experience Design Manager, HSBC
- **Rob Miles**, Offshore Safety Division, HSE
- **Dr Roland Barge**, Rolls Royce PLC
- **Dr Miles Watkins**, Aggregate Industries

Key points from these papers include:

- improvements in customer satisfaction consequent on the implementation of a human-centred design process
- the key role of standards and guidelines in enabling effective engineering practice
- the necessity of standards to demonstrate quality and consistency
- the need for more performance based standards for human and organisational factors such as safety management systems
- the extent to which human factors is an integral part of sustainable development
- the opportunity for the human factors community to integrate their thinking into business and industry by helping to shape the behavioural standards of the future

Panel debate objectives and format

The plenary speaker, Scott Steedman, will chair a debate between a panel drawn from the Symposium chair and speakers and members of the audience. The objectives of the debate will be to stimulate discussion with the audience over the opportunity for human factors to become embedded in business practice through voluntary codes, guidance and standards.

The debate will take the form of opening statements made by the panel in response to broad questions set by the chair, such as:

- How can we integrate human factors into business and industry through best practice codes and standards?

– What form should human factors take in codes, guidance and standards?
– How can we use standards to stimulate better behaviour in the workplace?
– How can ergonomists become better involved in the wider range of standards, in particular the newer 'third generation standards'?

The floor will then be opened for questions and/or comments that will be put to the panel members as appropriate.

The debate will end with a summary from the Chair of outcomes and potential actions.

PROMOTING
COLLABORATION

BRIDGING THE GAP: SHOULD THERE BE MORE COLLABORATION BETWEEN RESEARCHERS AND PRACTITIONERS?

Daniel P. Jenkins[1], Steven T. Shorrock[2,3] & Amy Z.Q. Chung[3]

[1]*DCA Design International, Warwick, UK*
[2]*Eurocontrol, Brétigny sur Orge, France*
[3]*The University of New South Wales, Sydney, Australia*

Most people within the IEHF community would acknowledge that there are differences between the challenges that researchers and practitioners face on a day-to-day basis. Furthermore, most would probably agree that there is much benefit from collaboration between the two groups. In this paper, we contend that more could be done to increase the level of collaboration between researchers and practitioners. A discussion session is proposed to discuss (1) in which areas collaboration is needed, and (2) the appropriate mechanism for supporting this collaboration.

Introduction

Most people who associate themselves with the discipline of human factors and ergonomics (HF/E) have similar goals, regardless of whether they are academic researchers or practitioners. These goals can be clustered into two main areas, adapted from the International Ergonomics Association's definition of HF/E:

1. The fundamental understanding of the interactions among humans and other elements of a system; and
2. The application of theoretical principles, data and methods to design in order to optimise human well-being and overall system performance.

The balance of focus will invariably depend on the individual and the demands of their employment. However, the stereotypical view is that those in academia are more focused on the former, and those working in industry are more interested in the latter. This difference in focus has contributed to a research-practice gap in HF/E (Meister, 1999; Waterson and Sell, 2006). The challenges that practitioners face often differ from the challenges faced by academics, especially in the application of research, where the barriers to research application are very different for those who work in universities and research organisations versus those who work in industry (Shorrock and Chung, 2010; Chung and Shorrock, 2011).

It is contended that the challenges facing academia are, comparatively at least, much better understood by the wider IEHF community. Many of these challenges are captured in academic reports and papers and disclosed to the world. Conversely, many of the challenges facing practitioners are not often discussed outside their respective organisations for commercial or political reasons.

A cursory flick through the ergonomics and human factors journals reveals a strong focus on proposals for new tools and techniques that allow the HF/E community to meet Objective 1. One only has to look at the contents page of one of the human factors methods books (e.g. Stanton et al, 2004; Kirwan and Ainsworth, 1992; Stanton et al, 2005; Sanders and McCormick, 1993; Wilson and Corlett, 2005) to understand that we have a wide range of tools at our disposal. However, when faced with the challenges of Objective 2, practitioners do not necessarily want new tools, theories or data as a priority.

Still, if HF/E is a discipline which comprises fundamental understanding as well as application in the real world, then it is important that research remains applicable. Practitioners need to remain up-to-date with substantial advances in thinking and researchers must demonstrate the application of their research. As such, the application of research has clear advantages to researchers, practitioners, policy makers, and the public.

Increasing collaboration, communication and networking between researchers and practitioners was the most cited suggestion by human factors and ergonomics practitioners in the largest study conducted on the research-practice relationship in HF/E (Chung and Shorrock, 2011). If this is to be realised, then two questions naturally follow.

1. In which aspects of application and practice is more collaboration needed?

There are several potential areas in which researchers and practitioners could cooperate. The ten next most cited suggestions made by HF/E professions for improved research application in Chung and Shorrock's (2010) study were:

- ensuring that research focus and methodology are relevant to the organisational environment;
- providing clearer implications/applications and more definitive conclusions in articles;
- increasing awareness of research among practitioners and policy-makers;
- seeking support from decision makers and stakeholders;
- applying research findings to real problems and organisational experience;
- reporting research in different media;
- report research in a more understandable, clear and readable manner;
- increasing research and publication among practitioners;
- improving availability/accessibility of research articles and providing better compilation of research; and
- understanding the value of research for practice.

There may well be other areas, which are further removed from the application of research and concern more the process of change, such as:

- communicating what HF/E is to decision makers;
- communicating its value in the design cycle or in management;
- combating resistance to change (such as impacts on cost, resource, and time); and
- working with other disciplines (sometimes conflicting) aims.

2. What is the best format for collaboration?

The research community have well-established mechanisms for sharing their work. These include:

- peer reviewed journals such as those affiliated with the IEHF (Ergonomics and Theoretical Issues in Ergonomic Science);
- conference papers;
- books;
- articles in the IEHF magazine 'The Ergonomist';
- blogs;
- microblogs (e.g. twitter); and
- discussion boards (e.g. LinkedIn).

Significant challenges facing practitioners are the availability of research articles and time to read research articles (Chung and Shorrock, 2011). Articles are currently spread in dozens of different journals, mostly behind paywalls, and it takes a significant investment in time to determine which are relevant and useful to a current project or area of work.

There are two ways of communicating research information to practitioners (and policy makers), by push and pull. Where practitioners have a specific question, they may 'pull' information from a search of journal papers, conference papers and books. However, many practitioners also want findings and new approaches and theories to be 'pushed' to them, for instance through conferences, workshops, seminars, articles in 'The Ergonomist', and social media such as blogs, microblogs (e.g. twitter), and discussion boards (e.g. LinkedIn).

However, research communication is not enough: more active collaboration is required to ensure that: 1) research questions are relevant in the first place, 2) there is access to organisations and key stakeholders for research and application, and 3) implications are clear.

The question remains, however, how ergonomics and human factors specialists (whether research- or application-focused) want to collaborate – assuming that there is sufficient motivation to collaborate in the first place.

Conclusions

The aim of this paper is to set the scene for the discussion, using this as a template, the aim will be to discuss (1) in which aspects of application and practice is collaboration is needed, and (2) the mechanisms for supporting this collaboration.

References

Chung, A.Z.Q. & Shorrock, S.T., 2011. The research-practice relationship in ergonomics and human factors – surveying and bridging the gap. *Ergonomics, 54* (5), 413–429.

Kirwan, B. & Ainsworth, L.K., 1992. *A guide to task analysis.* London: Taylor and Francis.

Meister, D., 1999. *The history of human factors and ergonomics.* Mahwah, NJ: Lawrence Erlbaum Associates.

Sanders, M. S., & McCormick, E. J. 1993. *Human factors in engineering and design* 7th ed. (New York: McGraw-Hill).

Shorrock, S. & Chung, A., 2010. Human factors research and practice. Part 2: Bridging the gap. *The Ergonomist, 477,* 4–5.

Stanton, N. A., Hedge, A., Salas, E., Hendrick, H. & Brookhaus, K. (eds) 2004. *Handbook of Human Factors and Ergonomics Methods.* (Taylor and Francis London).

Stanton, N. A., Salmon, P. M., Walker, G. H., Baber, C. & Jenkins, D. 2005. *Human Factors Methods: A Practical Guide for Engineering and Design.* (Ashgate: Aldershot).

Waterson, P. & Sell, R., 2006. Recurrent themes and developments in the history of the Ergonomics Society. *Ergonomics, 49* (8), 743–799.

Wilson, J.R. & Corlett, E.N., eds., 2005. *Evaluation of human work* 3rd ed. (CRC: Boca Raton).

TRANSPORT

VIDEO SUPPORT TOOLS FOR TRAINING IN MARITIME SIMULATORS

Sashidharan Komandur & Sathiya Kumar Renganayagalu

Aalesund University College, Norway

Training is vital for the safety of any complex offshore operation. Advanced simulators are now a standard fixture to train operators. Familiarization of the procedures is an important first step prior to hands-on training in the simulators. We explore the utility of including videos of experts in the instruction material for familiarization. We also explore the utility of videos layered with eye tracking data (of experts) for familiarization as it may increase specificity in the instruction methods.

Introduction

In the maritime domain proper training is vital as most operations are highly demanding and safety critical. This underlines the importance of training, which is why offshore industry uses advanced training tools like modern simulators. Familiarization of procedures is an important first step and many introductory courses are geared towards this. A study in the medical domain showed that experts and novices have very different view patterns that directly affect individual performance (*Law, 2004*). So we believe, recording experts' visual attentions and showing it to novices will enhance the training experience and accelerate their familiarization process. The first aim of this study is to assess the utility of including videos of experts in the instruction material during the familiarization of operational procedures. The second aim of this study is to assess the utility to the familiarization process by including eye tracking data in the video.

Methods

Subjects will be recruited from the pool of students who attend certificate introductory courses in anchor-handling operation for working at the Aalesund university college (familiarization of anchor-handling procedures). These participants can be classified as novices. Data from subjects will be collected only after receiving ethics approval from REK (Norway) and NSD (Norway). *Materials* required for this study primarily consists of a simulator for anchor handling operations at the Aalesund University College. The simulator in addition to realistic visualizations has a physical layout that is proportional to real ship bridge layouts in offshore supply vessels. In this experiment we have chosen the familiarization part of training where the trainer gives a brief about the simulator and its equipment in the simulator space itself. Video data of experts presented as a part of instruction material is captured using two video cameras. One camera (video 1) captures the entire ship bridge layout when the expert carries out an example of the operational procedure. The second camera (video 2) is inbuilt in the eye-tracker and provides the perspective of the expert (trainer) during the operational procedure. The tools used to capture expert's data are a standard off the shelf video camera and an eye tracker (ASL Mobile eyeglasses 2012).

Methods for the study is as follows: The current instruction process by the trainer during familiarization consists of oral instructions and walk through in the simulator. This is followed by hands-on trials by the students in the simulator. In this study subjects undertake 5 hands-on trials and based on the perceived outcome of these trials effectiveness of instruction method is determined. The subjects in this study will be divided into three groups and they will go through the instruction process with the following additional material as shown below:

1) Group 1 (Control condition): standard instruction
2) Group 2 (Case 1): standard instruction + video 1
3) Group 3 (Case 2): standard instruction + video 1 + video 2

After the instruction process subjects in all the groups will undertake 5 hands-on trials in the simulator and based on the perceived outcome of these trials effectiveness of instruction method is determined. The perceived learning outcome will be determined by a self-report adapted from the study Saus et al. 2010. The instructors will also judge the outcome subjectively. In addition the subjects will also wear an eye tracker (Tobii eyeglasses 2010) during the hands-on training, as we would like to explore if any of the eye tracking metrics (e.g. fixation count or duration on chosen areas of interest) can become measures of learning outcome. Appropriate statistical methods will be employed to test for significant differences in the perceived learning outcomes between the groups. **This is a work in progress and our preliminary studies indicate that there are substantial gains in the perceived learning outcome when videos with eye tracking data is provided as a part of instruction material.**

References

Benjamin Law, M. Stella Atkins, A.E. Kirkpatrick, Alan J. Lomax, Christine L. Mackenzie, 2004, *Eye Gaze Patterns Differentiate Novice and Experts in a Virtual Laparoscopic Surgery Training Environment*.
Evelyn-Rose Saus, Bjørn Helge Johnsen, Jarle Eid, 2010, *Perceived learning outcome: the relationship between experience, realism, and situation awareness during simulator training*, Int Marit Health 61, 4: 258–264.

PREFERRED OR ADOPTED TIME HEADWAY?
A DRIVING SIMULATOR STUDY

Magali Gouy[1], Cyriel Diels[2], Nick Reed[1],
Alan Stevens[1] & Gary Burnett[3]

[1]*TRL, UK*
[2]*Coventry School of Art and Design, UK*
[3]*University of Nottingham, UK*

The aim of this study was to disentangle two concepts that are often confounded in driving and ergonomics research: preferred and adopted time headway (THW). It is argued that preferred THW represents a range of THWs drivers feel safe and comfortable with. In contrast, the adopted THW is the THW that drivers indeed adopt in a certain driving situation, which depends on current situational factors (e.g. visibility, other drivers, traffic flow). Situational factors were varied in a simulator study, as participants were asked to follow a lead vehicle in three different traffic conditions. The preferred THW was assessed after each traffic condition by means of the psychophysical method of limits. Preferred and adopted THW were compared in the data analysis and results support the existence of two distinct constructs.

Introduction

When following a car, drivers are constrained by the speed of a lead vehicle (LV) in contrast to free driving where drivers are not constrained in this way (Vogel, 2002). Time headway (THW) is a commonly used parameter to estimate the criticality of a car-following situation and is defined as the elapsed time between the back of the LV passing a point on the roadway and the front of the following vehicle passing the same point (Evans, 1991). The adopted THW is an outcome of the interaction between individual characteristics and situational factors.

On the one hand, individual characteristics appear to have a persistent, durable influence on THWs. It has been reported in the literature that the "preferred" THW is related to drivers' braking performance and perceptual-motor skills (van Winsum, 1998; van Winsum & Brouwer, 1997; Van Winsum & Heino, 1996). Thus, drivers who maintained a shorter THW were more efficient in their control of braking, braked harder and adjusted the intensity of braking better to the criticality (as measured by time-to-collision, TTC) of the moment the LV started to decelerate compared with drivers who preferred longer THWs. Further, it is often reported in literature that personality factors such as sensation-seeking influence the choice of THW, whereby individuals scoring high on sensation-seeking tend to leave shorter gaps (for review see Jonah, 1997). Finally, individual differences have been reported in adopted THW due to drivers' attributes such as age and gender with older and female drivers leaving larger gaps (Evans & Wasielewski, 1983; Taieb-Maimon & Shinar, 2001).

On the other hand, situational factors can have a more transient influence on THWs. Many different factors have been found to influence the THW adopted by drivers such as time-on-task (Fuller, 1981), traffic flow (Postans & Wilson, 1983), characteristics of other vehicles on the road (de Waard et al., 2008), intoxication (Smiley & Brookhuis, 1987) and visibility of the road ahead (Van der Hulst et al., 1998).

The Risk Allostasis Theory (RAT) (Fuller, 2008) states that feelings of risk, as a parameter determining task difficulty, are the primary controller of driver behaviour and drivers seek to maintain a feeling of risk within a preferred range. Based on the RAT, it is assumed here that the preferred THW is a an abstract construct representing a range of THW drivers feel safe and comfortable with, based on their perceived capabilities to adjust the intensity of braking to the criticality of the situation. Within this range, drivers adopt a THW that is most appropriate within a given situation. Hence, a driver might keep a shorter THW when there is a high traffic flow on the road and a larger one in conditions of reduced visibility (e.g. at night).

Therefore, it seems that two constructs can be disentangled. On the one hand, preferred THW represents a range of THWs determined by individual characteristics such as drivers' perceived braking skills. In contrast, the adopted THW is the THW that drivers indeed adopt in a certain driving situation, which depends on current situational factors (e.g. visibility, other drivers, traffic flow).

The aim of this study was to disentangle the two distinct constructs aforementioned. Adopted THW is easily measurable by means of a conventional car-following study in contrast to preferred THW. In the present study, participants were asked to drive in three different simulated traffic conditions. The psychophysical method of limits was designed to measure thresholds in general and was applied in the present study to assess the threshold of drivers' preferred THW (Gouy et al., 2012b). The minimum preferred THW thus obtained were compared to the minimum adopted THW obtained from a simulator study. The hypotheses were that minimum adopted THW would be dependent on situational factors, whereas minimum preferred THW would remain constant throughout.

Method

Participants

A total of 42 participants (21 males, 21 females) were involved in the study. The drivers ranged in age from 20 to 64 years ($M = 35.9$ years; $SD = 11.3$ years). They had their driving licence for 17.5 years on average ($SD = 10.7$ years). The mean annual mileage was between 2000 and 35000 miles ($M = 10,369$ miles; $SD = 6,212$ miles). All participants were recruited from TRL's participant database of local members of the public and had previously participated in driving simulator studies.

Apparatus

The experimental study was performed in a low-level driving simulator that consisted of a table upon which a steering wheel and manual gearbox (Logitech G27) were mounted, offset to the right to replicate the typical UK driving set-up. Corresponding pedals (clutch, brake and accelerator) were located beneath the steering wheel under the table. A 55″ plasma screen (HITACHI 55PMA550) was placed behind the table. The driving simulation was generated by SCANeR Studio 1.1 software (OKTAL). The speed and engine tachometer were displayed to the driver at the bottom of the screen. The simulated road environment consisted of a typical three-lane UK motorway.

Procedure and design

At the beginning of each trial, participants signed a consent form and they were then asked to fill in a demographic questionnaire about their age and driving experience and received written instructions for the trial.

The study alternated the evaluation of the adopted THW (simulator drive) under three different traffic conditions described below with the evaluation of the preferred THW (psychophysical method of limits) taking place after each of the simulator drives. A repeated measures design was employed: each participant drove on the same simulated road under the two platoon conditions (THW03 and THW10) along with the baseline (BL) and underwent the evaluation of preferred THW after each of the simulated drive.

In the simulator drive, participants were asked to follow a LV with the instruction to remain in the same lane as the LV throughout. To motivate drivers to follow the LV and make the scenario realistic in spite of lane restrictions, participants were told to imagine that they are invited for a birthday party and don't know the route. A friend is invited for the same party and knows the route so that they can follow. Each participant performed three drives in total. In two of the drives, a platoon of vehicles was driving on the left (slow) lane maintaining either a short THW of 0.3 s (THW03) or a longer THW of 1.0 s (THW10). The LV was constantly driving next to the platoon at the same speed (110 kph = 68.3 mph). In a third, baseline (BL) condition, there were no other vehicles present on the road other than the LV. In all three drives, the ego car (vehicle driven by participants) and the LV drove in the middle lane with random traffic driving on the outer right ('fast') lane. Each drive lasted for 6 minutes, whereby the first minute of each drive was not included in the data analysis because it was considered as a run-in to the trial.

The measurement of preferred THW took place on the same route with the exception that there were no other cars present but the LV. At the very start of the drive, the ego vehicle accelerated automatically up to 70 mph and took over lateral and longitudinal control of the vehicle but participants were asked to keep their hands on the steering wheel as if they were driving normally. Based on the psychophysics method of limits (Fechner, 1889), participants were exposed to a set of increasing

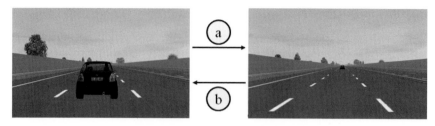

**Figure 1.   The psychophysical method of limits implemented for the assess-
ment of preferred THW includes the presentation of a set of (a) increasing
THWs starting with a very small THW (0.1 sec) and (b) decreasing THWs
starting with a large THW (2.5 sec).**

THWs, starting from a very short THW (0.1 s) (Figure 1a). After a THW was
presented for 5 s, the screen was blanked and participants were asked to respond
'yes' if they would normally adopt this THW or whether it was 'too short' or 'too
large'. Consecutive THWs increased with steps of 0.1 s The screen was blanked for
5 s, which was enough time for participants to respond. Afterwards, the incremented
THW was displayed for another 5.0 s. The presentation of THWs was stopped once
the preferred THW was reached. The same process was repeated with a set of
gradually decreasing THWs starting from a very large THW (2.5 sec) (Figure 1b).
The presentation of THWs was stopped at the point at which the THW no longer
represented drivers' preferred THW. The presentation of the set of increasing and the
set of decreasing distances was counterbalanced and the results from both sets were
averaged. As the result in each set represented a threshold, which was the lowest
THW that participants would accept, the output of the psychophysical method of
limits represents in fact a minimum preferred THW.

Prior to the experimental drives participants performed a familiarisation session in
particular to get used to the braking characteristics of the simulator vehicle.

Data

Concerning the objective data, driving performance data was recorded at a frequency of 20 Hz throughout each participant's drive. THW (s) was calculated as
follows: distance to the LV (m)/speed (m/s). The distance to the LV was measured
along the road from the front of the ego vehicle (vehicle driven by participants) to
the rear of the LV.

Results

The average minimum preferred THWs (Figure 2) measured in conditions BL
($M = 1.36$; $SD = .59$), THW03 ($M = 1.30$; $SD = .58$) and THW10 ($M = 1.28$;
$SD = .53$) varied significantly ($F(2, 82) = 3.26$, $p = 0.04$, $n_p^2 = .1$) but the F value

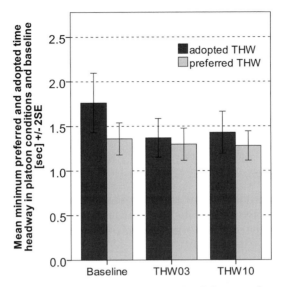

Figure 2. **Averaged minimum preferred and minimum adopted THW and standard error $(+/-2SE)$ in the three traffic conditions (BL, THW03, THW10) $(N = 42)$.**

is closed to the significant level and the effect is small (Cohen, 1992). Moreover, pair-wise comparisons using Bonferroni correction showed that these differences were not statistically significant.

The results of the averaged minimum adopted THW (Figure 2) showed an effect of the traffic condition in the expected direction: the mean THW was higher in the BL condition where there was no platoon present ($M = 1.76$; $SD = 1.08$). There was only a small difference between THW03 ($M = 2.04$; $SD = .99$) and THW10 ($M = 2.12$; $SD = .93$). There was a significant effect of platoon condition [$F(1.57, 64.43) = 10.6, p < .00, n_p^2 = .14$].

Discussion and conclusions

The aim of the study was to disentangle preferred and adopted THW. On the one hand the preferred THW represents a range of THWs determined by stable drivers' characteristics such as braking skills. The adopted THW, on the other hand, is the THW that drivers indeed adopt in a certain driving situation, which depends on transient situational factors (e.g. visibility, other drivers, traffic flow, workload, distraction, intention). Consequently, the lower threshold of preferred THW is supposed to be stable over time whereas its counterpart, the minimum adopted THW, is supposed to vary according to situational factors. In the present study, the variation in situational factors was induced by variation in traffic conditions within a simulated driving experience.

In line with the expectations, results showed that individual drivers selected a preferred THW that was consistent over time and conditions. As expected, the minimum adopted THW varied significantly across the different drives supporting the hypothesis that adopted THW depends on situational factors. The results support the idea raised in the introduction based on the Risk Allostasis Theory (RAT): preferred THW is an abstract construct representing a range within which a driver feels able to control the situation. Nevertheless, the lower threshold of this range is measurable by means of the psychophysical method of limits. Consequently, depending on the current situational factors (e.g. traffic), drivers will then decide about a THW they wish to keep (adopted THW). For instance, drivers will adopt a larger THW on an empty road compared to a driving situation within dense traffic.

However, it is still unclear how minimum THW is affected by drivers' braking performance and perceptual-motor skills and also by personality. The method can be employed in further studies to analyse the relation between drivers' personality or skills and their preferred THW.

Comparing the two parameters minimum adopted and preferred THW informs about any risk taken by drivers. The method of limit is therefore a promising research tool that can be implemented to find out when drivers are operating beyond their perceived capability and what the implications are. Also, drivers might have an erroneous idea of their braking capabilities. This has safety implication if drivers overestimate their braking capabilities and thus keep an inappropriate short THW on the road. In addition to its use in driver behavioural research, this method could be useful as a prediction tool for driver training to detect unsafe driver behaviour and to coach improvements in driving style.

Acknowledgement

The research leading to these results has received funding from the European Commission's Seventh Framework Program (FP7/2007-2013) under grant agreement n° 238833/ADAPTATION project, www.adaptation-itn.eu. We would like to thank Prof. Josef Krems for his advice in the development of the method. © Transport Research Laboratory 2012.

References

Cohen, J. 1992. A power primer. *Psychological Bulletin*, 112(1): 155.
de Waard, D., Kruizingaa, A., & Brookhuis, K. A. 2008. The consequences of an increase in heavy goods vehicles for passenger car drivers' mental workload and behaviour: A simulator study. *Accident Analysis and Prevention*, 40: 818–828.
Evans, L. 1991. *Traffic safety and the driver*. New York: Van Nostrand Reinhold Company.
Evans, L., & Wasielewski, P. 1983. Risky driving related to driver and vehicle characteristics. *Accident Analysis & Prevention*, 15(2): 121–136.

Fechner, G. T. 1889. *Elemente der psychophysik (Vol. 1).* Leipzig, Germany: Breitkopf and Härte.

Fuller, R. 1981. Determinants of time headway adopted by truck drivers. *Ergonomics,* 24(6): 463–474.

Fuller, R. 2008. What drives the driver? Surface tensions and hidden consensus. *Paper presented at the 4th International Conference on Traffic and Transport Psychology,* Washington, DC.

Gouy, M., Diels, C., Reed, N., Stevens, A., & Burnett, G. 2012. Preferred time headway assessment with the method of limits. In J.-F. P. P. V. Mora, & L. Mendoza (Eds.), *European Conference on Human Centered Design for Intelligent Transport Systems*: 87–93. Valencia: HUMANIST Publication.

Jonah, B. 1997. Sensation seeking and risky driving: a review and synthesis of the literature. *Accident Analysis & Prevention,* 29(5): 651–665.

Postans, R., & Wilson, W. 1983. Close-following on the motorway. Ergonomics, 26(4): 317–327.

Smiley, A., & Brookhuis, K. 1987. Alcohol, drugs and traffic safety. *Road users and traffic safety*: 83–105.

Taieb-Maimon, M., & Shinar, D. 2001. Minimum and comfortable driving headways: Reality versus perception. *Human Factors: The Journal of the Human Factors and Ergonomics Society,* 43(1): 159.

Van der Hulst, M., Rothengatter, T., & Meijman, T. 1998. Strategic adaptations to lack of preview in driving. Transportation Research Part F: *Psychology and Behaviour,* 1(1): 59–75.

van Winsum, W. 1998. Preferred time headway in car-following and individual differences in perceptual-motor skills. *Perceptual and Motor Skills,* 84: 863–873.

van Winsum, W., & Brouwer, W. 1997. Time headway in car following and operational performance during unexpected braking. *Perceptual and Motor Skills,* 84(3): 1247–1257.

Van Winsum, W., & Heino, A. 1996. Choice of time-headway in car-following and the role of time-to-collision information in braking. *Ergonomics,* 39(4): 579–592.

Vogel, K. 2002. What characterizes a "free vehicle" in an urban area? Transportation Research Part F: *Traffic Psychology and Behaviour,* 5(1): 15–29.

USER-ORIENTED INFORMATION SYSTEMS IN PUBLIC TRANSPORT

Stephan Hörold, Cindy Mayas & Heidi Krömker

Ilmenau University of Technology, Germany

The aim of this paper is the application of human-computer interaction methods to the field of passenger information in public transport. Appropriate methods are presented in order to analyse the characteristics and requirements of passenger information systems. Based on the analysis of users, tasks, environmental context, and systems, a framework is created for user-oriented development in public transport. This framework includes a set of methods, to analyse passenger information systems and a basic set of results.

Introduction

Public transport is an important part of transportation systems in cities and countries, moving people from one point to another, e.g. to get to work or to school and for leisure and tourism purposes. This role becomes more important considering climate changes and congestions of high frequented streets (Scholz, 2012). An increased quality of public transport can be achieved in different ways, for instance optimizing timetables, increasing comfort and services, or increasing the quality of passenger information systems.

In recent years, technological development of information systems, mobile devices, and mobile communication provides transport companies with a wide range of new possibilities. Passenger information systems can not only be static, stationary and collective, but dynamic, mobile and individual (Norbey et al., 2012). In the near future, systems will incorporate disturbances, real-time and prognostic data, and will provide personalized information. These new possibilities, as an addition to the existing ones, create new challenges for developers and transport companies.

This paper describes an approach for user-centered development of passenger information, based on the elements user, task, environmental context and system.

Elements of human-computer interaction

From an human factors point of view, increasing the quality of passenger information systems is related to an increase of usability. According to ISO 9241-11, the definition process of usability requirements depends on a clear understanding of users, tasks, environmental context, and systems (ISO 9241-11, 1998). These four elements describe the essential basis for usability requirements of a special system.

- The user is the person working and interacting with the system. Each user has individual capabilities and limitations, which influence the usage. The motivation of users to use a system, their expectations, and their goals may differ widely. (Cooper, 2007)
- Tasks are defined as those physical or cognitive activities necessary to accomplish a goal (ISO 9241-11, 1998). In reference to their goals and capabilities, users perform different tasks and subtasks. A task can be divided into more detailed subtasks up to single actions.
- The environmental context describes the physical and social context of a usage situation (ISO 9241-11, 1998). This context might include technical restrictions, ambient conditions, such as temperature or light, and organisational or cultural factors.
- Any equipment, which is necessary to fulfil a task, might influence the usability. This is why, hardware and software, as well as further working materials have to be considered a part of the system.

Characteristics of public transport

The process of identifying and defining users, tasks, environmental context, and systems in public transport has to consider the special preconditions of public transport. The adaptation of the elements of human-computer interaction to public transport results in the following characteristics.

Heterogeneous users

Usually, public transport systems are open to everyone. Due to this fact, the range of users includes nearly the entire population, visiting business travellers and tourists. Additionally, users with visual, hearing, mobility, or other impairments have to be considered. As a result, there is no definite sort of passenger respectively user, but rather a highly heterogeneous field of different passengers.

High amount of different tasks

While using public transport systems, users are confronted with a variety of physical and cognitive tasks. These tasks usually range from trip planning to finding stops and vehicles, and dealing with disturbances.

Varying environmental context along the journey

Conducting a journey with public transport includes changing from one environmental context to another. The travel chain (VDV, 2001) represents all stages along the journey, e.g. planning the journey at home, walking to the stop, travelling inside the vehicle, et cetera.

Different passenger information systems

Along with classic information material, such as paper charts and timetables, new systems, e.g. dynamic passenger information displays and mobile applications, complete the range of information systems. Developers of passenger information systems have to choose between these alternative technologies according to the information needs of the users.

As a consequence, the following challenges can be derived, in respect to a user-centred development of passenger information systems:

- Knowing who the users of the public transport are.
- Knowing which tasks are performed, when and where, along the journey.
- Knowing the environmental context, in which the user is using the passenger information system.
- Knowing which information is needed and which system suites the task best.

Methods

The development of passenger information creates special requirements for the selection of human-computer interaction methods. Especially, the involved experts from different fields of work, e.g. intelligent transport systems, informatics, and transport engineering, call for interdisciplinary, understandable results. Consequently, the selection process for appropriate methods has to consider the following challenges resulting from the domain of public transport:

- Experts from public transport often hold a ticket-oriented view of passengers.
- Public transport often consists of different systems, e.g. bus system, railway system, and underground system, with different characteristics.
- A wide range of stakeholders, from passengers to transport companies and politics, are involved in the development of passenger information systems.
- Regional differences in users, public transport systems and local circumstances, have to be integrated.

Based on these challenges and the criteria interdisciplinary comprehension, adaptability to public transport, and preparation effort, appropriate methods for analysing and describing users, tasks, environmental context, and systems were compared. The conducted analysis of human-computer interaction methods shows that the persona technique, the hierarchical task analysis (HTA), and the usability context analysis (UCA) meet the described challenges. Advantages and disadvantages of alternative methods, considered during the analysis of methods, are discussed later in this paper.

Personas for public transport

User-oriented processes in human-computer interaction provide a variety of methods to analyse and describe the users of a product or a service. In respect to an interdisciplinary development team, personas (Cooper, 2007) are suitable, as they can not only be adapted to local circumstances, but fulfil the role of an easily understood communication base. Additionally, personas can easily be used to empathize with the user, when analysing tasks and information needs.

Hierarchical task analysis for public transport

A journey with public transport consists of several tasks and subtasks, e.g. planning a trip, entering a vehicle, and finding the destination. The hierarchical task analysis (Annet, 2005) serves best for the analysis of complicated workflows with several different tasks and subtasks. In case of public transport with tasks in varying environments, the results can not only be grouped by tasks and subtasks, but by different phases along the travel chain.

Usability context analysis for public transport

At every stage of the travel chain, the environmental context can vary, due to local circumstances. A stop can be simply a sign at the road or a main train station with shops, restaurants, and an information centre. A vehicle can be crowded, with minimal space left for each passenger and a limited view to the next information system. For the development of passenger information systems, it is necessary to analyse the context and reveal typical characteristics. Human factors and ergonomics field of science has different methods for this task. The usability context analysis (Thomas & Bevan, 1996) provides a context questionnaire with questions on the physical, organizational, and technical environment. This questionnaire can be used to cover a wide range of factors, while keeping the results interdisciplinary understandable.

Selection and content of passenger information systems

Passenger information systems are characterized by technical parameters and information content. These systems can be differentiated by:

- Information characteristic: static vs. dynamic
- Mobility: stationary vs. mobile
- Addressees: collective vs. individual

Decisions regarding these three characteristics and the content can be made based on the results of the prior analysis and the information needs of the users. The framework for the identification of information needs in public transport (Hörold et al., 2012) can support the decision-making process and reveal needed information.

Application and results

The presented methods personas, hierarchical task analysis, and usability context analysis were applied to the German public transport system. These nationwide case studies analysed users, tasks, and the environmental context, in order to conclude usability requirements for different passenger information systems.

Persona descriptions

As a first result of the case study, a set of seven personas was created (Mayas et al., 2012). The analysis and clustering is based on a kickoff expert group, six further expert interviews with staff from two transportation companies and different scope of functions, two focus groups, and a survey with 145 passengers from different regions and ages. The revealed personas include passengers with different frequency and purpose of use. For instance, Maria, the power user, goes by public transport for all trips all day and Bernd, the ad-hoc user, who usually avoids public transport, and only relies on public transport when necessary. Several mobility impairments up to a slightly limited walking ability are also considered. In order to meet the high variety of impairments, an extension of the set of personas is required. The denotations of the revealed personas are shown in figure 1.

Tasks and subtasks

In a case study within German Public Transport (Hörold et al., 2012) the HTA was based on the results of an expert group which performed a cognitive walkthrough (Smith-Jackson, 2005) and a focus group with passengers. The results contain 16 tasks and 78 subtasks within the three phases: preparation, travel and dealing with disturbances. Figure 1 shows an extract of the results.

Context descriptions

The usability context analysis questionnaire was adapted to the field of public transport and a field study with different kinds of public transport companies was conducted. The study includes rural and urban areas and covers a wide range of different kinds of stops, vehicles, and surroundings. The results show that a definition of universal environmental context at a special stage of the travel chain is quite difficult. There are nearly no constant conditions within the environment at a particular stage of the travel chain. For example, the weather and atmospheric conditions, as well as facilities, number of other passengers, and available information systems vary, depending on date, time, vehicle, kind of stop, et cetera.

Information needs for passenger information systems

Within the case study the information needs for the set of seven personas were identified by experts using the framework for the identification of information needs in public transport (Hörold et al., 2012). The results consist of a persona-specific

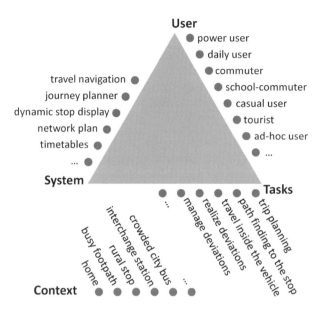

Figure 1.   Framework of users, tasks, systems, and context, as an adaption of the ABC-Modell (Frese & Brodbeck, 1989).

set of needed information, which can be mapped to tasks and subtasks, as well as stages of the travel chain.

Discussion

Although public transport is widely used, the application of human-computer inter-action methods, to increase the quality of passenger information systems, is not yet common practice. The described framework and case studies provide a base for fur-ther work in this area and give an example on how developments in public transport can be conducted, with more regard to the users' needs. Prior to the case studies, several analysing methods for users, tasks, environmental context, and systems were considered and compared, according to the interdisciplinary comprehension, the adaptability to public transport, and the preparation effort.

The persona technique is one of many methods to describe the user. One established alternative is the definition of user roles. But in reference to the area of public transport, the passengers' relations to the system in user roles are less differentiated and more formal (Constantine, 2002) than the passengers' goals, capabilities, and motivations in personas (Cooper, 1999). For this reason, personas are more suitable to describe the users of public transport. The effort for the persona construction can be compensated by the reuse of the data in the further development process.

The hierarchical task analysis is a suitable method for the identification and analyzing process of tasks in the area of public transport. It can be adapted to the special field of public transport and the level of detail can be adjusted depending on the purpose and requirements (Annett, 2005). The needed resources, especially time, depending on the level of detail, and experts for conducting the analysis, can make the application more difficult (Annett, 2005). These disadvantages can be reduced when the level of detail is appropriate and involved experts and stakeholders are chosen carefully. Especially in public transport companies, the knowledge of the travel chain and basic tasks already exist and can be used for further analysis.

Although the description of users via personas and the analysis of tasks and subtasks involve the description of the environmental context to some degree, the environmental context requires a more detailed description of the place. An established method, to gain a clear understanding of the environmental context, is described within the contextual design methodology (Beyer & Holtzblatt, 1998). Conducting contextual interviews would not only reveal the actual context, but the connections between context, users, tasks, and system. Considering that the diversity of the context along the stages of the travel chain is very high, this would mean a high demand for resources. Especially the need for experts conducting the interviews and analysing the environment would be very high. The applied questionnaire can be used with fewer resources and provides a structured tool that can be conducted without the need for experts in the field.

The identification of information needs can be done through different kinds of methods. Interviews, questionnaires, and focus groups are some of them. These methods have to deal with the problem that users might not know their real information needs or do not want to reveal it (Devadson & Pratap Lingam, 1997). The framework for the identification of information needs is based on user analysis, tasks, and available information and bypasses the mentioned problem.

Conclusion

The results of our case studies show that the described framework provides the necessary information for the development of passenger information systems and support the decision making process. The selected methods are flexible and can be adapted to local characteristics, available resources, and needed level of detail.

Especially for those public transport systems, which vary widely from German or European public transport systems, the acceptance and applicability of the suggested methods has to be proven in further studies. The results of the shown methods can be used, not only for the development of passenger information system, but for better understanding of users in general.

Acknowledgments

Part of this work was funded by the German Federal Ministry of Economy and Technology (BMWi) grant number 19P10003L within the IP-KOM-ÖV project. The project develops an interface standard for passenger information in German public transport.

References

Annet, J. 2005, Hierarchical Task Analysis (HTA). In N. Stanton. (ed.) *Handbook of Human Factors and Ergonomics Method*, (CRC Press, Boca Raton), 33-1–33-7

Beyer, H. and Holtzblatt, K. 1998, *Contextual design: defining customer-centered systems*, (Morgan Kaufmann Publ, San Francisco)

Constantine, L. 2006, Users, Roles, and Personas. In J. Pruitt, T. Adlin. (ed.) *Persona Lifecycle*, (Morgan Kaufmann Publishers, Amsterdam), 498–519

Cooper, A., Reimann, R. and Cronin, D. 2007, *About face 3: the essentials of interaction design*, (Wiley Publishing, Indianapolis)

Devadason, F. J. and Pratap Lingam, P. 1997, A Methodolody for the Identification of Information Needs. In *IFLA Journal*, Vol. 23, 41–51

DIN EN ISO 9241-11. 1998, *Ergonomic requirements for office work with visual display terminals (VDTs) – Part 11: Guidance on usability (German version)* (Beuth Verlag GmbH, Berlin)

Frese, M. and Brodbeck, F. C. 1989, *Computer in Büro und Verwaltung: psychologisches Wissen für die Praxis*, (Springer Verlag, Berlin)

Hörold, S., Mayas, C. and Krömker, H. 2012, Identifying the information needs of users in public transport. In N. Stanton. (ed.) *Advances in Human Aspects of Road and Rail Transportation*, (CRC Press, Boca Raton), 331–340

Mayas, C., Hörold, S. and Krömker, H. 2012, Meeting the Challenges of Individual Passenger Information with Personas. In N. Stanton. (ed.) *Advances in Human Aspects of Road and Rail Transportation*, (CRC Press, Boca Raton), 822–831

Norbey, M., Krömker, H., Hörold, S. and Mayas, C. 2012, 2022: Reisezeit – schöne Zeit! In G. Kempter. (ed.) *Technik für Menschen im nächsten Jahrzehnt: Beiträge zum Usability Day X*, (Pabst Science Publ, Lengerich), 33–41

Scholz, G. 2012, *IT-Systeme für Verkehrsunternehmen: Informationstechnik im öffentlichen Personenverkehr*, (dpunkt.verlag, Heidelberg)

Smith-Jackson, T. L. 2005, Cognitive Walk-Through Method (CWM). In N. Stanton. (ed.) *Handbook of Human Factors and Ergonomics Method*, (CRC Press, Boca Raton), 82-1–82-7

Thomas, C. and Bevan, N. 1996, "Usability context analysis: a practical guide". Retrieved September 25th, 2012, from: http://hdl.handle.net/2134/2652.

Verband Deutscher Verkehrsunternehmen (VDV). 2001, *Telematics in Public Transport in Germany*, (Alba Fachverlag, Düsseldorf)

TRAIN AUTOMATION AND CONTROL TECHNOLOGY – ERTMS FROM USERS' PERSPECTIVES

A. Buksh[1,2], S. Sharples[1], J.R. Wilson[1,2], G. Morrisroe[3] & B. Ryan[1]

[1] *Human Factors Research Group, University of Nottingham, UK*
[2] *Ergonomics Team, Network Rail, UK*
[3] *CCD Human Factors, Copenhagen, Denmark*

The European Train Control System (ETCS) as part of ERTMS (European Rail Traffic Management System) is a new automation and control system which has fairly recently been introduced into the UK rail system. This paper briefly reviews some of the potential human factors issues associated with driving with ERTMS, drawn from previous literature and rail standards and also from familiarisation activities. It then summarises and discusses some of these issues based on data collected from interviews with UK ERTMS drivers.

Introduction

The European Train Control System (ETCS) as part of ERTMS (European Rail Traffic Management System) is a new automation and control system which has been introduced into the UK rail system fairly recently. ERTMS operates in different levels and when driving in level two all movement authorities for the train are given via a planning area on a DMI (i.e. an in-cab digital display unit), and drivers are no longer using lineside signals. Therefore, ERTMS drivers receive their primary signaling information from inside, rather than outside the cab. Additionally, the ETCS provides speed profiles on the DMI, which the drivers must follow or the system will apply the brakes. Prior to the system applying the brakes, the system provides overspeed and warning alarms. Drivers are still required to monitor their environment outside the cab in order to detect and report any events or hazards, as well as monitoring parts of the infrastructure such as level crossings. Further information about ERTMS can be found on the UNIFE website (www.ertms-online.com).

UK train driving has been 'heads up' driving, where drivers are predominantly looking for information outside the cab and using their route knowledge to make decisions about speed and braking control. However, with the introduction of ERTMS, it is proposed that the driving philosophy and culture in the UK will change from 'heads up' to 'heads up, heads down' driving. Introducing ERTMS has also required changes in driving rules and lineside signage, which are also expected to influence long term driver route knowledge.

Prior to the implementation of ERTMS in the UK, there was a considerable amount of work discussing potential human factors issues with train driving under ERTMS. However, there has been little actual research into the effects of ERTMS on train driving in the UK since its implementation in 2009. Questions have been raised about its effect on driving strategies and styles, and on performance.

These questions have been highlighted by the recent Rail Accident Investigation Branch (RAIB) incident report (RAIB 2012) of the Llanbadarn automatic barrier crossing incident in June 2011. The incident occurred on an ERTMS route where a passenger train ran onto the level crossing when the crossing was raised and the indicator close to the crossing was flashing red. The report implicated in-cab signaling as one of several causal factors, and the driver reported that the he was observing the DMI at the same time he should have been observing the lineside indicator. One of the outcomes of the RAIB 2012 report was to recommend further human factors work investigating driving with ERTMS.

This paper briefly reviews some of the potential human factors issues associated with driving with ERTMS, drawn from both previous literature and rail standards and also from familiarisation activities undertaken by the first author. Secondly, it will summarise and discuss some of these issues based on data collected from interviews with UK ERTMS drivers.

Potential human factors issues

Automation

The introduction of the ETCS has increased the level of automation in train driving. Enhanced automation with ERTMS driving will ultimately *change the role of the driver*. The driver's role will consist of more monitoring tasks and anticipating intervention in case of any disruptions (Stoop et al. 2008). Therefore the driver's role and tasks need to be reassessed and how this will in turn impact on drivers' cognitive strategies and attention.

Supervisory control

Supervisory control systems represent situations in which automation and humans work together to accomplish a task (Sheridan 2002). The ETCS acts as a supervisory automated braking system which instructs any changes in speed and braking points. The use of automated subsystems means that the *role of the driver shifts towards supervisory control* and there will be greater use of cognitive functions such as goal monitoring, exception handling and recognising anomalies (Woods et al. 2010). However, in supervisory control, misdirected attention could cause goal conflicts to be missed or misprioritised (Woods et al. 2010) which could be a concern with ERTMS driving. Also, it is unclear how the changes in the driver's expertise and automating the driver's goal setting activity will affect driver control.

Route knowledge & situation awareness

Signals and route knowledge are essentially the main pieces of information that drivers use in the driving task in the UK. However, with the introduction of the ETCS, driver's route knowledge will be altered, in turn affecting that driver's mental model of driving. Real-time information regarding movement authorities and upcoming speed changes can support a driver to plan ahead. However the consistency of drivers' strategy selection and the automated algorithms underpinning the goal setting components could have *an impact on drivers' behaviours and their strategies* (Naweed et al. 2009). It has been suggested by Young et al. (2006) that enhanced level of information displayed on the DMI should improve a driver's situation awareness. However they further stated that this may not be the case, similar to automation in cars, if the driver's attention and awareness of the system is not maintained. Therefore it is unclear if driving with ERTMS improves a driver's situation awareness and if it has an impact on a driver's level of performance.

Allocation of attention

The driving task requires the driver to be able to allocate and shift attention in order to collect and integrate information, from both inside and outside the cab. Kecklund et al. (2011) conducted a risk analysis of ERTMS in the Swedish rail system and identified *information overload* and *divided attention* as major issues with driver-machine interaction. The ECTS also has some elements of attention control to help the driver's attention to be directed to the most critical information at the correct time to reduce information overload (Metzger et al 2012). Auditory signals and alarms are used to direct the driver's attention to important visual information. However there is also potential for these sounds to be a cause of distraction at crucial parts of the driving task.

Driving styles

Jansson et al. (2004) identified two different styles of train driving in the Swedish rail system; reactive and proactive. Drivers with the reactive driving style relied heavily on the feedback from the technical system. For example, with warning signals drivers with the proactive driving style were thinking ahead and focused on planning in order to prevent future warning signs. Jansson et al. (2004) proposed that the proactive driving style is more beneficial as drivers are more capable of anticipating and dealing with any irregularities. They further proposed that the in-cab 'information vacuum' could be a major contributing factor to drivers having reactive driving styles, as drivers are forced into a reactive mode of operation. The introduction of ERTMS could potentially force drivers from the more beneficial *proactive driving style to a reactive driving style*, however this needs to be further investigated.

Mental workload

A major concern with the introduction of ERTMS driving is *its potential impact on drivers' mental workloads*. It is crucial to understand the mental workload of drivers in the different levels of ERTMS as both overload and underload could lead to human error or reduce performance. Additionally, the introduction of automation could potentially introduce issues of *drivers being 'out of the loop'*. There are different levels of ERTMS, and tasks that a driver must carry out during transitions between levels (such as monitoring if ERTMS has been successfully achieved) could also lead to peaks in workload.

Methodology

Overview

A process of familiarisation was conducted at the start of the study to develop an initial understanding of ERTMS by reviewing standards and other documents from a range of sources. Also, informal interviews and discussions were held with various contacts in the rail industry. A participant-observer approach was taken by using various ERTMS simulators. Field observations were also made in four in-cab rides along an ERTMS route.

Interviews

Eighteen semi-structured interviews were conducted with ERTMS drivers. They consisted of seventeen males and one female, with an age range of 29-59 and an average age of 42.8. The drivers had been driving trains for an average of 7.38 years, which ranged from 2 to 22 years. Time spent driving with ERTMS varied from nine months to eighteen months. As well as ERTMS all drivers had previous experience with a variety of other signalling systems including radio electronic token block (RETB), coloured light signals and semaphore signals.

The semi-structured interview questions were piloted on a member of the Network Rail Ergonomics team and a member of the University's Human Factors Research Group. The questions chosen were influenced by the familiarisation process and observations. The interview consisted of 27 questions, which were sub-divided into four categories; general experience with driving with ERTMS, use of information from inside and outside the cab, cognitive experience with ERTMS driving and ERTMS driving in different scenarios. All interviews were arranged by a train operating company (TOC) and lasted approximately one hour.

Analysis

The interviews were recorded and then transcribed. The data was analysed using thematic analysis using a bottom-up approach. The data was initially analysed producing a set of themes and codes. The data was then re-analysed to ensure that the themes and codes matched the data. The six themes which emerged from the data are discussed in the next section.

Results & discussion

Adaptation to new technology

The majority of drivers were quite positive about ETCS and had adapted their driving to the new system. It was suggested *"that train driving is habit forming"* and that the introduction of ERTMS caused drivers to adjust some of their patterns of behaviour in driving. Trust in the new system appears to have increased with time as the drivers become more familiar and confident with the system. However, there were still some drivers who were not fully comfortable with the level of control of the new technology. On the whole drivers felt that the ETCS system was a safer system, however there were concerns about the increased time spent looking inside the cab at the DMI. The data also touched on the effects of ERTMS on driving culture. Drivers prided themselves to be *"professional, skilful, with responsibility of the job on their shoulders."* However, some suggested that using ERTMS might have removed some of their driving skills that they pride themselves in.

Style of driving

The data suggests that driving with ERTMS has shifted some drivers from being proactive drivers to be more reactive drivers, as they have more monitoring tasks and less anticipatory control, as proposed by Jansson et al. (2004). Drivers are using the brake and throttle more than previous types of driving as they have to brake and accelerate according to the system, and one driver described this type of driving to be *"twitchy."* Different driving styles have emerged, where some drivers try to utilise their route knowledge and judgement, within the parameters of the system, whilst others are more reliant on following the speed and braking information provided by the ETCS. In addition, it was reported that the alarms affect the way some drivers drive the train. Some drivers reported that they drove the train as they did previous to the implementation of ERTMS (within the parameters of the system), whilst other drivers drove the train slightly below the permitted speed in order to prevent overspeed or warning alarms, in order to reduce the level of noise in the cab. Train drivers also have to prioritise different pieces of information and there are occasions when the system is negatively impacting their prioritisation process, consistent with Woods et al (2010)'s proposal that supervisory control could cause goal conflicts to be missed or misprioritised. For example when drivers have to monitor crossings whilst waiting for a movement authority to extend.

There has also been a shift in control, as the ETCS has taken responsibility for some of the control from drivers in terms of braking, speed and enforcing set parameters on the drivers. Some drivers felt that they were unable to use their route knowledge and expertise to drive the train to the best of their capability and the system did not allow them to make "*positive decisions*." However, there was a minority of drivers who felt that they had the same level of control as with previous driving systems and the ETCS did not affect their driving.

Allocation of attention

Drivers reported that when driving with ERTMS they were spending varying amounts of time focusing on the DMI for information on speeds and movement authorities, which meant that they were spending less time monitoring outside the track. The majority of drivers stated that their focus of attention was now on the DMI, however this varied from occasional glances down to having to concentrate on the DMI for longer periods of time. Scenarios were identified were drivers had to examine the DMI for longer periods of time, as well as scenarios where drivers felt that their attention was split between the DMI and outside the cab. This is similar to Kecklund et al. (2011)'s finding of divided attention with ERTMS in the Swedish rail. Additionally, there are sound alerts and alarms to inform the driver of any changes on the DMI and this too could draw their attention away from the tracks to the DMI. The alerts were described by drivers to be both useful to attend to new information but also distracting.

Mental workload

From the interviews it was not clear if driving with ERTMS increased or decreased drivers' overall perceived mental workload, in comparison to their driving pre-ERTMS. However, what was clear is that there were periods of underload and overload. On the one hand, drivers reported increased mental workload due to having to monitor and control speed within narrower parameters, in comparison to previous driving, and by the knowledge that the system would intervene and apply the brakes if they overspeeded. A particular example of increased workload given was having to brake to the braking curve whilst going down a gradient heading towards a crossing. On the other hand, the data suggested that there are also periods of underload. These included periods of time where they did not have to reduce their speed and had movement authority for longer distances. It was also reported by some drivers that they could experience underload as their use of their own route knowledge had decreased. It is crucial to further investigate mental workload as both underload and overload could be detrimental to performance.

Route knowledge

Train drivers acquire 'route knowledge' from training and driving experience allowing them to anticipate, think ahead and be proactive drivers. The data suggests that ERTMS has affected drivers' route knowledge. For example, the braking points provided by the system may conflict with their own braking points which they have acquired with experience. However, other parts of route knowledge such as station braking knowledge; have not been affected. Drivers typically integrate several pieces of information such as braking points and weather conditions from their route knowledge when planning ahead. However this may be hindered if drivers have incomplete or conflicting route knowledge or unable to use their route knowledge, which could negatively impact on their situation awareness. This could potentially affect the driving style from being proactive to more reactive. As drivers are not using the entirety of their original route knowledge in ERTMS driving, when in abnormal working driver's may face difficulties. This is further discussed in the next sub-section.

Automation

It has been suggested that the introduction of ETCS has decreased the level of decision making by drivers. In particular, some drivers have reported that the system has taken over decisions regarding when to brake and the speed on approaches. Some drivers reported to attempt to brake close to where their braking points were within limits of the system and that the system did not effect their decision making, whilst other drivers reported to fully rely on the DMI to know when to start braking.

There is a small section of the route were ERTMS drivers have to transition back to coloured light signalling (ERTMS level NTC – national train control), where their movement authority is no longer given on the DMI but on signals out on the track; and the system is no longer monitoring the speeds. Again, drivers reported different experiences to this situation. Some drivers found that they had to pay more attention to remember where to brake, the various speed restrictions and the use of different lineside signage; whilst other drivers found that they had to pay equal or less attention whilst driving.

If there is a problem with the ETCS system, drivers must drive in 'degraded mode.' In this situation they are given movement authority by speaking to the signaller and must drive the train without the support of the ETCS system. All drivers found this situation difficult and experienced high workload, consistent with Bainbridge's "ironies of automation." One of the issues highlighted was that drivers' had to remember speeds and braking points that they did not normally need to recall. Additionally, they had to recall all the positions of the block markers, despite maybe never having to stop at them before in normal operation. Many drivers adopted the strategy of driving slower than the maximum degraded speed to ensure that they do not make any human errors.

Conclusions

The subjective data provided by the ERTMS drivers have provided an insight into the effects of ERTMS on train driving in the UK. However, the data also highlights, alongside the RAIB (2012) report, that further investigations must be conducted to further understand the effects of ERTMS on a human factors level. In particular, issues such as allocation of attention and mental workload need to be further investigated, in order to provide support and training for drivers but also to support future ERTMS projects. This study is the first in a programme of work and future studies will be conducted to address the impact of ERTMS on train driving in the UK.

References

Kecklund, L., Mowitz, A. & Dimgard, M. 2011. Human factors engineering in train cab design-prospects and problems. In P. C. Cacciabue, M. Hjalmdahl, A. Ludtke & C. Riccioli (eds) *Human modeling in assisted transportation* (Springer, Milan, Italy)

Jansson, A. Olsson, E. & Kecklund, L. 2004 Acting or reacting? A cognitive work analysis approach to the train driver task. In J. R. Wilson, B. Norris, T. Clarke & A. Mills (eds) *People and rail systems: human factors at the heart of the railway* (Ashgate pulishing limited, UK)

Metzger, U. & Vorderegger, J. 2012. Railroad. In M. Stein & P. Sandi (eds) *Information ergonomics: a theoretical approach and practical experience in transportation* (Springer, Berlin, Germany)

Naweed, A., Hockey, B. & Clarke, S. 2009 Enhanced information design for the high speed train displays: determining goal set operation under a supervisory automated braking system. In D. de Waard, J. Godthelo, F. L. Kooi & K.A. Brookhuis (eds) *Human factors, security and safety* (Shaker publishing, Maastricht, the Netherlands) 189–202

RAIB. 2012, "Rail Accident Report: Incident at Llanbadarn Auomatic Barrier Crosssing (Locally Monitored), near Aberystwyth, 19 June 2011." Retrieved 23rd September, 2012, from: http://www.raib.gov.uk/cms_resources.cfm?file=/120627_R112012_Llanbadarn.pdf

Sheridan, T. B. 2002 *Humans and Automation* (Wiley, New York)

Stoop, J. & Dekker, S. 2008. The ERTMS railway signaling system; deal on wheels? An inquiry into the safety architecture of the high speed train safety. In E. Hollnagel, F. Pieri, & E. Riguad (eds) *Proceeding of the third resilience engineering symposium* (Ecole de Mines Paris, Paris), 255–262

Woods, D. D. & Branlat, M. 2010. "Hollnagel's test: being 'in control' of highly interdependent multi-layered network systems." *Cognition, technology and work*, 12, 95–101

Young, M. S., Stanton, N. A. & Walker, G. H. 2006. "In loco intelligentia: human factors for the future European train driver." *International journal of industrial and systems engineering*, 1(4), 485–501

CROWD SOURCING OF PUBLIC TRANSPORT PROBLEMS

R.E. Sims, T. Ross & A.J. May

Loughborough Design School, Loughborough University, UK

Social sites on the World Wide Web allow increased sharing of ideas and problems. Such online networking includes the ability to report public transport problems and get support for campaigns to improve specific aspects of public transport. One site that offers such a facility provided access to its data on problems that were reported over a 13 month period. These were assessed and comparisons were made to see if those campaigns that got the most support were inherently different to the campaigns that received little support, and to investigate the types of problems that were reported online. Whilst definitive answers cannot be given, there are some suggestions that come out of the analysis and the overall impact on improving public transport accessibility and usability is discussed.

Introduction

The World Wide Web has grown exponentially since the first site was published in 1990. In 2011 it was estimated that 32.7% (that is 2.2 billion) of the global population were 'online' (Internet World Stats, 2012), having access to a computer at home/work/school or elsewhere that they could use to use the internet. Social networking sites such as Facebook, Twitter, etc. allow the interaction of people who might never have met face-to-face, but may share a common interest or aims. 'Crowd sourcing' (also known as user-generated content, community/social systems, or collective intelligence) is the use of data collected from multiple individuals, to develop shared outcomes (Doan et al., 2011). Crowd sourcing has been seen as a potential source of solutions to problems and a way of accessing ideas and innovations that might otherwise not be considered or discovered by standard design or innovation routes. FixMyTransport (www.fixmytransport.com) is one website that offers a platform for public transport users to report problems which are then automatically communicated to the relevant transport operator or local authority. FixMyTransport aims to make it easier for members of the public, and particularly those who are less confident about reporting problems, to send problem reports to the 'owner' of the issue. By enabling easier reporting of transport-related problems by a larger proportion of the population, it is expected that pressure can be brought to bear on transport operators to improve their services. In addition, the site will enable the public (and again particularly those less confident) to feel empowered to more actively engage in society by making their voice heard. Problems reported can include dirty seats, late trains, vandalised bus stops, poor signage etc. People are able to report issues as one-off 'problems', or they can post them as a 'campaign'. Campaigns can be supported by other people, who can opt to support the campaign and/or leave a message after the original posting.

The site can be accessed via computer or smart phone, enabling people to add problems as they occur, or to log on at a later point. The work detailed in this paper forms part of a larger project called Ideas in Transit (www.ideasintransit.org), with funding from the Engineering and Physical Sciences Research Council, the Technology Strategy Board and the Department for Transport. The project as a whole aimed to investigate many aspects of reducing car travel and promoting and supporting sustainable transport.

Aims

The aims of the work reported in this paper were to:

- Provide an overview of the campaigns that were reported on FixMyTransport in a 13 month period.
- To explore the characteristics of the campaigns that might influence the level of support it attracts (e.g. length of time since posted, transport mode, problem type, specific characteristics of the campaign post itself) to see if there was anything that could be said about the most successful campaigns when compared to campaigns with little support that might have influenced how successful they were (or not).

Methods

Access was provided to a database detailing all the campaigns and problems reported on FixMyTransport between 13th July 2011 and 20th August 2012. Simple counts were taken of how many campaigns had different numbers of supporters. It was found that there were 35 campaigns that had attracted 10 or more supporters. As 35 seemed a reasonable number of campaigns to investigate further within the time available, it was decided that these 35 would be designated the 'most supported' campaigns.

Two comparisons were then made between the 'most supported' campaigns and 35 campaigns that did not attract many or any supporters. Due to the large number of campaigns with 0 supporters or 1 supporter, the first comparison was made between 35 of these campaigns that were 'matched' to the 'most supported' campaigns as closely as possible according to the length of time since the campaign was posted, and mode of transport. The second comparison was made with 35 campaigns with 0 or 1 supporter that matched, as closely as possible, the problem that was described in the 'most supported' campaigns. Overviews of both the 'most supported' campaigns and the campaigns with 0 or 1 supporter were also generated.

Results

It was found that, within the timeframe given, there were 1935 campaigns started in total. The number of supporters for campaigns ranged from 0 to 256, but the majority of campaigns had 16 or fewer supporters, with 768 campaigns having only one supporter, and 576 having no supporters at the time of the analysis.

The types of problems that people started campaigns about varied widely. Table 1 shows the problems described in the campaigns with 10 or more supporters and the campaigns with 1 supporter. Percentages are used to take into account the different total sample sizes. There were 16 campaigns that only had 1 supporter where the problem described only occurred once: these are not in the table. In the campaigns with 10 or more supporters there were four that were only found once and no similar campaigns were found in the campaigns with 1 supporter. These were: bus stopping in advanced stop box for cycles at traffic lights, a ferry pier that was closing, a request to allow cycles on trams, and the location of the companion seat for accompanying a wheelchair user on trains. These are also not included in the table, to reduce the size of the table and make it easier to highlight the comparison and differences in percentages for the different types of problem in the campaigns with 1 supporter and the 'most supported' campaigns with 10 or more supporters.

In Table 1 'Timetable issues' include late running, slow running vehicles and vehicles leaving before the timetabled time. 'Routing' includes requests for services to different locations, unexpected changes of route on-journey, late changes of destination, and cancelled/missing services. 'Ticketing' includes problems with electronic payment, pricing issues, compensation requests and refund issues.

Each campaign is identified by a Problem ID number. These are issued successively, so lower numbers are campaigns that were posted longer ago than higher numbers. For the 'most supported' campaigns, in terms of length of time since campaign was posted and the number of supporters, it was found that: 20 of the campaigns had Problem IDs less than 1000, 7 have Problem ID between 1500 and 3000, 3 have Problem ID greater than 3000. This suggests that older campaigns are more likely to have gained more supporters. When looking at *all* campaigns with no supporters or only 1 supporter, however, there does not seem to be any relationship to length of time since the campaign was posted (Figure 1).

In terms of mode of transport and the 'most supported' campaigns it was found that train was the most frequent mode of transport with 26 out of 35 campaigns. There were 4 for bus travel, 4 for tram/metro travel and 1 for ferry travel. When looking at *all* campaigns with 0 or 1 supporter, bus issues were the subject of 849 campaigns compared to 379 campaigns for train issues.

The 35 'most supported' campaigns were matched as closely as possible by mode of transport and then as closely as possible to Problem ID (so length of time since posting) to 35 campaigns with 1 supporter. It was expected that these matched campaigns would be similar in terms of length of time since posting on the site and mode of transport they concerned. A brief content analysis was conducted of

Table 1. Percentages of each reported problem type for the 'most supported' campaigns and the campaigns with just 1 supporter (the totals might not be 100% due to some campaigns detailing multiple problems).

Problem type	% of sample (campaigns with 1 supporter, n = 768)	% of sample (campaigns with 10+ supporters, n = 35)
Poor driving/rude staff	12.1	2.9
Over-crowding	11.2	2.9
Info unavailable/out of date/requested	10.3	2.8
Ticketing	5.5	5.7
Shelter needed/damaged	3.8	2.9
Electronic display needed/inaccurate	3.0	2.9
Access	2	17.1
Ticket machines (lack of/problems	1.3	2.9
Entrance/exit gate issues	1.1	5.7
Announcement problems (volume/number)	1	2.9
Vehicles parked in bus stop	0.7	2.9
Signage problems	0.7	2.9
Cycle racks needed	0.4	11.4
Toilet issues	0.4	2.9
Body part trapped in closing door	0.4	2.9
Wi-fi request/problem	0.3	8.6
Footbridge needed/problems	0.3	2.8
Timetabling issues	23.9	
	20.1	

Problem type	% of sample (campaigns with 1 supporter, n = 768)	% of sample (campaigns with 10+ supporters, n = 35)
Failure of bus to pick up/put down	6.4	
Positioning of bus stop	2.2	
Old vehicles need replacing	1.8	
Heating on vehicles	1.7	
Station/stop improvements needed	1.6	
Broken down vehicle	1.3	
Stop/station needs cleaning	0.8	
Access to train cycle storage	0.7	
Car park issues	0.7	
Vehicle noise	0.7	
Internal vehicle layout (including luggage racks)	0.7	
Seating needed at stop/station	0.5	
Smoking on bus/at bus stop	0.5	
Leaking roof of vehicle/waiting area	0.4	
Integration between modes	0.3	
Seat reservation issues	0.3	
Grit needed/slippery station	0.3	
No refreshments on-board	0.3	
Station lighting	0.3	

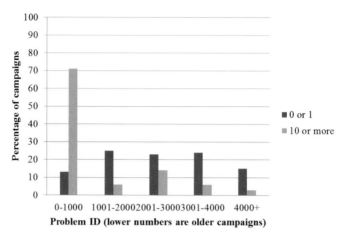

Figure 1.   Graph showing Problem ID against percentage of campaigns with 0 or 1 supporter compared to Problem ID for campaigns with more than 10 supporters.

Table 2.   Characteristics of campaign posts for campaigns with 1+ & 1 supporter, matched by mode of transport and Problem ID (n = 35 each).

	Characteristics	10+ supporters	1 supporter
Length of post	Short	26	22
	Medium	2	11
	Long	4	2
Type of event	one-off events/issues	6	5
	on-going issues	29	28
	mix one-off/on-going	0	2
	Offer solution(s)	18	14
Writing style	calm emotive	10	14
	not-calm emotive	1	10
	factual	24	11
	Evidence of research	5	2
	Mention consultation/petitions	3	0
	Mention legislation	6	0
	Use of exclamation marks	0	4
	Use of capital letters	0	1

the characteristics of the original campaign postings that were written on the site (Table 2).

When the 35 'most supported' campaigns were matched according to the type of problem being described in the campaign, it was not possible to find 35 matches in the campaigns with 1 supporter. In these instances the search was extended to include campaigns with 0 supporters. However, there were still some campaigns where no matching problem could be found. This resulted in 29 campaigns with 0

**Table 3.   Characteristics of campaign posts for 29 'most supported'
campaigns and 29 campaigns with 0/1 supporter, matched by problem type.**

	Characteristics	10+ supporters	0 or 1 supporter
Length of post	Short	25	19
	Medium	2	4
	Long	2	6
Type of event	one-off events/issues	3	4
	on-going issues	26	24
	mix one-off/on-going	0	1
	Offer solution(s)	18	13
Writing style	calm emotive	8	7
	not-calm emotive	1	4
	factual	20	18
	Evidence of research	5	2
	Mention consultation/petitions	3	0
	Mention legislation	6	1
	Use of exclamation marks	0	1
	Use of capital letters	0	3

or 1 supporter being matched to 29 of the 35 'most supported' campaigns. Table 3
details the analysis of the content of the initial campaign messages posted for each
of the 29 'most supported' campaigns and the matched campaigns with 0 or 1
supporter (matched by problem type).

Discussion and conclusions

It was found that the length of time since campaigns had been posted did not have an
obvious impact in the low-supported campaigns, but support increased with time for
the 'most supported' campaigns. There was a difference in prevalence of different
modes of transport in the 'most supported' and low-supported campaigns. It is
possible that this change arises from the fact that one train or train route will carry
more people than a single bus or bus route, so a significant issue on a train/train
route will impact on more people and garner more support. But overall there are
far more buses and bus routes than trains and train routes, so more people could be
impacted by individual problems when using buses than trains.

It is worth noting that the campaign with the most supporters had 256 supporters, with the next most supported campaign having only 41 supporters. The most
supported campaign linked into an organisation with other social media links, and
it is possible the campaign had been 'promoted' within the organisation, resulting in the increased support. In the comparison of 'most supported' campaigns
and low-supported campaigns matched on Problem ID and mode of transport it is
seen that there were differences between the amount of not-calm emotive language
('ranting') used (with more in the low-supported campaigns), whilst there was not
much difference between the numbers of campaigns offering solutions or being

one-off versus on-going issues. In the comparison of 'most supported' campaigns with low-supported campaigns matched on problem types, there was slightly more evidence of research/other petitions or consultation/legislation being mentioned, and less use of capital letters, exclamation marks and 'not calm' emotive language in the 'most supported' campaigns. There were also more medium/long campaign posts in the '0 or 1 supporter' matched group. In terms of the types of problems, the campaigns that gathered the most support were not necessarily the same as those problems which were found most frequently in the campaigns that only gained one supporter.

It is possible that short, to-the-point campaign posts, that avoid getting overly emotive and present the facts against a background of research/other support gain the most support. This has implications for people looking to start successful campaigns in the future, and for such websites in terms of the suggestions they make to contributors looking to make connections to other people and develop successful campaigns and networks with others with related interests. For campaigns that concern related issues but only have 1 supporter, if there was a way of joining these together to make a 'meta campaign', the number of instances of these problems being reported would result in them becoming part of the 'most supported' group of campaigns, with more than 10 supporters.

The work conducted here is only a very brief overview of the available data, but is an interesting first step into a relatively new area of exploration and within the context of the larger project to investigate ways to support and promote public transport usage, amongst other things. With increased time it would be possible to conduct a more rigorous qualitative analysis of all the data within the campaigns, and to draw more definitive conclusions regarding the content of more successful versus less successful campaigns. At the time of writing, in-depth questionnaires and interviews were being conducted with a sample of people who posted problems and campaigns on the site, to explore the 'type' of people who make complaints in this way, whether they had complained via alternative methods including talking to the driver/members of staff, and what impact they thought making the complaint online had. Initial results suggest that respondents generally had a positive response to raising their issues from the transport providers and found the experience of 'complaining' online to be positive. Crowd sourcing sites are keen to maximise the impact that their sites have and the benefit for the contributors to those sites, and for people working to improve public transport and uptake of public transport then any information that can help increase the effectiveness of crowd sourcing data and campaigns is of potential benefit.

References

Doan, A., Ramakrishnan, R., & Halevy, A.Y. 2011, Crowd sourcing systems of the World Wide Web. *Communications of the ACM, 54* (**4**), pp. 86–96.
Internet World Stats, 2012, www.internetworldstats.com (accessed September 2012)

FATIGUE IN THE MARITIME AND ROAD HAULAGE INDUSTRIES

Andrew P. Smith & Paul H. Allen

School of Psychology, Cardiff University, UK

This paper reports a comparison of fatigue in seafarers and lorry drivers. The seafarers data came from the Cardiff Seafarers' Fatigue Research Programme and this was compared with a small survey of lorry drivers. Results showed that reported fatigue was similar in the offshore and onshore samples. The reported fatigue scores from lorry drivers were similar to the crew of ferries. These groups also reported reduced health compared to other sectors. The present results probably reflect the effects of frequent port-turnarounds and deliveries/drop-offs. The importance of these risk factors for fatigue requires further investigation.

Introduction

It is widely recognised that fatigue at work has an effect on health and safety. This topic has been extensively studied in onshore industries and attention has focused on the safety critical tasks involved in transport operations. Fatigue at sea has received less attention (Allen, Wadsworth and Smith, 2007, 2008) but clearly has consequences for the health and safety of the crew and passengers. It also impacts on the environment, as seen when oil tankers run aground. Inefficiency due to fatigue also has huge economic costs as 90% of goods are transported by sea. The Cardiff Seafarers' Fatigue Research Programme (Smith, Wadsworth and Allen, 2006; Smith, 2007) investigated risk factors for fatigue, perceived levels of fatigue and health and safety outcomes of fatigue. The results showed that there are many risk factors for fatigue and it is inappropriate to focus on one issue such as working hours. Quite often manning levels and working hours are appropriate for plain sailing but are found wanting when the extra demands of port operations come into play. Research on fatigue in road haulage often focuses on hours driving without considering other activities such as loading/unloading. The aim of the present study was to compare fatigue in seafarers and road haulage. Port-turnarounds were often associated with increased fatigue in seafarers and the present study examined whether deliveries/drop-offs in road transport also led to greater fatigue than driving per se.

Development of a fatigue questionnaire for the road haulage industry

A questionnaire was designed containing all key outcome measures included in the seafarer survey whilst also addressing issues unique to the road haulage industry. The questionnaire was therefore based around the following sections:

Some basic information about you: Basic demographic questions including gender, age and level of education.
About your main job:
Questions about the type of lorry driven, type of employment and experience in the industry.
Your hours of work in your main job:
Questions about hours of work as a road haulier, the type of work involved in, miles covered and length/frequency of rest periods.
Additional work
Questions about any work conducted additional to road haulage.
Regulations to control working hours:
Questions addressing knowledge of regulations related to working hours.
Rest facilities
Questions assessing the acceptability of rest facilities for haulage drivers.
Fatigue at work:
Questions assessing experience of fatigue at work, fatigue related accidents and mitigating use of caffeine and naps.
About your work:
Standard questions assessing the work environment, exposure to negative work factors, job demands and support in the workplace.
Your sleep and your health:
Questions addressing typical quantity and quality of sleep as well as frequency of visits to the GP, job stress and smoking/drinking habits.
Accidents and Injuries:
Questions about recent visits to hospital and accidents experienced.
General Health and Well-being:
Standardised measures of health and well-being including the Cognitive Failures Questionnaire (CFQ), General Health Questionnaire (GHQ), Profile of Fatigue Related Symptoms scales (PFRS) and Short-Form Health Survey (SF-36).

All road haulage specific questions were designed in conjunction with a representative of the Road Haulage Association (RHA) in order to address those issues most important to the industry. A pilot survey was carried out to assess the suitability of the questionnaire and to refine any new items.

Samples

Road Haulage Sample: A total sample of 81 questionnaires were collected.

Demographics: Only 2.5% of the sample were female (n = 2) and 81.5% (n = 66) were married. Most of the sample had either completed some secondary education and left school before 16 (41.8%, n = 33) or reached GCSE/O-Level standard (40.5%, n = 32). Only 1 respondent was non-British (1.2%) and most drivers earned between £20,000 and £30,000 a year (68.4%, n = 54). The average age of the sample was 47.4 years old.

Type of haulage work: The type of vehicles driven by respondents are shown in table 1 below (vehicle descriptions taken from the Driving Standards Agency (DSA) website).

All but one of the sample were employed by a company (98.7%, n = 78) with an average company size of 513 (S.D. = 718.535, Range 9–3500). The type of work respondents reported being involved in is shown in table 2.

The average number of deliveries or pick-ups in a day for respondents who answered 'B' (see table 2) was 5.34 (S.D. = 4.7, Range = 2–20) and for respondents who answered 'C' was 11.79 (S.D = 7.3, Range = 3–25). On average respondents reported driving 228.7 miles in an average day (S.D. = 124.8, Range = 4–650) and 1124.1 miles in an average week (S.D. = 663.4, Range = 40–4000).

Comparison with seafarers

The data from over 1800 seafarers were compared with the lorry drivers (details of this sample are given in Wadsworth et al., 2008). First of all the road haulage

Table 1.   Type of haulage work.

Category	Type of vehicle	%	n
C1	A medium sized lorry with a maximum authorised mass (MAM) of at least four tonnes, capable of 80 km/h (50 mph).	2.7	2
C1+E	A drawbar outfit made from a combination of a category C1 vehicle towing a trailer of at least two tonnes MAM, with a combined length of at least eight metres in length and capable of 80 km/h (50 mph).	29.7	22
C	A rigid goods vehicle (not an articulated tractor unit) with a MAM of at least 10 tonnes, at least seven metres and at most 12 metres in length and capable of 80 km/h (50 mph).	6.8	5
C+E	(1) An articulated lorry with a MAM of at least 18 tonnes, of at least 12 metres in length and capable of 80 km/h (50 mph). (2) A drawbar outfit made from a combination of a category C vehicle with a trailer of at least four metres platform length, of at least four tonnes MAM, suitably braked and coupled, with a combined weight of at least 18 tonnes and a combined length of at least 12 metres, capable of 80 km/h (50 mph).	56.8	42
Other	Other type of vehicle not listed	4.1	3

Table 2.   Type of driving.

	Descriptions of work	%	n
A	"I drive long distances, only stopping along the way if I need to rest"	25.3	20
B	"I drive long distance but stop to make a number of deliveries along the way"	43.0	34
C	"I drive locally but have to make many deliveries during a typical working day"	22.8	18
D	Other	8.9	7

Table 3.   Fatigue at work and after work.

		Mean	N	SE	F	P
Fatigue AT work	Ferries	3.82	635	0.03	8.30	<0.001
	Tankers	3.49	303	0.05		
	Offshore support	3.65	494	0.04		
	Containers	3.42	113	0.08		
	Road Haulage	3.75	80	0.11		
Fatigue AFTER work	Ferries	2.51	641	0.02	5.08	<0.001
	Tankers	2.38	302	0.03		
	Offshore support	2.36	504	0.03		
	Containers	2.35	115	0.05		
	Road Haulage	2.45	80	0.07		

(on all scales a higher score indicates higher fatigue)

and seafaring samples were compared in terms of fatigue at work and fatigue after work. No differences were found between the groups.

The seafaring sample was then divided into broad vessel categories in order to assess how the road haulage respondents would compare with more specific sectors (see table 3 above). The road haulage sample were found to report fatigue levels most similar to seafarers working on ferries. Whilst finding similar levels of fatigue does not necessarily indicate similarity of cause, the fact that road haulage and ferry workers both operate according to strict time-ordered schedules with daily deliveries/port-calls and high demands on vigilance is nevertheless worthy of note. Further suggestion that deliveries may cause fatigue in road haulage drivers comes from a trend revealed in the results of another question, as shown in table 4 below. A clear trend can be seen with lorry drivers who reported the fewest deliveries (group A) also reporting the lowest average levels of fatigue.

Further analyses looked at reported health. Again, the road haulage and ferry groups were significantly worse than the other sectors on reported physical health, vitality and social functioning (see Table 5).

Table 4.   Levels of fatigue by type of work conducted (Road Haulage sample only).

	Fatigue at work Mean *(N, S.E)*	Fatigue after work Mean *(N, S.E)*
A: 'I drive long distances, only stopping along the way if I need to rest'	3.33 *(20, 0.26)*	2.18 *(20, 0.14)*
B: 'I drive long distances but stop to make a number of deliveries along the way'	3.99 *(34, 0.15)*	2.56 *(34, 0.11)*
C: 'I drive locally but have to make many deliveries during a typical work day'	3.89 *(18, 0.18)*	2.48 *(18, 0.13)*

(on all scales a higher score indicates higher fatigue)

Table 5.   Reported health of seafarers and road haulage groups.

		Mean	N	SE	F	P
SF-36 Physical Functioning	Ferries	85.00	672	0.55	3.14	<0.01
	Tankers	91.92	302	0.55		
	Offshore Support	90.61	505	0.59		
	Containers	88.98	108	1.47		
	Road Haulage	85.75	80	2.10		
SF-36 Vitality	Ferries	58.31	635	0.73	9.96	<0.001
	Tankers	64.22	295	1.06		
	Offshore Support	64.13	505	0.80		
	Containers	64.86	111	1.74		
	Road Haulage	55.25	80	2.45		
SF-36 Social Functioning	Ferries	79.75	642	0.88	4.88	<0.001
	Tankers	85.07	303	1.18		
	Offshore Support	84.26	505	0.96		
	Containers	83.72	109	2.24		
	Road Haulage	76.41	80	3.22		

(On all scales a lower score indicates poorer health status)

Discussion

There has been considerable research on transport fatigue although less than 5% of the studies have considered the maritime sector. Research has shown that there are many causes of fatigue and that the effects of different risks factors are additive. Working hours have often been the focus of studies of fatigue and yet it is clearly the case that they are not the only factors. Our prior research (Allen et al., 2005) has shown that fatigue is relatively low when the ships are sailing but is greatly increased when entering port, in port or leaving port. The present study confirmed this finding and showed that ships with frequent port-turnarounds often have the most fatigued crew.

The study also showed a similar pattern of results in the road haulage sector. Those drivers who regularly dropped off or changed loads reported greater fatigue than those who spent most of their time driving. In both sectors reported health was also influenced by the frequency of turn-arounds/drop-offs. These findings require replication and extension. However, these preliminary findings suggest that loading and unloading may lead to increased fatigue even in well regulated transport sectors. This suggests that fatigue levels should be audited rather than assuming that they will be safe provided legal working hours are adhered to.

Previous research on seafarers has shown that frequent turn-arounds are associated with both physical and mental fatigue. They are also associated with the perception that they increase the risk of critical incidents, both at the individual and ship level (e.g. slips, trips and falls; collisions). Similar analyses now need to carried out in the road transport sector.

At one level the solution is quite clear, namely to allow longer for turn-arounds. However, this is often not possible due to commercial constraints. Educational campaigns pointing out the fatigue induced by these working practices are desirable and these can be linked to training on fatigue prevention and management.

Conclusion

Transport fatigue is an established health and safety issue. The present paper describes an initial direct comparison between fatigue in the maritime and road haulage industries. Sectors with frequent port-turnarounds/drop-offs/deliverables reported the most fatigue and health problems. Future research must evaluate methods of preventing or managing this fatigue.

References

Allen, P., Wellens, B., McNamara, R. and Smith, A. 2005, It's not all plain sailing. Port turn-arounds and Seafarers' fatigue: A case study. In P. Bust and P. T. McCabe. (eds). *Contemporary Ergonomics 2005*, (Taylor & Francis), 563–567.

Allen, P. H., Wadsworth, E. J. and Smith, A. P. 2007, The prevention and management of seafarers' fatigue: A review. *International Maritime Health* 58: 167–177.

Allen, P. H., Wadsworth, E. J. and Smith, A. P. 2008, Seafarers' fatigue: a review of the recent literature. *International Maritime Health* 59, 1/4: 81–92.

Smith, A. 2007, Adequate crewing and seafarers' fatigue: The International Perspective. *International Transport Federation: London*. www.itfglobal.org/files/seealsodocs/3193/ITF%20FATIGUE%20REPORT%20final.pdf.

Smith, A., Allen, P. and Wadsworth, E. 2006, Seafarer fatigue: the Cardiff Research Programme. *MCA: Southampton*. http://www.seafarersfatigue.com.

Wadsworth, E. J., Allen, P. H., McNamara, R. L., Wellens, B. T. and Smith, A. P. 2008, Fatigue and health in a seafaring population. *OccupationalMedicine* 58: 198–204.

WHAT DO WE TELL DRIVERS ABOUT FATIGUE MANAGEMENT?

Ann Williamson[1], Rena Friswell[1],
Raphael Grzebieta[1] & Jake Olivier[2]

[1]*Transport and Road Safety Research, The University*
of New South Wales, Sydney, Australia
[2]*School of Mathematics and Statistics, The University*
of New South Wales, Sydney, Australia

Fatigue management strategies on road and in the workplace involve advice on the need to take breaks when tired. This approach assumes people can assess fatigue effects on performance but this assumption is questionable. The aim of this study was to investigate whether we have access to information about current fatigue state and levels of drowsiness, and the implications for detecting changes in driving performance and the likelihood of crashes in a driving simulator with tired drivers. The results reveal that drivers can report increasing fatigue especially across the earlier part of the drive. Importantly, drivers can detect the likelihood of falling asleep prior to crashing indicating that they can make an informed decision to drive or not drive when tired. This has implications for safety advice to drivers.

Introduction

It is well-known that fatigue affects our ability to perform. Fatigue is a particular problem while driving due to the demands for continuous attention while at the wheel and because a temporary lowering attention can have potentially disastrous consequences (Brown, 2001). Fatigue produces a gradual withdrawal of attention from road and traffic demands which is involuntary and almost impossible to resist (Brown, 1994). The driving task itself may increase the likelihood of a driver experiencing fatigue, especially when driving involves prolonged exposure to monotonous conditions such as on country roads or motorways. Driving is also vulnerable to the effects of fatigue from other sources such as sleep deprivation or circadian influences (Brown, 1994). Fatigue affects driving performance in a range of ways including poorer steering control, speed tracking, visual search and attentional selectivity (McDonald, 1984).

Fatigue is acknowledged to account for around 20 percent of crashes in Australia and the UK (RTA, 2001; Jackson et al., 2011). Furthermore, fatigue-related crashes are more likely to be fatal (Bunn et al., 2005). It is also acknowledged that fatigue is a hazard for performance in the workplace, but much less is known about the size, nature or causes of the problem for workplace safety.

Managing fatigue is not a simple matter for road or workplace safety. While there is considerable evidence about the conditions likely to produce fatigue and increase the likelihood of falling asleep at the wheel or in the workplace, the causes are multifactorial, the relationships between the factors are not well-understood and individual variability is high. Unlike other road or work safety problems, there are no clear exposure limits and fatigue management approaches take the form of guidance rather than prescribing specific actions through regulation. Fatigue management strategies on the road and in the workplace involve suggestions of driving or working limits and advise people to take breaks when they feel tired. A major assumption inherent in this advisory approach is that drivers have access to information about their levels of fatigue and drowsiness and are able to make the decision to stop and rest before their performance is sufficiently adversely affected that their risk of crashing becomes too high. There is considerable debate about the validity of this assumption.

The research on this issue suggests that people can detect decreasing alertness and increasing fatigue and sleepiness. Many studies have shown the expected decreases in alertness and increases in self-rated fatigue and sleepiness when sleep deprived (Dinges, et al., 1997), when required to work at vulnerable times in the circadian rhythm (Monk, 1991) or for prolonged periods without a break (Rosa and Colligan, 1988). One study (Nilsson, et al., 1997), showed that under simulation conditions drivers can make a judgment about when they should stop driving due to fatigue, apparently based on their physical symptoms. Interestingly, however, driver ratings of fatigue at the time they stopped driving were very similar, no matter whether the drive had been for only 40 minutes or for as long as 180 minutes. These results call into question the current approaches to fatigue management that are based on advisory limits on the length of drive time. Rather they suggest that encouraging drivers to respond to fatigue state may be more useful. The trouble is, however, that the relationship between subjective feelings of fatigue/drowsiness and changes in performance is not clear. We do not know when performance effects begin for a fatigued person and whether fatigued people have the capacity to detect the effect of these changes in state on performance. These are critical questions for safety. It is not enough simply to be aware of changes in alertness or feelings of fatigue and sleepiness. Drivers need to be able to detect and, importantly, to respond to changes which have impact on their capacity to drive safely.

Unfortunately, the evidence on the relationship between changes in alertness and sleepiness-related states and performance effects is equivocal. There is some evidence that increasing self-reported sleepiness is related to poorer performance in driving tasks. For example, Reyner and Horne (1998) showed in a driving simulator that increasing subjective sleepiness was significantly associated with an increase in the number of safety-related incidents and Horne and Baulk (2004) also found that subjective sleepiness, EEG-recorded sleepiness and lane deviations in a driving simulator were highly correlated. On the other hand some studies have shown that self-rated alertness or fatigue is significantly correlated with self-rated performance but that there changes are not well-correlated with actual performance (Dorrian, et al., 2000; Dorrian, et al., 2003). In addition, some on-road studies found no association between self-assessed fatigue and a number of non-driving performance measures (Williamson, Feyer and Friswell, 2000) or a set of driving-related performance measures (Belz, Robinson and Casali, 2004). Further research is needed to clarify when performance effects begin to occur and become noticeable for a fatigued person.

The link between detecting fatigue and sleepiness and detecting when these experiences might lead to falling asleep and potentially to crashing is also unclear. Horne and Reyner (1999) found that drivers underestimated the probability of falling asleep when sleepy and seemed to underestimate their likelihood of crashing. Even partially sleep-deprived people whose task was to sit quietly in a darkened room and predict when they were going to fall asleep found it difficult to do so (Kaplan, Itoi and Dement, 2008). There is evidence that people overestimate the time they take to fall asleep and they can be in the early stages of sleep without being aware of it (Baker, Maloney and Driver, 1999). Furthermore, it seems that people can be in the early stages of sleep and still be responding to external stimuli. For example, Ogilvie, Wilkinson and Allison (1989) showed that some individuals were able to continue responding to a simple reaction time task despite EEG recordings showing that they were in Stage 1 sleep.

In summary, it seems that drivers can detect that they are increasingly becoming fatigued or drowsy, but may be less able to detect when these changes signal the onset of sleep, or the likelihood of crashing. The aim of this study is to investigate the extent to which we have access to information about our current fatigue state and levels of drowsiness, and their implications for detection of changes in driving performance and the likelihood of crashes. This information will show whether relying on subjective fatigue states will provide a valid estimate of safety-relevant driving performance effects or whether other indicators would be more useful.

Method

Study design

This was a driving simulator study involving three conditions. Drivers did a two hour drive on a monotonous simulated country driving route and were asked to make prompted or unprompted judgments about their sleepiness and likelihood of crashing in different ways depending on the study condition:

- Condition 1 (Unprompted, button press) – participants made a button press on the steering wheel if they thought it likely they would crash in the next few minutes.
- Condition 2 (Prompted, ratings only) – participants were asked to make judgments about their level of sleepiness, and likelihood of falling asleep and likelihood of crashing in the next few minutes at a prompt (audible beep) occurring approximately every 200 seconds throughout the drive.

- Condition 3 (Prompted, button press) – participants were asked to make the same ratings of sleepiness and falling asleep in response to the same prompts as for Condition 2, but were also asked to make a button press at any time when they thought they were likely to crash in the next few minutes.

The aim of this using prompted and unprompted conditions was to look at the effect of regular reminders of the need to assess crash risk (prompting) in contrast to the more realistic situation of leaving drivers to make the judgment with no reminding. Driving performance and crashes were monitored continuously throughout the drive.

Driver sleepiness was enhanced by testing in the period coinciding with the mid-afternoon circadian low-point (all testing commenced at 14:30 hrs) and further enhancing afternoon sleepiness by shortened sleep the night before to five hours. Sleep duration was validated using wrist actigraphs, which measure body movement indicative of waking or sleeping. All participants were given a two hour practice drive on a day preceding the test day in order to reduce practice effects on performance and to enhance the monotony of driving during the test session.

Participants:

Each condition involved separate groups of 30 drivers who were normal night sleepers. They were recruited through advertising around the University and the local community. Participants were randomly allocated to study condition.

Measures:

- Simulator – A STISIM Drive personal computer based interactive driving simulator was used. This is a very widely used and validated simulator. The drive for this study was two hours long and designed to be monotonous. It involved variable speed limits to which drivers were asked to comply.
- Sleepiness detection – The 9-point Karolinska Sleepiness Scale was used at each prompt. This a validated (Akerstedt and Gillberg, 1990) and widely used subjective measure of sleepiness, where 1 = extremely alert, 2 = very alert, 3 = alert, 4 = rather alert, 5 = neither, 6 = some signs of sleepiness, 7 = sleepy, but no effort to stay awake, 8 = sleepy, but some effort to stay awake, 9 = very sleepy, great effort to stay awake.
- Likelihood of falling asleep – Every 200secs study participants were asked to answer the question 'What is the likelihood of you falling asleep during the next few minutes', with the response options of A = very unlikely, B = likely, C = possibly, D = unlikely, E = very unlikely.
- Driving performance – measures were taken of lateral deviation in lane, speed and speed limit compliance, headway of following distance to vehicles in front and reaction time to target signals. The number of crashes were also counted.
- Objective ratings of drowsiness – Drowsiness of drivers in the simulator were measured using Optalert which monitors eye and eyelid movements using an

infrared reflectance method using transducers housed in a light spectacle frame (Johns, Tucker and Chapman, 2006) which has been validated in a range of settings including on-road (Howard, Clarke et al., 2006; Stephan, Hosking et al., 2006).

- Questionnaire containing questions on demographic characteristics, sleep history (including the Epworth Sleepiness scale) and recent activities leading up to the simulator drive.

Results

The first stage of the results of the analysis of this study have been completed. Participants were all experienced male and female drivers (83.5% had >10 years driving experience, 56.8% female, average age 45.8 yrs), who drove at least once per week. None had any evidence of sleep problems. Drivers had 4.24 hrs sleep on average in the night before the drive and rated their sleep quality as only moderate (57.9 on a 100 point semantic differential scale).

Driver ratings were standardised to z scores and plotted. Sleepiness ratings increased linearly across the first half of the drive then the rate of increase slowed markedly. Driver ratings of the likelihood of falling asleep and of crashing showed a similar trajectory to sleepiness ratings. There were no significant differences in the ratings of drivers in Conditions 2 and 3 so the results were combined. The majority of drivers reached the highest ratings of sleepiness, falling asleep and crash likelihood at some stage in the drive.

Around one-third of drivers had at least one crash, nearly half crossed the centreline and almost all crossed the road edge at least once. Most crashes occurred in the first half of the drive. There were no significant differences between drivers who crashed and those who did not on any of the demographic or personal characteristics or on recent activities prior to the drive. Drivers who crashed, however, had higher scores overall for all ratings.

Survival analysis by Cox regression of the relationship between ratings and crashes showed that ratings of falling asleep predicted the first off-road crash ($X^2_{(1)} = 6.24$, $p < 0.01$). For drivers who rated falling asleep in the next few minutes as possible, likely or very likely the odds of crashing were 4.3 times higher than those who rated themselves as unlikely to crash. Neither sleepiness ratings nor ratings of likelihood of crashing predicted crashes.

Discussion

The results of this study demonstrated that drivers are able to report increasing sleepiness levels, likelihood of falling asleep and of crashing across a drive when they would be expected to be fatigued due to reduced sleep and the time of day.

Drivers who experienced high levels of sleepiness, higher likelihood of falling asleep and of crashing were more likely to crash. Most significantly, for the aims of this study, the results showed that drivers who rated the likelihood of falling asleep in the next few minutes as very likely, likely or even possible were more than four times more likely to crash. This shows that drivers are clearly aware of increasing experiences relating to fatigue and sleepiness and are able to detect when these experiences might lead to falling asleep. The fact that drivers are aware of these changes makes it possible for them to respond earlier to avoid adverse road safety consequences of crashing.

In order to find out how drivers were feeling during the drive, it was necessary to ask them to make ratings of their current sleepiness level and likelihood of falling asleep or crashing. This was a potential limitation of the study design as the act of making ratings or made drivers more aware of safer driving which may have enhanced driver performance leading to lower numbers of crashes. It was possible to test this hypothesis as the study included a condition requiring no ratings and one in which only sleepiness-related ratings were made. The results showed no effects of making ratings or not on the number of crashes. It is also of note that only sleep-related and crash likelihood ratings differentiated drivers who crashed from those who did not so indicating that crashing was unlikely to be due to other factors like age, gender, driving experience, amount of sleep or recent activities.

The results of this study help to direct much-needed policy on fatigue risk management. While we know a considerable amount about what causes fatigue and can make some predictions about when it might occur, our predictions are not perfect. This means that it is not sufficient to simply tell drivers that they should not drive during vulnerable periods such as the middle of the night, or when they haven't had enough sleep or have been driving for too long. Such prescriptions need more research before they could be implemented as limits that will have the desired effect of keeping tired drivers off the road and allowing alert drivers to drive. Furthermore, these one-size-fits-all approaches are unlikely to be suitable for every driver and on every occasion.

In the meantime, in the absence of evidence, drivers are advised to take a break from driving when tired and to sleep or nap before fatigue begins to affect their driving skills. This approach assumes that drivers have access to information about their levels of fatigue and are able to make the decision to stop and rest before their performance is sufficiently adversely affected that their risk of crashing becomes too high. As discussed above, the validity of this assumption is extremely questionable. This research is significant because it provides the much-needed evidence on which innovative fatigue policy can be developed for road safety, rather than the current make-do approach. The research shows that drivers are able to help themselves to manage driver fatigue. Drivers have access to information about sleepiness and falling asleep that should inform judgments about when to drive at all and when to take breaks from driving. This means that deciding to drive when fatigued or sleepy is a road safety judgment just like the decision to exceed the speed limit or drive after drinking alcohol. We need to encourage and motivate drivers to avoid driving when fatigued and sleepy as this experience is an important indicator that they can use to maintain their own safety on the road and that of others.

References

Baker, F.C., Maloney, S. & Driver, H.S. (1999). A comparison of selective esti-mates of sleep with objective polysomnographic data in healthy men and women. *Journal of Psychosomatic Research*, **47**(4), 335–341.

Belz, S.M., Robinson, G.S. & Casali, J.G. (2004). Temporal separation and self-rating of alertness as indicators of driver fatigue in commercial motor vehicle operators. *Human Factors*, **46**(1), 154–169.

Brown, I. (1994). Driver fatigue. *Human Factors,* **36**(2), 298–314.

Brown, I. (2001). Coping with driver fatigue: Is the long journey nearly over? In P. Hancock and P. Desmond (Eds.) *Stress, Workload and Fatigue.* New Jersey: Lawrence Erlbaum. pp. 596–606.

Bunn, T.L., Slavova, S., Struttmann, T.W. & Browning, S.R. (2005). Sleepiness/fatigue and distraction/inattention as factors for fatal versus nonfatal commercial motor vehicle driver injuries. *Accident Analysis & Prevention*, **37**, 862–869.

Dinges, D.F., Pack, F., Williams, K., Gillen, K.A., Powell, J.W., Ott, G.E., Aptowicz, C. & Pack, A.I. (1997). Cumulative sleepiness, mood disturbance, and psychomotor vigilance performance decrements during a week of sleep restricted to 4–5 hours per night. Sleep, 20(4), 267–277.

Dorrian, J., Lamond, N. & Dawson, D. (2000). The ability to self-monitor performance when fatigued. *Journal of Sleep Research,* **9**, 137–144.

Dorrian, J., Lamond, N., Holmes, A.L., Burgess, H.J., Roach, G.D., Fletcher, A. & Dawson, D. (2003). The ability to self-monitor performance during a week of simulated night shifts. *Sleep,* **26**(7), 871–877.

Horne, J.A. & Baulk, S.D. (2004). Awareness of sleepiness when driving. *Psychophysiology,* **41**, 161–165.

Horne, J. & Reyner, L. (1999). Vehicle accidents related to sleep: A review. *Occupational and Environmental Medicine,* **56**, 289–294.

Howard, M., Clarke, C., Gullo, M., Johns, M., Swan, P., Pierce, R. & Kennedy, G. Evaluation of two eyelid closure monitors for drowsiness detection. *Sleep and Biological Rhythms*, 2006, 4 (Suppl 1): A13.

Jackson, P., Hilditch, C., Holmes, A., Reed, N., Merat N. & Smith, L. (2011). Fatigue and Road Safety: A Critical Review of Recent Evidence. Department for Transport (www.dft.gov.uk/pgr/roadsafety/research/rsrr

Johns, M.W., Tucker, A.J., Chapman, R.J., Michael, N.J., Beale, C.A. & Stephens, M.N. A New Scale of Drowsiness Based on Multiple Characteristics of Eye and Eyelid Movements: The Johns Drowsiness Scale. *Sleep and Biological Rhythms*, 2006, 4 (Suppl 1): A37.

Kaplan, K.A., Itoi, A. & Dement, W.C. (2008). Awareness of sleepiness and abil-ity to predict sleep onset: Can drivers avoid falling asleep a the wheel? *Sleep Medicine*, 9, 71–79.

McDonald, N. (1984). *Fatigue, safety and the truck driver.* London: Taylor & Francis.

Monk, T. H. (1991). Circadian aspects of subjective sleepiness: A behavioural messenger? *Sleep, sleepiness and performance.* T. H. Monk. Chichester, John Wiley & Sons: 39–63.

Nilsson, T., Nelson, T.M. & Carlson, D. (1997). Development of fatigue symptomsduring simulated driving. *Accident Analysis & Prevention,* **29**(4), 479–488.

Ogilvie, R.D., Wilkinson, R.T. & Allison, S. (1989). The detection of sleep onset: Behavioural, physiological, and subjective convergence. *Sleep,* **12**(5), 458–474.

Reyner, L.A. & Horne, J.A. (1998). Falling asleep whilst driving: Are drivers aware of prior sleepiness? *International Journal of Legal Medicine,* **111**, 120–123.

Roads and Traffic Authority (2001). *Driver fatigue: Problem definition and countermeasure summary.* Sydney.

Rosa, R.R. & Colligan, M.J. (1988). Long workdays versus restdays: Assessing fatigue and alertness with a portable performance battery. *Human Factors* **30**(3): 305–317.

Stephan, K., Hosking, S., Regan, M., Verdoom, A., Young, K. & Haworth, N. (2006). The Relationship between Driving Performance and the Johns Drowsiness Scale as Measured by the Optalert System. *Monash University Accident Research Centre: Independent Report.*

Williamson, A.M., Feyer, A.-M., Friswell, R. & Finlay-Brown, S., (2000). Demonstration project for fatigue management programs in the road transport industry – summary of findings, *Federal Australian Transportation Safety Bureau report* CR192, 1–20.

AGEING WORKERS

DESIGN FOR HEALTHY AGEING

**Elaine-Yolande Gosling[1], Diane Gyi[1],
Roger A. Haslam[1] & Alistair Gibb[2]**

[1]*Loughborough Design School, Loughborough
University, UK*
[2]*Department of Civil and Building Engineering,
Loughborough University, UK*

The "Organiser for Working Late" (OWL), an online design resource co-designed with industry, will be introduced during a workshop at EHF2013. OWL encourages communication between managers and workers when exploring the design of the workplace environment, equipment and systems people interact with, and need to use, as they age. To provide the right guidance to help people manage the effects of ageing it is essential to understand their user needs with respect to ageing and design (Williams, 2012a). This research forms part of The New Dynamics of Ageing (NDA) research programme which is funded by the ESRC, BBSRC, AHRC, EPSRC and MRC (Williams, 2011).

OWL considers the role of good design and ergonomics in healthy working. Contained within OWL are design stories and activities which have been directly informed by the results from the research project. These will be explored and the activities will be used in this workshop.

Industrial collaboration was essential for the development of OWL (Williams, 2012b). For the first phase of the research a questionnaire survey ($n = 719$ from 21 companies) and triangulation interviews ($n = 25$ from 5 companies) were conducted. Through this first phase, industrial collaboration was gained for the duration of the project with four companies; construction (domestic), construction (commercial), manufacturing (cement) and welfare (animal). With these companies ergonomics observations and in-depth interviews ($n = 32$), focus groups ($n = 33$) and OWL evaluation interviews ($n = 7$) were conducted. Each phase of the research was used to inform, develop and guide the next. For each phase of the research there were no right or wrong answers, as with the activities and discussions during this workshop.

During the workshop participants will be introduced and invited to use the activities and the downloadable materials in OWL. These were developed to encourage managers to understand the design needs and requirements of their workers; environment, equipment and job design, as they age.

The workshop session will be presented in three parts. Firstly an introduction to OWL which will demonstrate its content; information, design stories and activities which include the downloadable '@ work' cards (20 minutes). Secondly there will be two group exercises (30 minutes). Initially, participants will be asked to think about their own jobs and how they might impact on their health at work and healthy ageing; this exercise will use the 'body @ work' cards. In addition, participants will be asked to provide examples of how design at work has helped or hindered their ability to work healthily, using the 'body @ work' cards. Next, again in groups, participants will be provided with case studies from the collaborator companies. These discussions will focus on the health risks and health needs of the workers, when thinking about the design of their jobs; workplace and equipment. Participants will also be asked to consider how best the workers from the case studies, could be encouraged to think more about their health in relation to their workplace design. These discussions will be facilitated by using the full suite of materials within OWL. Finally, participants will be asked how they feel OWL could be used within their respective industries and/or companies (10 minutes).

The general aim is to gain an insight into the usability and application of OWL with the participants, as well as to publicise the new design resource, OWL. The feedback from users will be used to inform future iterations of OWL. Participants will be provided with their own 'OWL pack' to take away and use in their respective workplaces.

References

Williams E Y., Gibb A., Gyi D E., Haslam R. 2011, Constructive ageing: a survey of workers in the construction industry, *CIB W099, Washington, USA*, 24–28 August 2011.

Williams E Y., Gibb A., Gyi D E., Haslam R. 2012a, Building Healthy Construction Workers. In Anderson, M. (ed.) *Contemporary Ergonomics and Human Factors 2012; proceedings of the International Conference on Ergonomics & Human Factors 2012, Blackpool, UK, 16–19 April 2012*, London, Taylor & Francis pp. 69–70.

Williams E Y., Gyi D E., Haslam R., Gibb A. 2012b, Facilitating Good Ergonomics: workplace design and wellbeing, *International Society for Occupational Ergonomics and Safety (ISOE), XXIVth Annual International Occupational Ergonomics And Safety Conference 2012, Fort Lauderdale, Florida, USA*, 7–8 June 2012. pp. 127–133

SYSTEMS APPROACH

THE ROLE OF ERGONOMICS IN THE DESIGN OF FUTURE CITIES

Andree Woodcock

Coventry University, UK

Recruiting students onto ergonomics courses continues to be difficult despite clear business cases being presented for the importance of ergonomics as a driver in organisational effectiveness and safety and its crucial role in securing inclusivity and usability in product design. Growth requires concerted efforts to make our discipline relevant to contemporary society. One way of achieving this is presented, through linking ergonomics to citizen centred design of future cities.

Introduction

This paper stems from the assertion that collectively, and with adequate resources humankind has the ability to solve many of the wicked problems (Rittel, 1973) or grand challenges faced at local, national and international levels. Whilst research and development investment has addressed piecemeal solutions, insufficient attention has been given to the inclusive implementation of these. Ergonomics/human factors can address this gap by extending the principles of user centred and participatory design to the creation of inclusive cities and by ensuring that all citizens are included in changes (transitions).

Current global 'grand challenges' include ageing, sustainability, digital and economic divisions. Common across these is the need for substantial individual and collective behavioural change. Governments and international agencies create and agree international targets, which drive cycles of research and development. In these, a shared vision is created with the expectation that (frequently technological) solutions will be created which fulfill the requirements of, or lead to, the envisioned future. However, in reality, R and D funding streams favour short term solutions, are heavily technology led, have limited opportunities for fundamental research, user centred design, long term evaluation and do not produce integrated systems.

Using transport as an example, in 2007 the International Panel on Climate Change reported that an 80% cut in greenhouse gas emissions was needed from developed countries in order to reduce their impact on global warming. In order for the UK to achieve this target by 2050, the transport sector, currently responsible for a third of carbon dioxide emissions, will need to make significant reductions in transport emissions (RAE, 2010). This may be achieved through a combination of technological and behavioural solutions supported by policy making which changes the nature of mobility and the driving task. In this sector, automotive ergonomics is

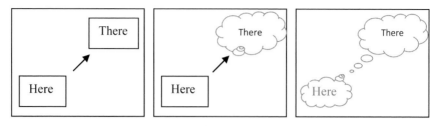

Figure 1.   Transitional societies (1a, 1b and 1c).

valued in terms of making vehicles safer, more comfortable, usable and inclusive and predicting the behavioural demands of autonomous vehicles etc.

Such solution led approaches may not achieve the level of changes demanded for a fundamental, step change. Ergonomics can appear to be limited (and unattractive to potential students) if it services technological innovation. Taking a user centred approach, which puts the citizen and their needs first (Maslow, 1943) can be used to drive the necessary societal innovation and restructuring. The unrecognised, crosscutting challenge is how can all sectors of society move from 'the here', to the shared vision of 'there'. Figure 1a represents the 'ideal path'. In Figure 1b, the shared vision of the future state becomes more complicated as it is recognised that proposed solutions only partly fulfil requirements and the notional future state is changed (in the light of evolving knowledge and understanding). Bridging the gap requires the development of inclusive implementation strategies which will ensure that citizens are able to transit in the direction needed, are supported in this and are not disadvantaged by their (in)actions.

Figure 1c, recognizes that the initial conceptualization of the present state may have been erroneous and that there is no plan to move from 'here' to 'there'. Ergonomics can provide a richer understanding of the here (i.e. through understanding of people's characteristics and motivations), primarily applied to the design of products and systems which contribute to the 'there'. Grand challenges require that societies (and their embodiment in cities), not just products, be redesigned. With its focus on UCD, (particularly macro and participatory ergonomics e.g. Haines et al, 2002) health and well being, user centred design principles and standards, ergonomics could make a significant contribution to the development of inclusive, transgenerational societies from the bottom up. As a discipline, the problem for us is whether we would be welcomed by those already engaged in re-engineering the future, where we can add value, and our willingness to work in cross disciplinary teams, e.g. with the design community through the Living Lab Network (http://www.openlivinglabs.eu/).

The transition gap to be addressed and its implications

The transition gap will always be problematic and as such non inclusive; not all people share the same goal, are willing or able to move at the same rate; solutions may not be incluive, speed of uptake of innovation is not constant, lack of planning means individuals are left to move forward on their own.

Using transport as an example, if the overall goal is to reduce carbon emissions, we may ask whether people believe the evidence on global warming, whether they see their own actions as contributing to it, and whether they believe that they should/could take direct actions to reduce their carbon emissions (e.g. Kellstedt, Zahran and Vedlitz, 2008; Leiserowitz, Maibach and Roser-Renouf, 2010). Those that are convinced, may require incentives and regulations before they change their behaviour, or they may simply be unable to make changes (due to cost). The design of obesogenic environments, long working hours and unsafe pedestrian routeways make it difficult for people to adopt more active lifestyles to reduce obesity, even when they are convinced of the need to do so. Moral dilemmas and conflicts are created when people believe that their behaviour may contribute to a problem, but they will not change their behaviour because of the personal costs incurred.

The poor design of the initial solutuons may be a barrier to behavioural change. Looking at the need to encourage independence in old age, there are significant issues in the degree to which user centred design approaches have been used to inform assistive technology products (Woodcock et al, 2013). For example, they might be designed to accommodate decline or disability in one function, but not associated ones; the context of use may not have considered, i.e. the home environment and personal circumstances of the end user; and there is a failure to consider higher level needs (Maslow, 1943). This means that potentially useful products which could increase independence and quality of life are not being adopted as quickly as was expected.

A further problem lies in designing for the early adopter. It may be argued that most technologies conform to the adoption pattern depicted by the Technology Adoption Lifecycle Model (described in Fenn, Raskino and Gammage, 2009; Rogers, 1983; Moore and Benbasat, 1991). Early adopters 'spread the message' to the early majority. This explains the importance given to successful early demonstrators (e.g. electric vehicles) in leading technology acceptance and converting people to adopters at a faster rate. Early adopters are key users, but they are atypical ones (e.g. Janson, 2011). They may have higher levels of education, income and higher status occupations than later adopters; they may be more technology savvy and willing to put up with inconvenience. In concentrating on this group, the late majority and the laggards may be disadvantaged and their characteristics ignored. Given that these groups account for 48% of the potential users, this is not an insignificant issue. Lantry (2012) recognizes the growing exclusion of underclasses in future 'creative cities'.

To summarize, it is clear that in transitional societies, some people will not share the vision, or want to be part of it and that some are excluded by poor design. In the conclusions of Rio + 20, it was stated that the rapid development of science and technology has brought unprecedented prosperity, but also challenges to societies, sometimes resulting in serious threats. According to Article 27 of the Universal

Declaration of Human Rights, "Everyone has the right … to participate and to share in scientific advancement and its benefits" UNESCO (2012).

The fulfillment of this article lies in inclusive design, but not simply at product level. A systems approach is need which allows all sectors of society to benefit from scientific advances, even during transitional periods. The lack of systems thinking is repeatedly exhibited in the move to a digital society, where people with, for example, poor levels of literacy and education, disabilities, and lacking technological know-how cannot access on-line services or benefit from them e.g. access to social benefits, on line shopping, on-line complaint forms.

Therefore a new avenue for ergonomics is proposed, specifically to address the problems faced by people and societies in making transitions, ensuring that through 'good design and human factors' products, services and cities are designed to be as inclusive as possible and that processes are embedded in the creation of future societies to ensure inclusivity. This requires bringing together areas such as transition management (e.g. Sterrenberg et al, 2012), socio technical design (e.g. Carayon, 2007), macroergonomics (e.g. Imada, 2002), user centred (e.g. Jokela et al, 2003) and persuasive design (e.g. Lockton, Harrison and Stanton, 2008).

The design of inclusive societies provides a clear route to realize the ambitions of ergonomics as expressed by the IEA's, Goal C *'to enhance the contribution of the ergonomics discipline to global society'*, through Objective C2 *'mobilise ergonomics profession to address major global challenges'* (http://www.iea.cc/ 04_project/Strategic%20Plan%20Goals%20and%20Objectives.html). With reference back to Figure 1, ergonomics as a discipline can be applied in the ways outlined below.

- Here: Ensuring user consultation and representation in design of infrastructure, services, products and buildings; keeping the citizen at the centre of the design, needs and requirements analysis
- Transition: ensuring inclusivity in change management, understanding skills and knowledge gaps, informing upskilling and mechanisms to support people making or effected by change, evaluation of product and sevice design
- There: Designing future cities: infrastructure, services, products and buildings, behaviour.

The systems approach proposed would design for each of the three states through long term planning, with a view to not only designing integrated future cities or communities which are fit for purpose and citizen based, taking into account latest technological developments, but which also encourage different behaviours.

The Hexagon Spindle Model (Woodcock, Benedyk and Woolner, 2009) developed to encourage a pupil centred approach to the design of classroom environments and products for children on the autistic spectrum, has been used here to show how a refocusing of attention of ergonomics away from the work environment could be used to set out the factors which need to be considered in the design of future cities. Significantly and usefully in this context is the recognition of factors which influence human system or product interaction (Figure 2). Additionally, and perhaps somewhat crudely in its current representation, the model acknowledges that all interactions, at any level are composed of a set of tasks over time, and that for an efficient and enjoyable experience each of these should be optimized. For example, a journey is composed of many elements, planning and preparation, travel to the transport gateway (e.g. bus stop), the vehicle assisted part of the journey, and reaching the final destination.

The next section illustrates how a systems approach is currently being used to create more inclusive cities and communities.

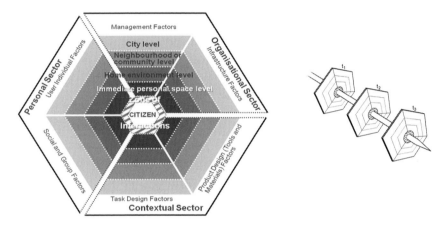

Figure 2.   Hexagon Spindle model.

Merocable in Medellin

The user centred approach undertaken by the developers of a new transport system for Medellin was reported in Atkinson et al (2012). Here a technological innovation was used to drive and act as a catalyst for societal and economic transformation in the poorest districts of Medellin. A participatory approach was adopted many years in advance of the build, which recognised the opportunities the Metrocable would provide to support behavioural change, community cohesion, confidence, education and health provision, economic growth (directly through jobs association with construction) and indirectly through the growth of small businesses around the metro station etc. In terms of the H-S model, the task design factors related to the need to increase and enhance the mobility of vast amounts of commuters into and out of the city in a comfortable, enjoyable and efficient manner. The product was the Metrocable system itself, the operation of which required the development of new infrastructure and management processes. User issues were considered in terms of the design of the service and the cable car. New ways of behaving towards each other and with regard to public transport were developed alongside the system, based on respectful, helpful behaviour. Each element of the journey was considered.

However, more work needs to be undertaken in terms of enabling people to get from their front door to the stations, and in improving disabled access with other stations.

Design of houses for adults with cognitive disabilities

Steele and Ahrentzen (2010) have adopted a user centred approach to the design of houses for adults with autism and other cognitive disabilities. An understanding of the need for predictability in everyday life has been matched with technological innovations (e.g. telemonitoring) and home design (e.g., motorised windows, adjustable sinks). The day has been broken down into stages, and the design of each task or environment where activities take place optimised Importantly this has been taken one stage further, into community design based around residents' needs such as ensuring safety and security, minimizing sensory overload, allowing opportunities for controlling social interaction, fostering health and well being ensuring accessibility and support in the surrounding neighbourhood and achieving affordability and durability. In designing a community which is safe and pleasurable for the most vulnerable in society, a pleasant environment has been created for other residents.

Focusing on eco-design, the global village network (http://www.gen-europe.org/) provides other examples of citizen based approaches to sustainable communities, where sustainable also includes economic and inclusive elements.

Conclusions

The speed of technological development is increasing and is led by the need to deal with grand challenges. An underlying need remains to ensure that changes, new societies and cities are made in as inclusive a way as possible which does not lose sight of the citizen, their needs and requirements. The case studies show that this is beginning to happen, but perhaps without involvement of ergonomists. There is still a clear role for ergonomics to transfer the knowledge, tools and techniques accrued in different environments to the design of people centred cities. In this way the discipline will be perceived as relevant to contemporary issues and leading in shaping the future.

The transition gap identified at the start of the paper, is starting to be considered e.g. in apps that use gaming elements to compare performance, or make a game out of knowledge transfer and in the consideration of the interfaces between gated communities and their surrounding environments. Transitional ergonomics could be used:

- To develop new knowledge about human characteristics and capabilities and context of use to inform the development of products and systems which lead to and support change
- To understand the barriers to societal change, at individual, group, organisational and societal levels and work towards removing these

- To develop strategies to help individuals, organisations (etc) to cope with the change process
- To understand the extent and effect of legacy systems and how their effects can be ameliorated
- To develop new business models which support societal change
- To provide effective user consultation processes for the development of new products, systems and societies.

References

Ahrentzen, S. and Steele, K. 2010, *Advancing Full Spectrum Housing: Designing for Adults with Autism Spectrum Disorders.* Tempe, AZ: Arizona Board of Regents. Retrieved 1st October 2012 from http://stardust.asu.edu/research_resources/detail.php?id=60

Atkinson, A., Atkinson, P., Osmond, J. and Woodcock, A. 2012, The impact of an integrated transport system; an ethnographic study of Medellin, Colombia; *1st International Conference on Human Factors in Transportation*, San Francisco, July 2012

Carayon, P. 2007, Human factors and ergonomics in health care and patient safety, *IX International Ergonomics Congress of SEMAC*; Mexico City, Mexico

Fenn, J., Raskino M. and Gammage, B. 2009, *Gartner's Hyper Cycle Special Report for 2009*. Retrieved 1st October 2012 from http://www.bnlcp.org.uk/sites/bnlcp/files/report/Gartners_hype_cycle_special_2009.pdf

Haines, H., Wilson, J.R., Vink, P. and Koningsveld, E. 2002, "Validating a framework for participatory ergonomics (the PEF)." *Ergonomics*, 45(4), pp. 309–327

Imada, A.S. 2002, "A macroergonomic approach to reducing work-related injuries." In: Hendrick H.W., Kleiner B.M. (eds) *Macroergonomics: Theory, Methods and Applications*. Lawrence Erlbaum Associates, pp. 151–172

Jokela, T., Iivari, N., Matero, J. and Karukka, M. 2003, The standard of user-centered design and the standard definition of usability: analyzing ISO 13407 against ISO 9241-11, Proceedings of the Latin American conference on Human-computer interaction, CLIHC'03, pp. 53–60.

Kellstedt, P.M., Zahran, S. and Vedlitz, A. 2008, "Personal Efficacy, the Information Environment, and Attitudes Toward Global Warming and Climate Change in the United States." *Risk Analysis*, 28, pp. 113–126

Leiserowitz, A., Maibach, E., and Roser-Renouf, C. 2010, *Climate change in the American Mind: Americans' global warming beliefs and attitudes in January 2010.* Retrieved 1st October 2012 from http://e360.yale.edu/images/digest/AmericansGlobalWarmingBeliefs2010.pdf Yale University and George Mason University. New Haven

Lockton, D., Harrison, D. and Stanton, N. 2008, Design with Intent: Persuasive Technology in a Wider Context, in H. Oinas-Kukkonen et al. (Eds.): *PERSUASIVE 2008*, Springer-Verlag Berlin Heidelberg, pp. 274–278

Maslow, A.H. 1943, "A theory of human motivation." *Psychological Review*, 50(4), 370–96

Moore G.C. and Benbasat, I. 1991, "Development of an instrument to measure the perception of adopting an information technology innovation" *Information Systems Research* 2, 192–223

RAE, 2010. *Electric Vehicles: charged with potential.* London: Royal Academy of Engineering, ISBN 1-903496-56-X

Rittel, H. 1973, "Dilemmas in a General Theory of Planning." *Policy Sciences*, pp. 155–169

Rogers E.M. 1983, *Diffusion of Innovations*, 3rd edition. Free Press, New York

UNESCO. 2012, The future we want must be an ethical one, Retrieved 1st Ocober 2012 from http://www.unesco.org/new/en/social-and-human-sciences/ themes/global-environmental-change/sv1/news/the_future_we_want_must_be_ an_ethical_one/

Woodcock, A., Ward, G., Unwin, G., Osmond, J., Fielden, S. and Ray, S. 2013, "Unlocking the potential of the younger older consumer." *Contemporary Ergonomics*, this volume

Woodcock, A., Woolner, A. and Benedyk, R. 2009. "Applying the Hexagon Spindle Model to the design of school environments for children with autistic spectrum disorders." *Work: A Journal of Prevention, Assessment and Rehabilitation* 32 (3), 249–260

THE NEED TO UNDERSTAND SYSTEMS OF SYSTEMS

C.E. Siemieniuch, M.A. Sinclair & M.J. de C. Henshaw

School of Electronic, Electrical & Systems Engineering
Loughborough University, Loughborough, U.K.

Typically, systems are constructed of subsystems, and are hierarchical in nature. Systems of systems (SoS) differ, in that they are composed of federated systems, each with an individual owner, and usually held together by a network of contracts. SoS can last longer than their component systems (and vice versa), leading to important issues of evolution, emergence, resilience and agility. Those that last longer do so because of their importance for society – energy, healthcare, policing and justice, food and agriculture are examples. This paper explores the importance of SoS in society, outlines gaps in our systems of systems knowledge of behaviour and engineering, and outlines potential systems ergonomics/human factors contributions to filling the gaps.

Introduction

Systems of Systems (SoS) have been a part of our society since prehistoric times, without recognition. Once human society moved from a hunter-gatherer lifestyle into agriculture and fixed abodes, systems of systems of some duration must have emerged, and are still around today. However, only in the last 50 years or so have their roles in society been recognised as social artefacts needing engineering and operation. This has been brought more into evidence as the benefits and associated difficulties and dependencies of information technology have pervaded our society, and have become the basic infrastructure underpinning the operations and behaviours of SoS.

SoS are now recognised as fundamental to our society, supplying many critical functions, such as education, energy, and healthcare. They are also fundamental to agriculture and the food supply chains that support us, as well as the industries that deliver the built environment around us, and the civilisation that we currently enjoy. In fact, so pervasive are they, and so immersed within their operations are we, that as the adage expresses it, 'we cannot see the wood for the trees.'. It could be argued that SoS become apparent only when there is a failure; we give some examples below. However, a brief commentary on the differences between systems (with which all ergonomists are familiar) and SoS (where there is less widespread, formal recognition) is beneficial.

Characteristics of SoS compared to systems

Typically, systems are constructed of hierarchies of subsystems. The collection of subsystems has a single hierarchy for control, in order to meet the goals of the system as a whole. By contrast, SoS differ in that they are composed of federated systems, each with an individual owner, and usually held together by a network of contracts. Occasionally, they may be hierarchical. SoS may last longer than their component systems (particularly for long-lasting SoS that are critical for society), leading to issues of evolution, emergence, resilience and agility.

There are shorter-lived SoS as well, that may be in existence for defined periods of a couple of years, or less. In these, the components tend to be established, and unlikely to change. For these, efficient operation of the SoS depends critically on management of the interfaces between the systems and organisations.

(Maier 1998) has characterized the fundamental characteristics of SoS well:

- The elements of the SoS are themselves sufficiently complex to be considered systems
- Operating together the SoS produces functions and fulfils purposes not produced or fulfilled by the systems alone
- The elements possess operational independence. Each element fulfils useful purposes whether or nor connected to the assemblage. If disconnected each system continues to fulfil useful purposes
- The elements possess managerial independence. Each element is managed mainly for its own purposes rather than the purposes of the collective.

The corollaries of this are:

- SoS architecture is more about the interfaces between the systems, not the systems themselves.
- The absence of a control hierarchy implies that co-operation and collaboration must be negotiated, and then maintained by some form of contract, or other agreement. This is part of the architecting process.
- Because each organisation may have intellectual property rights (IPRs) to protect, and may have entered confidentiality agreements with other organisations, there will be limits and delays to the information flows necessary for fully-efficient operations of the SoS
- The characteristics and corollaries above imply that the SoS will demonstrate unexpected, or emergent, behaviour; consequently, resilience and agility are critical attributes of the SoS, necessary for its long-term survival. Furthermore, the longer life expectancy of SoS also implies that the operational environment of the SoS will change, perhaps abruptly, again emphasising the importance of ensuring resilience and agility in the SoS.

Lurking in these characteristics and corollaries are some systems ergonomics issues; firstly, the need for negotiations and secondly the possibility, often probability, of emergent behaviour point to the need for first-, second-, and third-order learning over the lifetime of the SoS. The latter two are where humans predominate. Thirdly, the need for trustworthy, ethical behaviour of the component systems within the SoS (to compensate for limitations on information flows, among other aspects), allied to the demands for good governance (for legal and commercial reasons) again point to the involvement of humans within the SoS (Siemieniuch and Sinclair 2004; Henshaw, Siemieniuch et al. 2010). These are in addition to the usual arguments for the involvement of ergonomics in the design of individual systems. Finally, the more obvious implication of the characteristics is that resilience and agility are key issues, and ergonomics knowledge on organisational design, roles and jobs and the provision of resources will be very important in addressing these issues.

Examples of the need for resilience and agility

An example is Toyota, 1997 (Sheffi 2005). Aisin Seiki made P-valves for brake systems for Toyota and supplied 99% of all Toyota models. The main factory caught fire; 506 machines were destroyed. Toyota's alternate supplier making 1%, was unable to ramp up production. Toyota at this time was running at 115% of normal production, as a commercial response to impending legislation.

Toyota had only a few hours' stock of valves, with trucks on the road carrying another 2 day's capacity. Aisin salvaged some tools, replaced others and was in production in 2 weeks, making 10% of requirements, 60% after 6 weeks, and 100% after 2 months. Both Aisin and Toyota, in their respective *keiretsus*, asked for short-term help. 22 organisations from the Aisin *keiretsu* and 36 from the Toyota *keiretsu* replied. Within 5 days, Aisin had made available blueprints and process expertise and production had been allocated. Notably, Denso, a major Toyota supplier, outsourced their own production to free up tools and processes to produce these P-valves (as did others), and helped to develop alternative processes for the valves using different precision tools in other smaller suppliers. Within 2 days some valves were delivered by these alternate suppliers; within 9 days of the fire, all Toyota plants were functioning as normal again.

During this period, neither financial nor legal negotiation took place, nor was pressure applied to Aisin Seiki to prioritise Toyota over other customers. However, Aisin eventually covered the direct costs of the suppliers – labour, equipment, materials, and Toyota gave their Tier 1 suppliers 1% of their respective sales to Toyota for the January-March quarter in appreciation.

For contrast, consider the Nokia & Ericsson example (Sanchez-Garcia and Nunez-Zavala 2010); a fire in a chip fabrication plant in the USA in 2000 that serviced both Nokia and Ericsson caused a shortage of a vital component for a range of phones. Nokia reacted fast, creating a task force to source the world for replacements, using their buying power (in those days). Ericsson was slower to react, and more

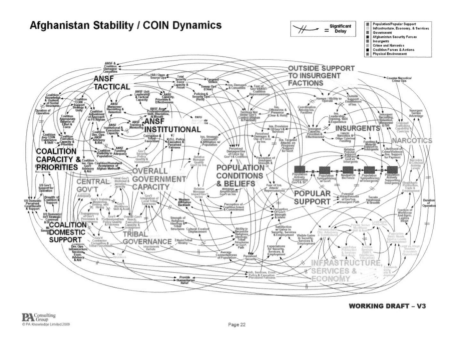

Figure 1.   An extreme example of a SoS; NATO ISAF in Afghanistan.

consensual in approach, and wound up starved of chips. The outcome was that Ericsson faced significant loss of market share, engineered a merger with Sony, and later decided to leave the mobile market completely.

There are many differences in these two examples; differences in approach, in the sharing knowledge, and in relationships. If the relationship is defined as a formal contract, then it becomes easier to define the boundary of a SoS; it is the network of contracts, even though this may include competitors as in the Nokia/Ericsson example. However, if we use a looser definition such as the cross-share-holdings model of a *keiretsu*, then the SoS is much larger.

A third example (Collins 2012) brings other lessons. In 2009, during Christmas, customers in a suburb of London lost their supply of gas and electricity. This was caused by a burst water main enabling water to enter an adjacent gas main, and blocking it. Because the repair would take time, the gas company distributed 3000 electric heaters and cookers to customers. The extra load on the electricity grid kept tripping the circuit breakers, leading to requests to ration their use, and a cold Christmas for many. Close proximity can create SoS relationships, too.

Finally, fig. 1 presents an extreme example of a SoS; no comment is necessary.

There are many lessons to be learnt about the design, implementation and operation of SoS over their lifetimes, and because of the deep, complex, long-lasting involvement of humans in them, there is a strong role for Ergonomics and Human Factors. We discuss some of the implications in the next section.

Capability gaps in addressing systems of systems

A recent study has explored the state of knowledge regarding SoS, and identifying some future needs (T-AREA-SoS; https://www.tareasos.eu). Some capability gaps that this study reports are listed below, with a brief ergonomics commentary. It will be noted that these gaps are related, and overlap.

- SoS Measurement. Just defining the boundaries is difficult, let alone measuring effectiveness or resilience. While there are a number of measures that can be applied to technological aspects of the SoS, there are few for the organisational aspects, particularly for organisational trade-off decisions.
- Modelling & Simulation for SoS. The complexities of many SoS preclude easy assimilation by humans; modelling and simulation is important to support this. We need extensions of modelling ontologies to cover the non-linearities of human involvement, and techniques for creating and managing distributed, heterogeneous, black-box models of organisations. An important issue is the evaluation of the dependability of such models.
- Verification & Validation for dynamic SoS. Especially for safety- and security-critical SoS (e.g. energy, food), V&V is important. Because of the continual evolution of SoS, V&V must be continuous and 'good enough' too, and should consider the dynamic problems of changing personnel, roles and even organisations within the SoS. This is particularly important when autonomous systems are included, given that they are impoverished with resect to ethical behaviour and responsibility (Asaro 2011; Asaro 2012).
- Prototyping of SoS. Necessarily, this will involve simulation, and at affordable cost. There is a trade-off between the level of detail, the cost, and the extent to which parts of the SoS can be left to the ingenuity and resilience to be expected of the humans occupying roles in the SoS.
- Prediction and analysis of emergent behaviour, both beneficial and maladroit. Emergent behaviour is very likely; SoS have all the trappings of complex adaptive systems, or 'wicked problems'; predicting the range, frequency and severity of this behaviour is not easy (Taleb 2008). Distributed situation awareness, strong communications, appropriate roles and other ergonomics skills, are relevant to this, but need elaboration for the SoS domain.
- Integration of corporate and engineering governance for enterprise SoS. Because of their size, importance and geographical reach across multiple jurisdictions, the governance of SoS is a clear issue. The implications for competition policy, the potential effects on huge numbers of people, and the capabilities of SoS to act at a distance (Asaro 2011), show that there is a need for responsible, ethical behaviour by the SoS as a whole as well as for the component systems. Necessarily, this will extend to regulatory bodies at national and regulatory co-operation at international levels.
- Dynamic composition of SoS. This concerns organisational architectures and sustainability of the SoS as it evolves, from sub-organisational levels to the SoS itself, and the maintenance of consistency between the levels as well. There are many lessons that could be extracted from ergonomics professionals working

in organisational design that could make a significant improvement to current knowledge, particularly for distributed situation awareness, authority over resources, response to mundane variation, etc.

- Agility through reconfiguration in dynamic SoS. Avoiding the argument about which subsumes which for agility and resilience, both of these attributes are delivered mainly by human action. For both of these, there is a need to understand the effects of constraints due to the contract network, the sources of complexity and emergence, and how these may be redistributed to alleviate their effect, thereby improving both of the attributes.

- Evolution and migration of legacy systems. This is a perennial issue, SoS or not. The problem is usually the loss or lack of documentation, loss of expertise, and lack of support for hardware. The net consequence of these can be the sterilization of the legacy system's adaptability; it can become a limited, monolithic, geographically constrained system performing only a limited set of functions, and kept operational by the power of prayer. Assimilation of these systems into the SoS requires special attention, and considerable human involvement at all times, both to regenerate the missing knowledge and to exercise resilience and agility when the inevitable system failure occurs. It indicates the need for knowledge conservation across the SoS; one or two organisations failing to do this may place the SoS at risk.

- Secure SoS implementations. This covers all levels of the network; humans, applications and physical levels (NCOIC 2011). It is still the case that networked control systems in important industries are easily attacked (e.g. Stuxnet), however secure the rest of the system is. Furthermore, separately from attacks, the lack of security internal to the network may mean that action at a distance can inadvertently occur (Asaro 2011; Asaro 2012), perhaps as a result of unintended human lbehaviour. It is the distributed control interaction of humans and technology which needs attention.

- Assuredly safe SoS implementations. We now certify systems and SoS for operational use. Given the evolutionary nature of SoS and the complexity that can characterize an SoS, this paradigm may need replacement by 'safety by continuous assurance'. This paradigm needs exploration.

- Economic resilience. This refers to the overall behaviour of the SoS. The problem is to design and then sustain the SoS in a changing environment. While there are technological aspects to this, the management issues are the most important, those of the disposition of roles, and assurance of resilience in this disposition, allied to the provision of sufficient, appropriate resources to enable resilience and agility to be demonstrated. This is partly understood for systems, but the complexities of this for SoS are not well understood; furthermore, there is a shortage of tools to carry out such activities; for example versions of system dynamics for SoS, trade-space analyses for organisations, and for distributed simulations. An important consideration is the potential size and global scope of SoS; they can stretch far beyond national boundaries, and within a national boundary could become dominant – 'too big to fail' – and act in anti-competitive ways. For this reason, it is important that regulatory bodies be involved; in the words of a senior manager in a major supermarket chain, 'to keep the bastards honest'.

Addressing the capability gaps

It is evident that the gaps listed above have large overlaps in content, and with cyclical relationships (for example, SoS measurement will require extensions to ontologies, but these ontologies will require some measurement of real-life SoS to provide the constructs and to validate them. This will require new tools with measures …). The prevalence of these relationships among the gaps indicates that a coherent, parallel, investigation is needed and that this investigation will require the participation of real-life SoS in a variety of domains to provide the necessary data to energise the knowledge development processes.

Consequently, we need to ensure that there is a community of practice involved in the development of understanding about SoS, with the strong participation of industry and the public sectors, and with representation from business leaders and legal, economics, industrial systems, politics and government, organisational design, and IT professions (and probably others). Furthermore, the engagement of practitioners as well as academics will be required.

A coherent programme of work by these groups will be needed as well. Perhaps the best model for this is the Framework Programmes of the European Union (which could also provide the scale of funding required for this coherent programme). Fortunately, some thinking about is happening in the European Commission already. Unfortunately, the obvious tie-up between the EPSRC and SERC funding bodies in the UK necessary to support a national effort may first require some clearance of the bureaucratic hurdles that currently exist.

Looking at the gaps, it is also clear that there is strong content of ergonomics topics, albeit perhaps not enough ergonomics knowledge within them covering the particular characteristics of SoS. It is important that the members of the IEHF are involved in these discussions; one of the real strength of the profession is the capability of its academics and practitioners to think sensibly and effectively across a wide range of disciplines. However, becoming involved in the planning of the research enterprise outlined above is essential, and we could perhaps look to the senior members of the Institute to become involved.

Finally, what of the skills, tools and expertise required of the ergonomics/human factors community? There is certainly an awareness gap; there are few papers in the literature about the issues above, and an education process is necessary, both in CPD and in academic programmes. The essential information-gleaning tools required do not need much change; we are after all looking at interface issues, and the majority of tools and techniques we use are appropriate. The big lack is in the melding of this information into larger schemes of understanding, and here we do fall short. Since this is an area of considerable interest and profit both to the IT&C profession and to management and cognate professions in general, we may be sure that tools will be developed, though not necessarily in ways appropriate to our interests. An example might be another EU project, COMPASS (http://www.compass-research.eu).

Conclusion

This paper outlines the importance of SoS in society, indicates gaps in our knowledge about them and their lifecycles, and points to the need for a big, coherent research effort to understand them better. That we need this understanding can be seen in the required transition to a sustainable, carbon-free energy supply. Consider the truly huge investments made over the past century or so to create the infrastructures and deliver energy for human use – electricity, gas, hydrocarbons, etc. – and the limited time in which to replace these networks, and it is likely that only by the efficient and effective utilisation of SoS that we will be able to make this transition. We need to develop our knowledge of SoS.

References

Asaro, P. M. 2011. "Remote-control crimes." *IEEE Robotics & Automation* 18(1): 68–71.

Asaro, P. M. 2012. "A body to kick, but no soul to damn: perspectives on robotics". In P. Lin, K. Abney and G. A. Bekey eds. *Robot ethics*, Cambridge, MA, MIT Press: 169–186.

Collins, B. 2012. *"Building resilience into complex systems"*. Presentation at Loughborough University, 18 May 2012.

Henshaw, M. J. d., C. E. Siemieniuch, et al. 2010. *"Aiding designers, operators and regulators to deal with legal and ethical considerations in the design and use of lethal autonomous systems"*. International Conference on Emerging Security Technolgies (ROBOSEC), Canterbury UK.

Maier, M. W. 1998. "Architecting principles for systems-of-systems." *Systems Engineering* 1(4): 267–284.

NCOIC 2011. NCOIC *"Interoperability Framework 2.1"*. Washington, DC, Network Centric Operations Industry Consortium.

Sanchez-Garcia, J. and M. Nunez-Zavala 2010. *"Nokia y Ericsson"*, IPADE, Universidad Panamericana, Aguascalientes, Mexico.

Sheffi, Y. 2005. *"The resilient enterprise: overcoming vulnerability for competitive advantage"*. Boston, MA, MIT Press.

Siemieniuch, C. E. and M. A. Sinclair 2004. "Process ownership and the long-term assurance of occupational safety: creating the foundations for a safety culture". *Contemporary Ergonomics 2004*. P. T. McCabe. CRC Press, Boca Raton, Florida, USA: 245–249.

Taleb, N. N. 2008. *"The black swan: the impact of the highly improbable"*. London, Penguin.

HIERARCHICAL SYSTEM DESCRIPTION (HSD) USING MODAF AND ISO 26800

Mike Tainsh

Lockheed Martin, Ampthill, Bedford, UK

MODAF provides a systems engineering framework for the representation of requirements based on a series of viewpoints. It provides the top level system description for UK MoD's Human Factors Integration Programmes (HFIPs). This paper reports how this approach can be supplemented using the concepts from ISO 26800 to provide a technique for documenting parts of the HFIPs design work. Central to this technique is the concept of integration/fitting which is seen in two parts: organisational and technical. The Hierarchical System Description (HSD) technique is found to support and benefit both parts by enabling traceability and transparency of design decisions based on human factors principles.

Introduction

MODAF

The UK MoD's Human Factors Integration (HFI) technical guidance is contained mainly within DEFSTAN 00-250. This starts with a consideration of UK MoD's Architectural Framework i.e. MODAF which states that all HFI activities should be seen within that systems architectural framework. MODAF provides a framework for handling the requirements that flow down into the human factors design and implementation.

There are five architectural viewpoints: strategic, operational, system, technical and acquisition. Human factors engineering activities (which are the parts of the HFIP of interest here) belong within the operational viewpoint. MODAF states: "The Operational Viewpoint describes the tasks and activities, operational elements and information exchanges required to conduct business and operational activities." The implications of the viewpoints for all the HFI domains have been described by the MoD HFIDTC (HFIDTC, 2008).

There are three MODAF viewpoints which are specific to human factors engineering although others may be relevant. The three are OV4 (Organisational Relationships Chart), OV5 (Operational Activity Model) and OV6 which addresses Operational Rules, State Transitions, and Event-Trace Descriptions. These provide a description of the top level requirements from which roles, tasks, equipment designs and other system characteristics will be derived.

ISO26800

ISO 26800:2011 outlines a general approach that can be taken by ergonomists working in a systems context, the principles that they may apply and the concepts that they may use. The principles, approach and concepts are all described at a high level. They refer to the broader aspects of working rather than the details of solving particular problems. The principles include fitting the User's characteristics to tasks and environment, including equipment design characteristics. The user, task, equipment and environment are seen as elements of a work system. However, the concept of fitting, as the concept of integration in ISO 15288, appears somewhat imprecise. Hence there is no way of knowing, from this source alone, how to achieve the fit between system elements, or when the fit or integration is achieved.

Aim

The aim of this paper is to show that by bringing together the comprehensive systems requirements approach from MODAF with the statement of ergonomics principles from ISO26800, a Hierarchical System Description (HSD) has been developed which supports system integration, design decision traceability and transparency, and implementation.

Hierarchical System Description (HSD)

Design decisions and integration

This paper presents a technique referred to as HSD for use within HFIPs, or similar programmes, as a means for supporting principled design decisions, lowering the risk of failures to integration, and hence implementation. The concept of integration is central to this technique.

The HSD approach employs a description technique that traces contributions from all the contributing systems engineering disciplines (including ergonomics/human factors) within the Project Team (PT), and includes the Users.

HSD is a management tool. It has been designed for use in ergonomics work such as HFI, or similar, programmes where it is necessary to demonstrate to all the PT's systems engineering professionals, users, managers, customers and others that:

(a) The hierarchy of processes that lead to a design decision being made can be traced back to the original statement of MODAF or other expression of requirements;
(b) The integration/matching processes associated with ergonomics design decisions can be demonstrated – they are principled designs in terms of ISO 26800;
(c) Various named groups have participated in the ergonomics activities – there is transparency of the participation and contributions of all stakeholders.

Hierarchy of descriptions

ISO 26800 starts from a statement of design principles. It states which ergonomics systems elements should be fitted to which others to create a well designed work system.

Following the guidance which can be inferred from ISO 26800, it is possible to propose that the HSD documentation should be constructed in six levels:

(a) Statement of Requirements and appropriate Principles – as described in MODAF and ISO 26800
(b) Elements of system as described in MODAF and developed into a recognisable (e.g. man-machine system) from the point of view of all stakeholders
(c) High level consideration of system design at the level of job characteristics
(d) Development of system design at the level of task characteristics
(e) Final level of detailed design addressing activities and the detailed specification of the equipment that supports them
(f) Empirical test specifications for acceptance trials. This will combine detailed specifications of ergonomics aspects.

These descriptions are proposed to enable design judgements to be made which fit the user characteristics, the equipment, the tasks and the environmental conditions to match the requirement for an integrated work system.

1. Description of Requirements and Principles

This may form part of a policy paper for a project. The description will comply with the requirements as expressed within MODAF or other expressions of requirements. The description will be divided into a series of Viewpoints which will include statements on how elements of the system should be connected. A system element may be such as hardware, software or Users.

2. Elements of the System

The elements of system will include User(s), tasks, equipment and environment, in accordance with the MODAF Glossary. ISO 26800 employs the word "components". In general these elements/components can be represented as shown in Diagram 1. This is likely to be a brief statement but it provides an essential link between a statement of requirements and a consideration of design descriptions.

3. High Level System Design

The initial level of design consideration is at the level of a job description. Here, initial consideration is given to "Work Groups" of users, and the coordinated sets of tasks and activities that may take several hours or more. The approach here is to create an initial description of the system that covers the whole of a job to be carried out by the users. This is a vital stage of the design, even though it is coarse grain, as variations on this design may become increasingly difficulty/costly later in the design process. The design can be represented in a set of five columns as in Table 1. This table is constructed for the case of an interactive system within

Environment

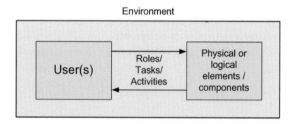

Diagram 1.   Representation of system elements (MODAF) or Work System ISO 26800.

a military vehicle (DEFSTAN 23-09). In this case there are two Users (1 and 2), working at generic interactive tasks (A and B) with 3 activities for each.

This type of management document make three points clear: firstly it is essential to involve the engineering Subject Matter Experts (SMEs) to describe the equipment and environmental characteristics; secondly, it is essential to involve User SMEs to describe the User characteristics and tasks, while there may be additional constraints such as performance or other requirements. The various groups of SMEs may make decisions on the basis of available evidence. The ergonomist will need to understand all these contributions and the requirement from ISO 26800 that the columns of the above table should match, or in ISO 15288 terms "be integrated".

4. Design and Development at the Level of Task Characteristics

This stage of the work takes Table 1 and develops it to a point where tasks of the order of minutes are described in terms of activities that take second.

5. Design and Development at the Level of Activity Characteristics

This takes the design down to sub-activities that may take only seconds or less. This is the smallest possible unit of time that is required but also carries out all the checks across the whole of the design to ensure that there is consistency in the application of design conventions and similar standards.

6. Test Validation and Acceptance Documentation

The test validation description will place all of the above in the context of detailed test specification which will ensure that integration is complete with all of the other work and other systems that comprise the overall product. At this point the principled designs are tested by empirical evidence prior to acceptance.

Integration of design information

The single most important opportunity that stems from a consideration of ISO 26800 is the development of a technique to achieve integration of the columns of principled design decision information within Table 1. ISO 26800 is imprecise in this area.

Table 1.   Hypothetical, example of High level System Description (HSD), using five sets of HFI characteristics, based on MODAF and ISO 26800.

User Characteristics	Equipment Characteristics	Characteristics of Job/Tasks	Performance Characteristics/ Criteria	Environmental Characteristics/ Requirements
User 1, Physical and Mental Characteristics (Set 1)	*[diagram: Design of Screen and Controls for User 1]*	1 Task A Activity A1 Activity A2 Activity A3	Level of Performance A Error Rate A Comfort Rating A /Ease of Reach A	Temperature/ Humidity Spatial Arrangement Noise/ Vibration
User 2, Physical and Mental Characteristics (Set 2)	*[diagram: Design of Screen and Controls for User 2]*	2 Task B Activity B1 Activity B2 Activity B3	Level of Performance B Error Rate B Comfort Rating B/ Ease of Reach B	Temperature/ Humidity Spatial Arrangement Noise/ Vibration

The need for a detailed consideration of how design information is integrated is vital as various engineering groups of SMEs will be contributing to different parts of the Table.

There are at least two aspects to integration:

(a) Consideration of the technical content of the HSD table, where design options have to be represented prior to consideration, and the impact of one set of design considerations on another must be assessed and then amended until the impacts are acceptable.

(b) Integration can be seen as the product of the resolution of conflicting technical positions between the various professional, User and management groups within the PT. These differences need to be resolved through evidence-based discussion, including trade off studies and the assessment of design options by the SMEs.

Table 2.   The outcome of row characteristics impacting on column characteristics.

	User Characteristics	Equipment Design Features	Task Characteristics	Performance Criteria	Environmental Requirements
User Characteristics		Change design	Change task	Change criteria	Change requirements
Equipment Design features	Change user/ train/select		Change task	Change criteria	Change requirements
Task characteristics	Change user/ train/select	Change design		Change criteria	Change requirements
Performance Criteria	Change user/ train/select	Change design	Change task		Change requirements
Environmental Requirements	Change user/ train/select	Change design	Change task	Change criteria	

Resolving integration issues

Firstly, the technical issues require consideration: these are the impacts of one set of proposed system characteristics upon another. As an example, it is assumed that five sets of vehicle characteristics as shown in Table 1 are under consideration.

There are two sets of impacts for each combination of five characteristics. These can be represented as in Table 2 where the complete set of comparisons and their outcomes are shown.

In the case of Table 2, it is the impact of the items in the rows upon the columns that is entered into the cell of the table. For example, when considering row one and column two the impact of the "User on the Design" is assessed. Later when considering row two and column one the impact of the "Design on the User" is assessed. In the first case the PT might consider changing of the design while in the second case, one might change the User (by selection, training or other means).

Secondly those who have responsibility for the design of the elements within the system will need to look at the outcome of Table 2 and check for conflict of expert judgments or consensus.

It is rare in large organisations for a single group to have total responsibility for such a table. It is common practice that different professional groups have responsibility for manpower and selection, training, design, health and safety. Where both groups conclude there are conflicts, there is a need to resolve them. Where both groups see consensus there may be an opportunity to improve beyond system requirements at that stage.

Consequences and benefits of using HSD

Current experience of applying HSD

Lockheed Martin is involved in two major UK MoD equipment acquisition programmes. These are for WARRIOR armoured vehicle, and SV SCOUT.

The current experience of this technique is derived entirely with these two major vehicle development programmes. HSD has helped in explaining to Users how requirements, Use cases, task descriptions and systems engineering design can be integrated to the benefit of the Users. HSD has also enabled contributions from various systems engineering disciplines to understand how their contribution impacts on User tasks and performance levels.

Design opportunities, and risk

HSD enables the discussion within project teams to move from conflict to the consideration of opportunities. By placing design discussions and conflicts in a hierarchical context where it is simultaneously possible to consider other systems engineering, human factors/ergonomics, and User considerations, there is increased scope for considering design opportunities.

HSD emphasises the contributions of the various engineering disciplines as well as that of human factors/engineering. The discussions on WARRIOR and SCOUT have aimed to encourage opportunities within the project for proposing improvements in design and performance, rather than simply meeting requirements via a risk free process.

Acknowledgements

My sincere thanks go to all my system engineering and human factors/ergonomics colleagues at Lockheed Martin, Ampthill who have supported and encouraged me when writing this paper.

References

HFI DTC, Systems Engineering and Assessment (2008), *The Human View Handbook for MODAF*, MOD, UK.

ISO 15288: 2001, *Information Technology – Life Cycle Management – System Life Cycle Processes*.

ISO 26800:2011, *Ergonomics – General Approach, Principles and Concepts.* DEFSTAN 23-09 (2010), *Generic Vehicle Architecture*, MoD, UK.

DEFSTAN 00-250 (Parts 0, 1, 2, 3, 4), (2008) *Human Factors for Designers of Systems*, MoD, UK.

MODAF Architectural Framework, (2004–8), MoD, UK.

CROWD BEHAVIOUR

HAZARD EXPERIENCE AND RISK PERCEPTION AMONG SPECIAL CONSTABLES

James I. Morgan

Psychology Research Group, Sheffield Hallam University, UK

Despite injury rates that are more than double that of other industries, to date, no research has examined risk perception in Police Service personnel. Equivocal findings across other high-risk industries suggest that, for frontline workers, risk perception is not homogenous. Risk appraisals of those closest to hazards are dependent on domain- or role-specific cognitive, social, and cultural factors. In turn, these are shaped by situational features specific to a particular work environment. The present study examined the magnitude and correlates of risk perception among Special Constables (SCs). SCs reported high levels of risk perception, associated with greater hazard exposure. Theoretical and practical implications are discussed.

Introduction

Detailed police accident and injury statistics are not readily available. However, the UK Health and Safety Executive (HSE) have published summary data for injuries to Police Service employees for the five years preceding, and including, 2009/10 (HSE, 2011). These data appear to show that members of the Police Service are at considerably greater risk of injury compared to workers from other industries. For instance, in 2008/09 the overall RIDDOR injury rate per 100,000 for Police Service employees was 1224.8 compared to an all-other-industry average of 502.2. The data also appears to show that Police Service employees experience qualitatively different risks to those faced by other safety-critical workers. Whilst slips, trips and falls are often cited as the most common cause of injury, even in high-risk domains (see Morgan, Jones, & Harris, 2013), it seems that, for the police, the most frequent injuries are not self-inflicted. According to HSE statistics the majority of injuries to police employees are the result of acts of violence against them, including physical assault, or are sustained during the handling of offenders (during arrest, custody processing, or cell transfer).

Given the unique and unpredictable nature of police operations, and the seemingly high injury risk, it is perhaps surprising that, to date, no studies have assessed risk perception among police personnel. It is intuitive to assume that frontline workers in high-risk environments hold perceptions that are in line with the 'objective risk'. To the contrary, findings from research conducted in various high-risk industries suggest that worker reactions to risk are not homogenous. It appears that there are both inter- and intra-organisational differences in the magnitude of risk perceived. For example, Ostberg (1980) found that across six groups of forestry personnel those workers on the frontline, closest to the hazard (e.g., supervisory staff and workmen), underestimated risk while members of more distal groups (e.g., senior managers and safety officers) tended to overestimate risk. In contrast, when Weyman and Clarke (2003) replicated Ostberg's (1980) study, they found a reverse pattern of results in groups of deep coal mine workers, with frontline workers perceiving more risk than senior managers. One explanation for these mixed findings is that levels of objective risk differ across organisations and job-roles. Objective risk cannot be easily determined, so this account is difficult to refute. In any case, it fails to explain differences in frontline worker's risk appraisal across high-risk industries, where risk for frontline workers should be comparably high. An alternative explanation is that perceptions of risk are dependent on domain- or role-specific cognitive, social, and cultural factors (Weyman & Clarke, 2003). In turn, these factors are shaped by situational features of the work environment, including production pressure (Hopkins, 1999; Ostberg, 1980), hazard knowledge and utility of controls (Hopkins, 1999; Rundmo, 1992), and exposure to hazards, and their consequences (Rundmo, 1992; 1996).

The present paper presents the preliminary findings from research exploring the potential correlates of risk perception in frontline Police Service personnel. Of initial interest was whether subjective risk appraisals of a specific group of frontline officers, namely Special Constables, are in line with the apparent high 'objective risk' (according to HSE injury rates). Also of interest was whether job experience (tenure) and hazard experience (risk exposure and previous injury frequency) were associated with these risk perceptions.

Method

Design

The study adopted a correlational design. The variables included were tenure, (serious and periodic) risk exposure, previous injury frequency, and risk perception.

Participants

A total of 49 Special Constables (SCs) volunteered via a Police Service contact to participate in the study. Special Constables are part-time unpaid volunteers who work alongside regular Police Constables (PCs). They carry out the same duties, wear the same uniform, are based at the same police stations, and have the same powers in law, including the power of arrest. Duties include neighbourhood policing and patrols, tackling anti-social behaviour, policing large events, and dealing with crimes such as burglaries, assaults, and criminal damage. The sample comprised 22 females and 27 males, aged between 19 and 52 years (M = 29 years, SD = 8.09 years). Tenure ranged from 3 months to 16 years (M = 3.27 years, SD = 3.08 years).

Materials

All measures were contained in a single questionnaire. At the beginning of the questionnaire SCs were asked to provide demographic information including their age, gender, and tenure.

The level of risk exposure at work was determined using two one-item sub-scales. The first of these measured serious risk exposure using an open-ended question, "how many times have you been in a situation at work where your life has been at risk?" The second sub-scale, periodic risk exposure, was assessed using the item, "on a monthly average how many times would you say you have been in a risky situation?" The 6-point response scale for this item ranged from Never (1) to Five or more times a month (6). The correlation between the two sub-scales was weak and non-significant ($rs = .22$), and as such they were not combined into a composite measure of risk exposure.

Previous injury frequency was measured using the open-ended question, "how many times have you been injured at work?"

Estimates of personal risk at work were established by assessing participants' level of agreement with the statement, "my job role means putting myself at risk." Responses were made using a 5-point scale ranging from Strongly Disagree (1) to Strongly Agree (5).

Procedure

Having consented to take part, SCs completed a paper-based questionnaire booklet, containing all of the measures described above, during a rest period at their police station. At the end of the questionnaire participants were thanked for their involvement and were debriefed.

Results

Data analysis

Data were not normally distributed. Accordingly, nonparametric Spearman's rank order correlation coefficient (Spearman's rho) analysis was performed to determine the strength and direction of variable associations. Table 1 shows all intercorrelations and descriptive statistics (median scores [*MDs*] and interquartile ranges [*IQRs*]).

Descriptive statistics

As can be seen in Table 1, the median score for injury frequency was zero. Inspection of the raw data revealed that very few Special Constables had been injured at work (11 of the 49 participants). However, because tenure was also low ($MD = 2$, $IQR = 4$), and some officers were injured on more than one occasion (up to 10 times), the injury rate cannot be accurately described as nil. The medium to strong positive correlation between tenure and injury frequency ($rs[49] = .34$, $p = .02$) indicates that there is a greater risk of injury for those that stay in the job for a longer period. Risk perception scores (measured on a 5-point scale) imply that Special

Table 1.   Intercorrelations and descriptive statistics (MDs and IQRs on the diagonal. *p < 0.05, **p < 0.01).

	1	2	3	4	5
1. Tenure	2.00 (4.00)				
2. Serious Risk Exposure	.48**	1.00 (3.50)			
3. Periodic Risk Exposure	.08	.22	2.00 (1.00)		
4. Previous Injury Frequency	.34*	.39**	.26	0.00 (0.00)	
5. Personal Risk Perception	.10	.11	.37**	.12	4.00 (1.00)

Constables recognise personal risk ($MD = 4$, $IQR = 1$) in their work. Despite this perception of risk, both periodic risk exposure ($MD = 2$, $IQR = 1$) and serious risk exposure ($MD = 1$, $IQR = 3.5$) appeared low.

Correlates of risk perception

Risk perception was positively correlated with periodic risk exposure ($rs[49] = .37$, $p = .01$) suggesting that a greater exposure to risk (per month), the more personal risk perceived. Risk perception was not correlated with tenure ($rs[49] = .10$, $p = .48$), serious risk exposure ($rs[49] = .11$, $p = .47$), or previous injury frequency ($rs[49] = .12$, $p = .42$).

Discussion

The reported study represents an initial exploratory investigation of frontline police officer's perceptions of risk at work, and the potential correlates. The results show that, in line with Weyman and Clarke (2003), those closest to the hazard, in this case Special Constables, perceive high risk. Although self-reported exposure to hazards and their consequences (in the form of previous injury rates) appeared low, the correlation between periodic risk exposure and risk perception implies that hazard familiarity may play a role in the perception of high risk. This finding is in contrast to Ostberg's (1980), who found that frontline forestry workers underestimated risk, despite their close proximity to the hazards in question (those involved in the production process). A similar familiarity effect has been observed in other domains. In aviation research, for example, studies have found that pilots with more hazard experience of flying into difficult meteorological conditions perceive less risk in doing so (e.g., Pauley, O'Hare, Mullen, and Wiggins, 2008). More broadly, sociocultural research has revealed an attenuative effect of residential proximity on risk judgments about high hazard installations (e.g., Walker, Simmons, Wynne, and Urwin, 1998).

The apparent equivocal influence of hazard proximity, or familiarity, on risk perception across contexts reflects the heterogeneity of the construct, and lends support to the proposition that it is cognitively, socially, and culturally defined. While it may be difficult to ascertain the specific cultural or social influences across high-risk work environments it may be possible to explain the underestimation and overestimation of risk in terms of cognitive bias. While underestimations can be attributed to an elevated notion of personal control and optimism (Linville, Fischer, and Fischhoff, 1993; McKenna, 1993; Weinstein, 1980, 1982), high risk estimates, such as those reported in the present study, can be interpreted in terms of cognitive availability in heuristic processing (see Weyman and Clarke, 2003). The premise of the availability and simulation heuristic is that it is the ease in which risk outcomes come to mind, that guides the appraisal of that risk. If risk information is readily available (e.g., via previous hazard experience), and mental simulations are lucid, the outcome in question becomes more subjectively probable (Kahneman and Tversky, 1982). Evidence from research on the hypothetical risk judgments of rail engineers supports this explanation. Morgan, Jones, and Harris (2013) found that workers who created better simulations while reading a risk scenario, made safer decisions. While risk perception was not measured directly it is not unreasonable to suggest that 'goodness of simulation' heightened the subjective probability of loss associated with the risky choice, and thus, made the safe option more attractive.

Conclusion

The results of the present study add to a history of mixed findings concerning the degree of risk perceived by operatives working close to hazards in safety-critical environments. These equivocal findings have a number of implications for those responsible for safety assurance. Firstly, it should be recognised that the way in which risk is processed and negotiated is dependent on a complex interplay of situational and individual factors. The intuitive assumption that hazard exposure is sufficient to generate accurate risk appraisals should be avoided, as should the temptation to attribute accidents or injuries to ignorance, lack of insight, or appreciation of risk, a common managerial view (see DeJoy, 1985).

In the present study, and in the majority of previous research on risk estimation, it was not possible to compare reported risk estimates with an accurate measure of objective risk, derived from detailed injury statistics for the population in question. However, Special Constables rated their risk as high, which is in line with the general Police Service injury statistics published by the HSE. While the perception of high risk was associated with hazard exposure, and specifically attributable to availability bias, the potential contribution of other variables should not be overlooked. In order to pinpoint the other predictors of risk perception in Special Constables in future research it may be beneficial to explore the reasons for why risk is not underestimated in this population, when it is in others (e.g., Forestry workers, Pilots, etc.). It is suggested that the influence of operative disposition should not be discounted (see Rundmo and Sjoberg, 1998; Zajonc, 1980), as well as worker's perceptions of control (see Harris, 1996; Klein and Helweg-Larsen, 2001; Weinstein and Klein, 1996). These factors may be of particularly importance for police officers who face unpredictable, and sometimes, uncontrollable work situations.

Given the high levels of risk perceived by Special Constables in the present study and the demanding nature of police work in general, it is recommended that the psychological well-being of these operatives should also be considered. In a recent study of Norwegian offshore petroleum workers, Nielsen, Mearns, Matthiesen, and Eid (2011) found that greater risk perception was associated with lower levels of job satisfaction. However, the researchers found that this effect diminished when workers perceived their safety climate as positive. These findings suggest that if high levels of risk perception add to the demand of police work, confidence in the effectiveness of safety procedures may buffer any negative effects.

Although the present study provides sufficient foundation for further work investigating the potential predictors or consequences of risk perception in Police Service personnel, there is a noteworthy caveat. It is possible that because single-item self-report scales were utilised, the reported correlation between periodic risk exposure and risk perception was a consequence of common-method variance. Anecdotal evidence gathered via verbal exchanges with Special Constables suggests that this is not the case. However, in order to help overcome this potential problem in future work, it may be beneficial to consider alternatives to self-report methods of measuring risk perception (e.g., see Morgan & Garthwaite, 2012).

Acknowledgements

The author would like to thank Special Sergeant Trudie Gray for her role in data collection, and all of the Special Constables who kindly gave up their time to take part.

References

DeJoy, D. M. 1985, Attributional processes in hazard control management in industry, *Journal of Safety Research, 16*, 61–71.

Harris, P. 1996, Sufficient grounds for optimism? The relationship between perceived controllability and optimistic bias, *Journal of Social and Clinical Psychology, 15*, 9–52.

HSE 2011. http://www.hse.gov.uk/services/police/statistics.htm (October 2012, date last accessed).

Hopkins, A. 1999, *Managing major hazards – The lessons of the Moura Mine disaster* (London: Allen & Unwin).

Kahneman, D. and Tversky, A. 1982, The simulation heuristic. In D. Kahneman, P. Slovic, and A. Tversky (eds.), *Judgement under uncertainty: Heuristics and Biases* (New York: Cambridge University Press), 201–208.

Klein, C. and Helweg-Larsen, M. 2002, Perceived control and the optimistic bias: A meta-analytic review, *Psychology and Health, 17*, 437–446.

Linville, P. W., Fischer, G. W., and Fischhoff, B. 1993, Aids risk perceptions and decision biases. In J. B. Prior and G. D. Reeder (eds.), *The social psychology of risk perception* (Hillsdale, NJ: Erlbaum), 5–38.

McKenna, F. P. 1993, It won't happen to me: Unrealistic optimism of illusion of control? *British Journal of Psychology, 84*, 39–50.

Morgan, J.I. and Garthwaite, J. 2012, Differentiating the effects of negative state on optimism and the implicit perception of everyday injury risk. In M. Anderson (ed.) *Contemporary Ergonomics and Human Factors 2012*, (Taylor & Francis, London), 181–188.

Morgan, J.I., Jones, F.A., and Harris, P.R. 2013, Direct and indirect effects of mood on risk decision making in safety-critical workers, *Accident Analysis and Prevention, 50*, 472–482.

Nielsen, M.B., Mearns, K., Matthiesen, S.B., and Eid, J. 2011, Using the Job Demands–Resources model to investigate risk perception, safety climate and job satisfaction in safety critical organizations, *Scandinavian Journal of Psychology, 52*, 465–475.

Ostberg, O. 1980, Risk perception and work behaviour in forestry – Implications for accident prevention policy, *Accident Analysis and Prevention, 12*, 189–200.

Pauley, K. A., O'Hare, D., Mullen, N. W., and Wiggins, M. 2008, Implicit perceptions of risk and anxiety and pilot involvement in hazardous events, *Human Factors*, 50(5), 723–733.

Rundmo, T. 1992, Risk perception and safety on offshore petroleum platform – Part 1: Perception of risk, *Safety Science, 15*, 39–52.

Rundmo, T. 1996, Associations between risk perception and safety, *Safety Science*, 24(3), 197–209.

Rundmo, T., and Sjoberg, L. 1998, Risk perception by offshore oil personnel during bad weather conditions, *Risk Analysis, 18*, 111–118.

Spielberger, C. D. 1983, *Manual for the State-Trait Anxiety Inventory* (Consulting Psychologists Press, Palo Alto).

Walker, G., Simmons, P., Wynne, B., and Urwin, A. 1998, *Public perception of risks associated with major accident hazards* (Contract Research Rep. No. 194). Sheffield, United Kingdom: Health and Safety Executive, HSE Books.

Weinstein, N. D. 1980, Unrealistic optimism about future life events, *Journal of Personality and Social Psychology*, 39(5), 806–820.

Weinstein, N.D. 1982, Unrealistic optimism about the susceptibility to health problems, *Journal of Behavioral Medicine, 5*, 441–460.

Weinstein, N.D. and Klein, W.M. 1996, Unrealistic optimism: present and future. *Journal of Social and Clinical Psychology, 15*, 1–8.

Weyman, A.K., Clarke, D.D. 2003, Investigating the influence of organisational role on perception of risk in deep coal mines, *Journal of Applied Psychology 88(3)*, 404–412.

Zajonc, D. 1980, Feeling and thinking: Preferences need no inferences, *American Psychologist, 35*, 151–175.

CROWD SATISFACTION AT SPORTING EVENTS

Victoria L. Kendrick, Roger A. Haslam &
Patrick E. Waterson

Loughborough Design School, Loughborough University, UK

This paper presents findings from in depth semi-structured interviews with event organisers and deliverers, investigating the organisation, coordination, and security of a variety of spectator sporting events. Safety was identified by those responsible for organising and delivering events as a key priority, with less attention given to user experience, crowd comfort and satisfaction. An evidence based description was developed to embody findings of the current study, illustrating the central issues that influence crowd satisfaction within spectator sporting events: anticipation, facilities, and planning (prior to the event); influences and monitoring (carried out during the event); and reflection (engaged in after the event).

Introduction

Despite the popularity of spectator sporting events, academic research examining how the organisation of crowd events can be enhanced remains relatively under-developed (Zhang et al., 2007; Johnson, 2008). Research addressing crowd events has largely focused on crowd safety (Zhen et al., 2008), pedestrian flow modelling (Smith et al., 2009), and event management over recent years (Getz, 2008): with substantial research around public order policing (Reicher et al., 2004; Drury & Stott, 2011), and hooliganism prevention (Stott et al., 2008). Consideration of the wellbeing of the crowd, particularly crowd satisfaction, comfort and performance has received less attention (Ryan et al., 2010; Machleit et al., 2000; Berlonghi, 1995). Moreover, the extent to which academic research findings influence the organisation of spectator sporting events is unclear.

Background

Prior research by the authors explored the user experience of crowds through focus groups, revealing differences in the factors affecting crowd satisfaction, with age and expectations (Kendrick & Haslam, 2010). However, venue design, event organisation, and safety and security concerns were found to highly affect crowd satisfaction, irrespective of group differences or crowd environments, showing the importance of these issues for all crowd events, for all crowd members. In light of the findings from the crowd participant focus groups, the current event organiser and event delivery interview study was undertaken, to explore the extent to which organiser actions meet the needs of the user.

This paper presents a subset of findings that form part of a larger study that used ethnography with spectator events of various descriptions, to explore the user experience of crowds. This included a case study of special events within a large UK university (Kendrick et al., 2012). The study presented here focused on the organisation, coordination, and security of a variety of spectator sporting events. The aim was to develop an evidence based description of important factors contributing towards crowd satisfaction.

Method

Semi-structured interviews were used to investigate the organisation of crowd events, including: approaches and processes used in the planning for crowd situations; attitudes and beliefs regarding crowd satisfaction, comfort, safety, and performance; and commitment to each (Robson, 1993). Interviewees were drawn from relevant stakeholder groups to achieve a structured convenience sample (Bryman, 2004a).

Sporting events encompassed the following crowd types: ambulatory (walking); spectator (watching an activity or event); expressive (emotional release, shouting, chanting); and limited movement (restricted movement) (Berlonghi, 1995).

A standardised interview question set was developed, with the same facilitator leading each interview (approximately 90 minutes each). Interview recordings were subsequently transcribed verbatim, and imported into the qualitative software tool, NVivo (Version 9.0) to enable systematic analysis (Hignett & Wilson, 2004).

Development of qualitative analysis involved hybrid thematic analysis of interview data, with data driven codes developed, and the identification of emergent overarching themes (Bryman, 2004b). Reliability was enhanced through the systematic review of the data by two independent researchers.

The study complied with ethical requirements of Loughborough University ethics committee: all interviewees were provided with information about the study and informed consent was obtained.

Results and discussion

Eighteen in depth stakeholder interviews were conducted (16 males; 2 females), comprising event organisers; health and safety officers; public and private security officers; and ground stewards. A variety of spectator sporting events were captured (indoor and outdoor), including various: football, rugby, handball, ice-hockey, and athletics events.

Eleven common themes emerged from the data:

- Health and safety,
- Public order,

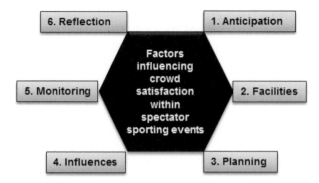

Figure 1.   Factors influencing crowd satisfaction within spectator sporting events.

- Communication,
- Physical environment,
- Public relations,
- Crowd movement,
- Event capacity,
- Facilities,
- Satisfaction,
- Comfort,
- Crowd characteristics.

Safety was seen to be a high priority, due primarily to legal obligations and a desire to protect venue reputation. However, the comfort and satisfaction of the crowd participants often received less attention, with budget considerations cited as a key reason. Additionally, inadequate communication and management systems were in place to ensure compliance with internal procedures. Interviewees highlighted a lack of usable guidance available to assist with the organisation of special events.

Findings from the interviews are summarised in Figure 1, which illustrates six central issues suggested as influencing crowd comfort, performance, safety, and satisfaction with the organisation and delivery of spectator sporting events: anticipation, facilities, and planning (prior to the event); influences and monitoring (carried out during the event); and reflection (engaged after the event). These are explained further below.

Anticipation, facilities, and planning (prior to the event)

These are aspects which take place in advance of the event. Anticipation of the target audience, as well as communication and sharing of knowledge and experience within and between events was limited. There appeared to be a lack of information available to organisers involved in relatively small scale sporting events in particular. Moreover, findings highlight the importance of tailoring crowd planning guidance to different crowd situations, supporting previous research (Berlonghi, 1995; Lee & Hughes, 2007; Ryan et al., 2010).

Important facility provision was not always well linked to individual event needs, for example in relation to car parking:

> *"Tescos over the road ... supporters also park there ... next to the stadium they have a specific number of supporters allocated a specific parking bay number, and they must park there. The problem is that the side streets get clogged with traffic parked up. So there is not a good relationship with the local residents." (Security officer)*

Acceptance of ad hoc arrangements such as this suggests a lack of appreciation of user needs as a valid problem requiring attention. Such findings are in line with the underdeveloped literature in this area, with a limited evidence base of knowledge, and usable guidance for planning crowd events (Berlonghi, 1995; Ryan et al., 2010).

Also planning and attention to crowd user comfort, performance and satisfaction, were often based on *"personal judgment"* (*Event Organiser*); and influenced by budget considerations, indicating that financial considerations often take precedence over user comfort and satisfaction.

Event influences and monitoring (carried out during the event)

These relate to the event itself, and the need to adapt rapidly to changing circumstances. Influences including extreme weather were a major concern for those organising and delivering crowd events with, for example, one police officer describing:

> *"If there was a severe weather problem like ice and snow ... then the police would look at it with people from the football ground, and the referee in terms of whether it's safe to play the match. But in terms of the people getting to and from ... then the football ground and the police would look at that." (Police Sergeant)*

Monitoring capacity during events was an issue stressed during interviews, as highlighted during an interview with an event organiser involved in rugby and athletics events:

> *"And there will normally be at least two staff on the exits ... either to open it up massively if we've got to get people out, which in that kind of venue is very low risk, because it's obviously outside. I mean it's literally a field. I mean you could pull people into the rugby pitch if you really needed to get people away from a specific area." (Event Organiser)*

Reflection (after the event)

Reflection concerns issues that should be resolved following the event. The importance of gaining feedback from all crowd users for example (including all staff working at the event), and implementing the information into future events was recognised throughout interviews. Although post-event learning was recognised as important, this was not always undertaken in practice. One event organiser involved in football and athletics events explained:

> *"We do, I mean ... You see the problem is that I just don't have the time ... we got some good feedback on what they thought ... but without spending a long time correlating certain aspects ... there was only sort of bits we could pull out ... But yeah ... it's something that we need to do a lot better." (Event Organiser)*

The importance of gathering and utilising feedback after events is reflected in research on crowds, although this is predominant in the context of crowd safety rather than other aspects (Lee & Hughes, 2007).

Conclusion

This interview study has identified important factors (anticipation, facilities, planning, influences, monitoring, and reflection) that are important to crowd satisfaction within spectator sporting events. Achieving a positive, high-quality crowd experience is desirable for overall event success, and of benefit to all stakeholders. These findings have fed into further work by the authors, with the goal of developing practical guidance (in the form of a tool for event organiser to use during the planning of crowd events), to aid organisation and enhance the user experience of crowds.

References

Berlonghi, A. E. (1995). Understanding and planning for different spectator crowds. *Safety Science, 18*(4), 239–247.

Bryman, A. (2004a). *Social Research Methods.* Oxford: Oxford University Press. pp. 183–4.

Bryman, A. (2004b). *Social Research Methods.* Oxford: Oxford University Press. pp. 554–5.

Drury, J., & Stott, C. (2011). Contextualising the crowd in contemporary social science. *Contemporary Social Science, 6*(3), 275–288.

Getz, D. (2008). Event tourism: Definition, evolution, and research. *Tourism Management, 29*(3), 403–428.

Hignett, S., & Wilson, J. R. (2004). The role for qualitative methodology in ergonomics: A case study to explore theoretical issues. *Theoretical Issues in Ergonomics Science, 5*(6), 473–493.

Johnson, C. W. (2008). Using evacuation simulations for contingency planning to enhance the security and safety of the 2012 olympic venues. *Safety Science*, *46*(2), 302–322.

Kendrick, V. L., Haslam R. A., & Waterson P. E. (2012). Planning crowd events to achieve high participant satisfaction. *Work: A Journal of Prevention, Assessment and Rehabilitation*, 41(1), 3223–3226.

Kendrick, V. L., & Haslam, R. A. (2010). The user experience of crowds – A human factors challenge. *Proceedings of the Human Factors and Ergonomics Society Annual Meeting*, *54*(23), 2000–2004.

Lee, R. S. C., & Hughes, R. L. (2007). Minimisation of the risk of trampling in a crowd. *Mathematics and Computers in Simulation*, *74*(1), 29–37.

Machleit, K. A., Eroglu, S. A., & Mantel, S. P. (2000). Perceived retail crowding and shopping satisfaction: What modifies this relationship? *Journal of Consumer Psychology*, *9*(1), 29–42.

Reicher, S., Stott, C., Cronin, P., & Adang, O. (2004). An integrated approach to crowd psychology and public order policing. *Policing – An International Journal of Police Strategies & Management*, *27*(4), 558–572.

Robson, C. (1993). *The Real World Research – A Resource for Social Scientists and Practitioner-researchers*. Oxford: Blackwell Publications.

Ryan, C., Shih Shuo, Y. & Huan, T. (2010). Theme parks and a structural equation model of determinants of visitor satisfaction – Janfusan Fancyworld, Taiwan. *Journal of Vacation Marketing*, *16*(3), 185–199.

Smith, A., James, C., Jones, R., Langston, P., Lester, E., & Drury, J. (2009). Modelling contra-flow in crowd dynamics DEM simulation. *Safety Science*, *47*(3), 395–404.

Stott, C., Adang, O., Livingstone, A., & Schreiber, M. (2008). Tackling football hooliganism: A quantitative study of public order, policing and crowd psychology. *Psychology, Public Policy, and Law*, *14*(2), 115–141.

Zhang, Q., Liu, M., Wu, C., & Zhao, G. (2007). A stranded-crowd model (SCM) for performance-based design of stadium egress. *Building and Environment*, *42*(7), 2630–2636.

Zhen, W., Mao, L., & Yuan, Z. (2008). Analysis of trample disaster and a case study – Mihong bridge fatality in china in 2004. *Safety Science*, *46*(8), 1255–1270.

BIOMECHANICS

COMPARISON OF REQUIRED COEFFICIENT OF FRICTION FOR BOTH FEET FOR STRAIGHT WALKING

Wen-Ruey Chang, Simon Matz & Chien-Chi Chang

Liberty Mutual Research Institute for Safety, USA

The required coefficient of friction (RCOF) is critical in determining whether a slip incident might occur. When using RCOF, most researchers do not differentiate between the two feet of the same participant under the same walking condition. This paper presents a comparison of the RCOF for both feet of 50 participants under four walking conditions using a t-test. The results in the current study indicated that 78% of the RCOF data showed a statistically significant difference between the RCOF from two feet for the same participant under each walking condition. The results of the logistic regression analysis indicated that the walking speed was the only factor with a statistical significance ($p = 0.044$) in contributing to the outcome of the t-test.

Introduction

Costs for disabling workplace injuries in 2009 due to falls on the same level in the US were estimated to be approximately 7.94 billion US dollars or 15.8% of the total cost burden according to the data from the Liberty Mutual Safety Index (Liberty Mutual Research Institute for Safety, 2011).

Required coefficient of friction (RCOF) represents the friction needed at the shoe and floor interface under dry conditions to support human locomotion. The available coefficient of friction (ACOF) represents the friction that can be supported at the shoe and floor interface without a slip. A slip may occur when the RCOF for an activity exceeds the ACOF at the shoe and floor interface (Redfern et al., 2001).

Because every step during walking is slightly different from others due to differences in gait parameters, the RCOF is not a constant for each individual even under identical conditions, as demonstrated by Chang et al. (2008). Typically, the RCOF data of both feet from the same participant under the same walking condition are pooled together without any understanding of potential differences in the RCOF between the two feet.

The objective of the current study was to compare and relate the RCOF of both feet under the same walking conditions for 50 participants with the t-test and linear regression analysis. In addition, a logistic regression was used to identify factors that contributed to the outcome of the t-test. More details of the current study could be found in Chang et al. (2012).

Methods

Fifty participants, evenly divided by gender, took part in this experiment. The average and standard deviation of weight, height and age for all male participants were $80.9 \pm 14.07\,$kg, $172.7 \pm 8.46\,$cm, 45.4 ± 13.2 years, respectively, and for all female participants were $64.8 \pm 9.16\,$kg, $162.4 \pm 5.54\,$cm, 45.2 ± 13.9 years, respectively. The participants gave written informed consent and were screened to assure no active musculoskeletal disorders. The protocol was approved by an institutional review board.

Details of the experimental setup could be found in Chang et al. (2012). The walkway was kept under dry conditions throughout the data collection period. To reflect those of normal and hasty situations in daily life, the participants walked at self-selected normal and fast speeds. Two types of footwear, a leather loafer (a slip-on) and a sneaker (a trainer), were used. Two factorial design combinations of the walking speed and footwear type were randomized, yielding four different walking conditions: loafer-fast (LF), loafer-normal (LN), sneaker-fast (SF), sneaker-normal (SN). The sampling rate for the force plates was 1000 Hz. The methodology to identify the RCOF introduced by Chang et al. (2011) was used.

The data pool for each foot under each walking condition for each participant was extracted. The data from both feet under the same walking condition for each participant were compared with a t-test. Factors that determined whether the RCOF values of both feet under the same condition were different based on the t-test were identified by a logistic regression analysis with gender, age group, footwear type and walking speed as the independent variables. Age was divided into 3 groups: 18–25, 26–54, 55 and older. For each walking condition, a linear regression equation across 50 participants for the average RCOF between the right and left feet for each participant was calculated.

Results and discussion

A total of 30,968 successful strikes were collected. Detailed information about sample sizes, RCOF values and walking speed were reported by Chang et al. (2012).

In 156 cases out of 200 (78%), the t-test results indicated a statistically significant difference between the RCOF values from both left and right feet ($p < 0.05$). The walking speed was the only factor that contributed to the outcome of the t-test with a statistical significance ($p = 0.044$) according to the results of the logistic regression analysis. In the linear regression equation for the average RCOF between the right and left feet for each participant for each walking condition, the RCOF for the right and left feet were treated as independent and dependent variables, respectively. The coefficients for the RCOF right foot (slope), intercepts and adjusted R^2 values for the regression equations are shown in Table 1. The relationship between the average RCOF values of the right and left feet for each participant under the LF condition across 50 participants is shown in Figure 1. Information about variations of the RCOF such as standard deviation, skewness and kurtosis was eliminated in

Table 1. The slope, intercept and adjusted R^2 of the regression equation across 50 participants for the average required coefficient of friction between the right and left feet for each participant under each walking condition. The RCOF of the right and left feet were treated as independent and dependent variables, respectively.

Condition*	Slope	Intercept	Adjusted R^2
LF	0.779	0.0525	0.49
LN	0.738	0.0575	0.63
SF	0.799	0.0467	0.58
SN	0.704	0.0684	0.60

*L = Loafer, S = Sneaker, F = Fast speed, N = Normal speed.

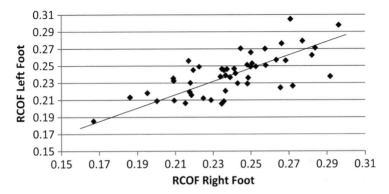

Figure 1. A graphic illustration of the relationship between the average RCOF values of the right and left feet for each participant across 50 participants under the LF condition.

averaging the RCOF of the same foot for the same participant and walking condition. Despite these eliminations, the right and left feet for each participant still had different average RCOF values as shown in Table 1 and Figure 1. More details of the results and discussions were reported by Chang et al. (2012).

In determining the possibility of slip incidents, the RCOF is usually compared with the ACOF. The results obtained in this study indicate that the RCOF values from both feet from the same participant under the same walking conditions were different in 78% of cases. When the RCOF data from both feet are different statistically, they should be kept separated. The comparison between the RCOF and ACOF should be applied to each foot separately. Then, the slip probability from both feet should be averaged to obtain the final probability for the individual.

Conclusions

This paper presents an investigation on the difference of the RCOF from both feet for level walking. The results of the t-test indicated that 78% of the RCOF data showed a statistically significant difference between two feet for the same participant under each walking condition. The logistic regression analysis indicated that the walking speed was the only factor with a statistical significance ($p = 0.044$) in contributing to the outcome of the t-test.

References

Chang, W. R., Chang, C. C., Matz, S. and Lesch, M. F., 2008, A methodology to quantify the stochastic distribution of friction coefficient required for level walking, *Applied Ergonomics*, 39 (6), 766–771.

Chang, W. R., Chang, C. C. and Matz, S., 2011, The effect of transverse shear force on the required coefficient of friction for level walking, *Human Factors*, 53 (5), 461–473.

Chang, W. R., Matz, S., and Chang, C. C., 2012, A comparison of required coefficient of friction for both feet in level walking, *Safety Science*, 50 (2), 240–243.

Liberty Mutual Research Institute for Safety, 2011, 2011 Workplace Safety Index, From Research to Reality, 14(3), available at http://www.libertymutual.com/researchinstitute.

Redfern, M. S., Cham, R., Gielo-Perczak, K., Grönqvist, R., Hirvonen, M., Lanshammar, H., Marpet, M., Pai, C. Y. C. and Powers, C., 2001, Biomechanics of slips, *Ergonomics*, 44 (13), 1138–1166.

BIOMECHANICAL ANALYSIS OF THE WALKING OF ENCUMBERED AND UNENCUMBERED MALES

Barbara May, James Shippen & Andree Woodcock

Coventry University, UK

This study investigates the biomechanical and postural tasks of able-bodied males undertaking unencumbered and encumbered walking journeys carrying a 5 kgs, 10 kgs and 20 kgs weighted holdall uni-laterally in one hand. The journeys were timed and analysed using 3-dimensional motion capture and muscle modelling techniques. Analysis indicated that the trunk bends towards the contralateral (unloaded) side to compensate for perturbation from the external load. Forces generated in trunk muscles were greatest on the con-tralateral side, whereas forces generated in the arm and shoulder muscles increased on the loaded side of the body. In the 20 kgs encumbered walk the deltoideus scapularis muscle exceeded its max-imal isometric force (nominal strength) which could lead to muscle fatigue and possible musculoskeletal injury.

Introduction

Travelling by air requires the traveller to frequently transfer from one mode of transport to another and walk considerable distances in the transit facility. The burden of travel is exacerbated by being encumbered with baggage whilst walking. The first assessment of airport walking distances from curbside to aircraft identified that considerable distances had to be walked (TRB, 2010). Distances traversed may not be within the capabilities of all passengers or be conducive to a positive journey experience. The definition of passenger walking distances within an airport has been defined as not only curbside to aircraft distances but also the distance between the check-in desk and the departure gate for departing passengers; the distance between the gate and the baggage delivery area for arriving passengers and the walking distance between the two gates to which two connecting flights have been assigned for transferring passengers.

The walking distance to the furthest gate at the following airports is reported as (AJC, 2010): Chicago O'Hare Airport 1,438 metres; New York, JFK Airport 2,371 metres; Los Angeles International Airport 2,023 metres; Atlanta Airport 816 metres. Whilst it is advantageous to minimise walking distances and the concomitant exer-tion experienced by the traveller, the tolerable maximum walking distance for a pedestrian over a specific route has not been defined.

Although minimising passenger walking distances might be a goal in the design of airport terminals, there are operational and economic constraints placed on airport managers which include regulatory practices, current airport configurations, gate assignment criteria and flight scheduling. The concept of 'designing for all' may be a morally desirable goal but it may not be a financially viable option. Airline managers have identified one solution to the economic dichotomy of having to maintain profits whilst appearing not to increase air fares in making ancillary revenue by charging fees for checked-in baggage. Higher baggage charges deliver hundreds of millions of pounds/dollars in ancillary revenue but the result has been that fewer bags are checked-in.US Airways calculated that only 30.2% of the 26 million passengers flying in a 6 month period in 2008 checked in their baggage (www.usairways.com). It is presumed therefore that the majority of passengers who do not check-in their bags with an airline carry them whilst transiting through the airport.

Research into bag carrying has focused on the physical effect on schoolchildren and military personnel of wearing backpacks (e.g. Shasmin et al, 2007; Birrell et al, 2007). Problems of carrying significant external loads include generalized fatigue and the development of stress fractures which may affect postural sway which in turn may increase the likelihood of falls and injury. Other research has focused on gait patterns and the position adopted for carrying bags i.e. on one shoulder, in the hand or in the crook of the arm (An et al, 2010; Crowe et al, 1993).

The purpose of this study was to compare the biomechanical ability of able-bodied males to undertake four walking journeys over a given distance at a self-paced speed, both unencumbered and encumbered with a weighted double-strapped holdall. The weighted holdall represented the 'hand-luggage' which is permitted to be taken on board commercial airlines e.g. 5 kgs on Virgin Atlantic and Thomson Airways; 10 kgs on Hawaiian Air and Air Canada and 20 kgs on British Airways, American, Continental, US Airways and Delta.

Although the participants were young able-bodied men it is acknowledged that the able-bodied young person is becoming less representative of the abilities and needs of the majority of people as populations are growing older with an associated increase in people with age-related disabilities.

Methodology

46 able-bodied males, aged between 19 and 27 years participated in the study. The age, weight and height of the participants were noted and their walking speed was recorded in each journey. A post-walk interview was also conducted. Three methodologies were utilised for the study.

Encumbered and unencumbered walking journeys

All participants undertook four walking journeys at a self-paced speed. Each journey measured 55 metres. The walking area was level and sufficiently wide to enable uninhibited gait patterns and excluded impedances such as uneven surfaces, obstacles or other persons. The first journey was walked without any encumbrance. The participants walked 3 more journeys, with 5, 10, and 20 kg weighted double-strapped holdall in one hand with the arm at full extension at the side of the body.

Motion data capture

3-dimensional movement analysis provides a visual interpretation and method for establishing normative gait parameters enabling the full body range of movement and flexibilities of individuals to be accurately measured. It also provides quantitative, objective data of a person's movements, including positions of segments of the body, position of centre of mass, joint angles, linear and angular velocities and linear and angular accelerations. A 12-camera Vicon optical tracking system together with twin AMTI ORS6-7 forceplates were used to analyse and compare the unencumbered gait of one participant chosen at random with the participant's gait whilst walking with a 20 kgs weight in the holdall held unilaterally by the double straps. The motion data was recorded at 100 frames per second and 12 channels of ground reaction data was recorded at 24,000 samples per second.

Muscle modelling

Muscle modelling provides an objective analysis of muscle forces, muscle activations and muscle timings and therefore it is possible to calculate the internal loads and torques which result from movement without using invasive methods. The 3-D motion data of the selected participant was used to undertake the whole body muscle analysis using BoB (Biomechanics of Bodies) muscle modelling software developed in-house (Shippen and May, 2010).

Data analysis

Encumbered and unencumbered walking journeys

After completing all the tasks, the participants were informed of their walking speed for each of the four journeys. During the post-walk interview the participants were asked to comment on any aspect of their journeys and the rationale for their walking speeds. In general, with the heavier loads the participants' concentration on achieving the task and walking speed increased.

The unencumbered walking journey and the three encumbered walking journeys were analysed using MatLab. This data is presented for the 46 participants.

The distribution of plots in figure 1 emphasises and agrees with the comments given by the participants. The close distribution of plots for the 5 kgs weighted holdall would suggest that the participants found this to be well within their capability. For the journey carrying the 10 kgs weighted holdall, five participants increased their speed of walking by 10% normalised to their unencumbered speed. However the greatest difference in the participants' walking speed was during the journey carrying the 20 kgs weighted holdall. Six participants walked considerably slower, which equates to their comments of "running out of energy during the heavy-weight walk": however the majority reported that they "consciously tried to keep to the same pace to keep the momentum going", which is borne out by the plots (figure 1) and four participants "walked quicker to get the job done".

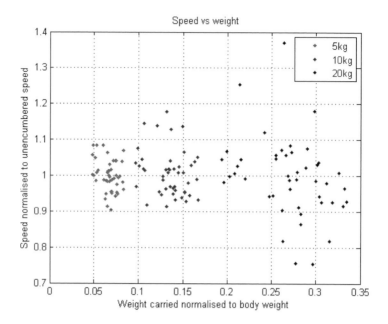

Figure 1.   Encumbered walking speed normalised to unencumbered walking speed, speed versus weight, and the weight carried normalised to body weight.

Motion data capture and muscle modelling

Figure 2 illustrates the trunk, neck and arm muscle activations of the participant. The skeleton on the left shows muscles during the unencumbered phase and the skeleton on the right during the walk carrying the 20 kgs weighted holdall in his right hand. For ease of visualisation, the colour of each muscle changes throughout activation from blue (inactive) through to green, yellow and red when the muscle is generating most force. The muscles also "bulge" to illustrate the muscles' maximum force compared to nominal strength.

Biomechanical analysis indicated a compensatory trunk inclination to counteract the characteristics of the 20 kgs weighted load and that the trunk bent towards the contralateral (unloaded) side to compensate for the perturbation from the external load (right skeleton). However the forces in the arm and shoulder were greatest on the right side of the body (the weighted holdall was carried in the right hand). Although whole body analysis was undertaken, it would not be feasible to report on all the 666 muscle units of the body; therefore four pertinent muscles were selected for specific examination – latissimus dorsi, erector spinae, deltoideus scapularis and sternocleidomastoideus.

Biomechanical analysis indicated a compensatory trunk inclination to counteract the characteristics of the 20 kgs weighted load and that the trunk bent towards the contralateral (unloaded) side to compensate for the perturbation from the external load (right skeleton). However the forces in the arm and shoulder were greatest on the right side of the body (the weighted holdall was carried in the right hand).

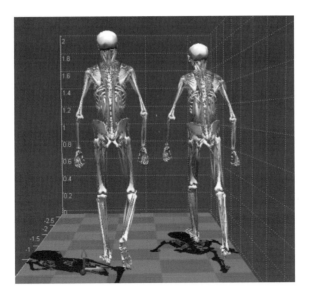

Figure 2. Muscles activation in posterior trunk – left skeleton = unencumbered, right skeleton = encumbered.

The latissimus dorsi muscle has its origin on the spinous processes of the lower six thoracic vertebrae, lumbar vertebrae, sacral vertebrae, supraspinal ligament and posterior part of the iliac crest through the lumbar (thoracolumbar) fascia, lower three or four ribs and the inferior angle of the scapula. It is inserted into the bottom of intertubercular (bicipital) groove i.e. the humerus just below the shoulder. It is responsible for extension, adduction, horizontal abduction and internal rotation of the shoulder joint and has a synergistic role in extension and lateral flexion of the lumbar spine. The maximal isometric force (nominal strength) which it can generate is 766 Newtons. During the encumbered walking journey carrying the 20 kgs weighted holdall the maximal isometric force generated in latissimus dorsi on the left side of the body peaked at 460 Newtons as compared with only 95 Newtons on the unencumbered walk (figure 3).

Erector spinae (sacrospinalis) is a complex of three sets of muscles: iliocostalis, longissimus and spinalis. The origin of this group is the medial and lateral sacral crests, the medial part of the iliac crests and the spinous processes and supraspinal ligament of lumbar and eleventh and twelfth thoracic vertebrae. Its function is the bilateral flexion of the cervical spine, extension of the cervical, lumbar and thoracic spine and rotation of the cervical spine. The muscle has twenty-eight origins and insertion points. The maximal isometric force for erector spinae(lt) is 579.6 Newtons. During the walking journey with the 20 kgs weighted holdall the maximal isometric force generated on the left side peaked at 440 Newtons compared to 150 Newtons on the unencumbered walk (figure 4).

The deltoideus scapularis muscle (figure 5) takes part in all movements of the upper arm. The origin of the anterior portion is the anterior border and superior

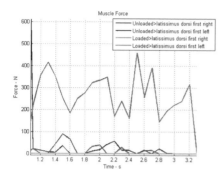

Figure 3. Latissimus dorsi, loaded and unloaded.

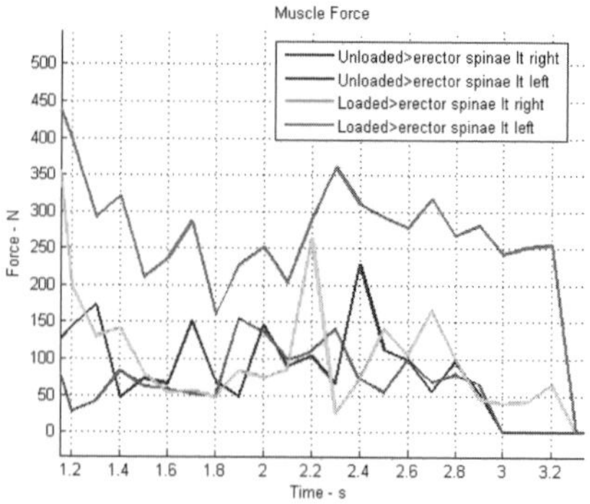

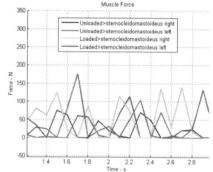

Figure 4. Erector spinae – loaded and unloaded.

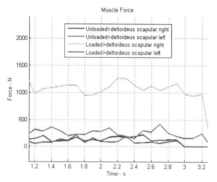

Figure 5. Deltoideus scapularis – loaded and unloaded.

Figure 6. Sternocleidomastoideus loaded and unloaded.

surface of the lateral third of the clavicle; the middle portion is the lateral border of the acromion process and the posterior portion is the lower border of the crest of the spine of the scapula. Its insertion is the deltoid tuberosity on the middle of the lateral surface of the shaft of the humerus. The maximal isometric force this muscle can generate is 885.6 Newtons. However during the encumbered walking journey with the 20 kgs weighted holdall, its maximal isometric force on the right side of the body was exceeded throughout the journey. This indicates that the muscle was operating at higher than its maximal isometric strength. It has been shown that muscles can generate forces in excess of their maximal isometric force for brief durations of time but it might be an indicator that the muscle could become injured through overuse.

The sternocleidomastoideus muscle (figure 6) has two origins – the sternal head on the manubrium of the sternum and the clavicular head on the medial part of clavicle. Its insertion is the mastoid process of the temporal bone and the lateral half of the superior nuchal line of the occipital bone. Its function is to bend the neck laterally, rotate the head, flex the neck, draw the head ventrally, elevate the chin and draw the sternum superiorly in deep inspiration. Its nominal strength is 168.4 Newtons.

During the encumbered walk both the left and right sternocleidomastoideus muscle were generating more force than during normal gait in order to maintain the head in an erect position. It is also possible that the participant was breathing more deeply during the exertion of carrying the 20 kgs weighted holdall which would increase the forces generated in this muscle to facilitate deep inhalation.

Discussion

Research identified that little attempt has been made to analyse the whole pedestrian journey and the needs of pedestrians in transport policy tend to be dealt with rather crudely, using either an "as the crow flies" buffer around transport access points or by representing the pedestrian as using road networks. Therefore transport strategy may be ineffective not only for the pedestrian but also in improving social inclusion for those people with differing capabilities e.g. the young, the elderly or disabled who face different impediments to movement and access.

There is growing evidence that products and services should be 'designed for all' and public transport and their transit environments, by their very nature, should be accessible for all who wish to use them. It is therefore suggested that any design concept or operating practice which reduces the burden of walking at transit facilities, especially over considerable distances, would be beneficial to all users and could provide a positive journey experience for the traveller and thereby increase the potential use of those facilities. Unfortunately carrying baggage throughout the considerable time and distances walked at transit facilities could produce not only a negative impact on the travellers' whole journey experience but also affect the travellers' health and well-being through an increased risk of musculoskeletal injuries caused by muscle fatigue. However if designers of, for example, airports are constrained by the need to design for the 5th–95th percentile users, this means that not only good design decisions are made but also poor design decisions are reinforced as they are not 'designed for all'.

Conclusion

The fastest unencumbered walking journey recorded was 28.6 seconds and the slowest speed of 58.8 seconds occurred during the 20 kgs encumbered walk. A comparison of participants' walking speeds identified the greatest difference in speed whilst carrying the 20 kgs weighted holdall. Carrying a holdall in one hand inhibited the free arm swing resulting in modifications to gait to compensate for the restricted arm movement together with a progressive decrease in stride width. The 3-dimensional motion capture and muscle modelling analysis identified that during the lateral carrying of the weighted holdall the trunk of the body was bent towards the unloaded side, the amplitude of trunk bend increased with load and there was also an increase in mediolateral centre of pressure displacement.

The results indicated that asymmetric load carrying not only induced postural compensation during walking journeys but also increased the forces generated in muscles. Muscles operating at or above their optimum nominal strength whilst carrying heavy baggage over considerable distances can become fatigued with the resultant risk of injuries to the musculoskeletal system. This has implications for the non-checking-in of baggage at an airport which may be detrimental to passengers' physical health and well-being as postural instability and generalized muscle fatigue caused by the carriage of heavy baggage over considerable distances could lead musculoskeletal injury.

Only able bodied males were included in this study. Future research on the effect of load carrying on the musculoskeletal system could include females, the elderly, the infirm, the young and the disabled. Further research could also investigate the effect on posture and muscle loading in response to the use of weighted wheeled luggage as compared to hand luggage. This data could be used to inform the design and development of transit facilities which are 'designed for all' which would not only provide a positive journey experience for all travelers but also increase revenue for operators as such facilities would be accessible and useable by everyone wishing to travel.

References

An, D-H., Yoon, J-Y., Yoo, W-G. and Kim, K-Y. 2010, Comparisons of the gait parameters of young Korean women carrying a single-strap bag. *Nursing and Health Sciences*, 12, 87–93

Atlanta Journal Constitution. 2010, *"Delta hikes baggage fees again"*

Birrell, S., Hooper, R. and Haslam, R. 2007, The effect of military load carriage on ground reaction forces. *Gait Posture*, 26 (4), pp. 611–614

Crowe, A., Schiereck, P. and Keessen, W. 1993, Gait adaptations of young adult females to hand-held loads determined from ground reaction forces. *Gait Posture*, 1, 154–160

Shasmin, H., Abu-Osman, N., Razali, R., Usman, J. and Wan-Abas, W. 2007, The Effect of Load Carriage Among Primary School Boys: A Preliminary Study. *Journal of Mechanics in Medicine and Biology*, 7 (3), pp. 265–274

Shippen, J. and May, B. 2010, Calculation of Muscle Loading and Joint Contact Forces in Irish Dance. *Journal of Dance Medicine and Science* 14, (1)

Transportation Research Board. 2010, *Airport Passenger Terminal Planning and Design: Guidebook*, Airport Cooperative Research Program, Federal Aviation Administration United States, Landum and Brown

DESIGN

SAFE DESIGN OF MOBILE CONSTRUCTION AND MINING EQUIPMENT

Tim Horberry[1,2] & Mike Bradley[1]

[1]*Engineering Design Centre, University of Cambridge, UK*
[2]*Minerals Industry Safety & Health Centre,*
University of Queensland, Australia

A design review of current mobile construction and mining equipment was undertaken using operator interviews, observations, an ergonomics audit checklist and a safe design workshop. The purpose was to identify any major ergonomics, usability or safety deficiencies associated with the equipment that may be considered in future design changes. Operators with different experience levels (novice, intermediate and expert) were the participants. This paper summarises the methods used and then outlines the key findings emerging, these include issues with the design of the cab workspace and the operator interface. The benefit of using operator-centred approaches for potentially improving the safe design of mobile construction and mining equipment is highlighted.

Introduction

Mobile construction and mining equipment includes heavy vehicles such as dump trucks, forklifts, backhoe loaders, bulldozers, rollers and mobile cranes (Horberry and Cooke, 2012). Such equipment are essential tools for materials handling, as well as being flexible vehicles for many freight movement tasks in mining, manufacturing, construction and elsewhere. They offer many benefits, such as improving work efficiency or reducing manual handling, but they can also pose a major occupational hazard, especially where used in close proximity to other 'pedestrian' workers (Larsson and Rechnitzer, 1994). Similarly, such equipment can pose safety and health hazards to operators and maintainers of the equipment, these hazards include access/egress injuries, noise induced hearing damage, vehicle fires and explosions, and loss of vehicle control (Horberry et al, 2010).

So whilst there are many benefits from such industrial mobile equipment, there are also some serious safety issues (Larsson and Rechnitzer, 1994). To combat this, a collection of design, engineering and administrative controls is therefore required: this often should include improved traffic engineering on sites, safer initial design, and, where necessary, the use of retrofitted vehicle technologies such as speed limiting systems or seatbelt interlocks to reduce risks to operators and pedestrian workers (Horberry et al, 2004). The scope of the research described here is upon the safe design of such equipment.

Figure 1.   Example of a Backhoe Loader.

This work specifically focused on one type of mobile equipment: a backhoe loader. This is a heavy vehicle used in many mining, quarrying, construction and urban engineering-related tasks such as building roadways, demolitions, transportation of materials or digging trenches. There are many different manufacturers of backhoe loaders worldwide, and for reasons of confidentiality the exact manufacturer used in this study will not be named. However, for virtually all manufacturers, a backhoe loader consists of a tractor fitted with a shovel/bucket on the front of the vehicle and a backhoe digging bucket on the back. Figure 1 shows an example of a backhoe loader (made by a different manufacturer to the one used for this research).

The overall purpose of this work was to identify, from a user-centred perspective, any major ergonomics, usability or safety deficiencies with the current design of this industrial vehicle. From this, possible design inadequacies could be identified and improvements thus proposed. Obtaining end-user opinions and perspectives in a structured and rigorous manner is an essential part of conducting a user-based design review (Horberry et al, 2010). Further, it would be inadequate from an ergonomics perspective to consider the equipment in isolation, without considering who is using the equipment, for what purpose they are using the equipment and in what environment the equipment is being used (Torma-Krajewski et al, 2009).

It should be noted that the design of many industrial vehicles do not often rad-ically alter when new models are introduced: that is, the actual operational and maintenance tasks are not likely to change too much (Horberry et al, 2010). Therefore, current end-user experience can be more effectively used to provide direct operational input into such small iterations compared to more radical design changes (e.g. automation, currently being widely introduced in Australian mining) where the new operational and maintenance tasks could be quite different.

Methods

Following initial inspections and observations of the equipment being used in a series of tasks, three operator-centred methods were used to gather the design review data.

Operator interviews

The participants were chosen to represent the range of experience levels that operators of such equipment might have. It was anticipated that using operators with differing experience levels may reveal different types of design issues: for example, basic equipment handling issues revealed by the least experienced operator, or persistent ergonomics deficiencies with the vehicle's controls and display revealed by the most experienced operator (Horberry et al, 2010).

Given the wide usage context of this equipment in many different domains, operators with a wide range of experience levels, from complete novices through to highly experienced workers, might therefore be using it. Although only a small sample, the research therefore interviewed three operators with vast differences in experience levels (although broadly representative of the wider user group):

- one novice (never operated a similar machine before), male and approximately 20 years old.
- one intermediate (experienced with similar equipment, but not this exact type of backhoe loader), male and approximately 30 years old.
- one experienced operator (over 15 years use with this exact equipment type), male and approximately 32 years old.

Participants were interviewed after they had performed a number of prescribed tasks with the backhoe loader that included reversing the machine, digging a trench and loading/dumping soil. The work was conducted under Cambridge University human ethics guidelines, and all the operators gave informed consent before being involved.

Human factors checklists

The main interview prompt device used was the Vehicle Ergonomics Audit Tool (VEAT) previously developed by one of the authors. The VEAT was not intentionally designed to be a comprehensive instrument to audit every operator-related aspect of a vehicle: instead it was used here merely as a structured tool to guide the interview questions. It was derived from several ergonomics checklists and 'classic' textbooks (e.g. Stevens, 1999; Grandjean, 1988) as well as more contemporary sources (e.g. Horberry et al, 2010). The VEAT had been previously used to assess the designs of ambulances, mobile mining equipment and police vehicles both in the UK and in Australia. It considered both the design of the equipment and the wider use context. It covered the following four broad areas:

1. Workspace-Physical Demands (e.g. operator posture, design of pedals and seating, and overall cab visibility)
2. Workspace-Cognitive Demands (e.g. design of warning lights and auditory alarms)
3. Environmental and Task Load (e.g. air conditioning and ventilation in the cab, and task scheduling issues)
4. General/Open questions (e.g. what operators considered to be the main ergonomic deficiencies of the backhoe loader)

Safe design

By means of a two hour workshop with the two human factors researchers, an experienced operator and a design engineer from the manufacturer of the backhoe loader, a 'Safe Design' process was used to reveal the potential difficulties with using the equipment to complete different tasks. 'Safe Design' is a general process that aims to eliminate Occupational Health and Safety hazards, or at least minimise potential risks, by systematically involving end-users and decision makers, and by employing hazard analysis/risk assessment methods throughout the full life cycle of the equipment (Australia Safety and Compensation Council, 2006; Hale, Kirwan, and Kjellén, 2007).

Two tasks were chosen to be investigated in the safe design workshop. These were common ones that had caused problems in the past to the experienced operator whilst he was gaining experience. These tasks were: levelling a roadway, and loading and dumping earth effectively. Building on the Safety in Design Ergonomics (SiDE) processes outlined by Horberry (2012), the safe design workshop went through each of these tasks and outlined the main steps in them. Then, for each step in the task, the workshop examined what possible things could go wrong, and how the design of the backhoe loader equipment could be improved.

Results

For reasons of space, only a selection of the obtained results is presented here: first the summary findings of one of the safe design workshops and then an amalgamation and summary of the key findings deriving from the interviews, observations and ergonomics audits. They should not be seen as a complete list of ergonomics issues experienced by all end users, as only three operators were involved in the interview stage, and a limited range of tasks were reviewed. Despite the restricted scope, it highlighted many ergonomic issues relevant to the design and operation of this type of mobile equipment that could potentially be improved in future models.

Table 1.   Sub-tasks identified in levelling a roadway.

Step	Task Description
1	Spreading and undertaking a basic levelling of the stone over the whole area – usually using the front bucket of the backhoe loader. The stone needs to be spread high, as in later stages it would compact.
2	Roll whole area solid with a roller. Keep going over and over until the roadway is compacted.
3	With ditching blade on the backhoe loader (5 Foot back blade with a big 'lip' on it), carve the stone off to get it compacted and more even. After using this blade, the operator would examine how level the roadway was (often done with a string or a hand held laser level). If still uneven then they would go over again with the blade.
4	Roll again with roller. This is to knit the top layer into the roadway. This final stage would usually occur without vibration if the previous stage was done well.
5	The roadway then is ready for concrete or tarmac.

Safe design workshop: levelling a roadway

The first task investigated in the safe design workshop was levelling a roadway. In addition to the operator skill requirements to perform the task, the size of stone used was a key determinant in the process, especially where large chunks were present. Usually it was done as a two person task. The second operator working outside the backhoe was equally important to set up/sort out the whole leveling process, and clear communication between these two was therefore vital.

Table 1 shows the main sub-steps in the task of levelling a roadway that were identified in the Safe Design workshop.

Step 3 was identified as being the key one for bringing stone to accuracy. As such, the workshop then explored where any design changes to the backhoe loader could assist with safe and efficient completion of this sub-task. The task was understood to be a fine motor control one, where the operators' skill in accurately using the vehicle's controls and display was vital.

In terms of some of the solutions proposed, an infrared beam fitted to the vehicle could help to identify any lumps and unevenness in the roadway. This would be quicker and potentially more accurate than setting strings. It has the additional safety benefit that the operator would not need to get out of the cab, so minimizing the likelihood of access/egress injuries or the severity of being struck by another vehicle. Additionally, better vehicle mirrors and improved cab visibility were proposed as ways to assist this predominantly visual task.

Interviews, observations and ergonomics audits

Table 2 below shows some of the other usability or safety deficiencies found with the equipment that may be considered in future design changes. For example, the labeling issues were revealed through the tasks that involved joystick usage

Table 2.   **Potential Design Deficiencies and Recommendations.**

Issue	Description
Visibility and communication	Visibility outside the cab is often an issue in larger mobile equipment. Regarding this backhoe loader, seeing the rear wheels was a problem. This was often worsened by the fairly small wing mirrors fitted as standard (which one operator replaced with larger ones). Side windows that open might be an advantage for communication with outside workers without having to open the large rear window.
Cab design	Although not explicitly measured, the cab workspace was generally adequate for most smaller operators; however, it may present potential problems for larger operators (e.g. above the 75th percentile male). This was supported by comments from the operators interviewed (e.g. hitting feet on the console floor when turning around to use the rear backhoe).
Seating	The seat was largely adequate, but additional support/padding/fixing would prevent it bottoming out or to prevent excessive vibration (especially when working to the rear).
Visual warnings	The in-cab visual warnings were not optimal and could be improved, especially to tell a less experienced operator exactly what the problem is (rather than just a warning triangle with an exclamation mark, for example). Similarly, an accurate fuel gauge (perhaps showing 'range to empty') could be useful.
Labels	Clearer labels for the operation of the joystick and other machine controls was recommended, as the current labels were not well-understood by operators they were aimed at (i.e. less experienced ones).
Emergencies	Better emergency evacuation procedures from the backhoe loader are recommended. The novice operator interviewed did not know how to isolate/deactivate the machine, so it could be left in an unsafe state if he/she needed to get out of the cab in an emergency.
Air conditioning/ ventilation	The air conditioning was largely adequate, but extra fan speeds (both faster and slower) were recommended, especially to help prevent the windows steaming up on a wet day.

and then subsequently identified by the application of the VEAT and the operator interviews.

Discussion

The type of backhoe loader investigated in this research has been in production and use for many years, with numerous design iterations over that period. Therefore, in many ways it is probably a fairly well-designed product for which it is unlikely that substantial ergonomics, usability or safe design issues would be revealed here. Despite this, one conclusion of this paper is that the Human Factors style approach used here (of user-centred safe design) has been shown to be a valuable way of generating additional data to assist in potential design improvements for future iterations of the equipment. This general approach may also allow the operator to function as the 'extended arm' of the designer when coping with abnormal operating conditions (Rasmussen and Goodstein, 1985). Further, this operator-centred design process might be of even more importance if new technologies (eg collision warning systems) and automation are installed into the vehicle, as their use is becoming more widespread in many industrial domains such as mining (Horberry et al, 2010).

In the longer term, further studies are certainly required to attempt to get more data from a wider sample of users and usage contexts to increase confidence in the prioritisation of issues to address in future design modifications. It should be noted that this research only focused on three male users (of different experience levels) on a fairly limited selection of tasks: it does not take into account ergonomic issues that would be experienced by females, other males of different sizes, ages or experience levels, nor of varying usage contexts or other tasks. Also, other approaches such as video recording operators whilst they are engaged in routine tasks with the backhoe loader might reveal additional usability, safety and ergonomics issues (Horberry and Cooke, 2012).

The research described here used a participatory ergonomics approach, which employed end-user feedback to assist in future designs of construction and mining equipment. In domains such as, mining, quarrying, and construction it is often difficult for equipment manufacturers get sufficient access to end-users (Horberry et al, 2010); Hale, Kirwan and Kjellén (2007) previously argued that the most important part of improving the safe design process was obtaining such operational input. It is therefore recommended that the structured and time-efficient end-user approaches that were used in this research are more widely employed with the design of construction and mining equipment (especially for new equipment and technologies). Similarly, during equipment procurement, it is important that potential equipment purchasers require manufacturers to demonstrate how well operational and maintenance risks have been addressed in the equipment design.

Acknowledgments

The paper authors would like to thank the support of the equipment manufacturer and colleagues at the University of Queensland (Australia) and University of Cambridge (UK). The paper was partly written with support of an EC Marie Curie Fellowship '*Safety in Design Ergonomics*' (project number 268162) held by the first author at the Engineering Design Centre, University of Cambridge, UK.

References

Australian Safety and Compensation Council. 2006, *"Guidance on the principles of safe design for work."* Retrieved 17th September 2012 from: http://safeworkaustralia.gov.au/AboutSafeWorkAustralia/WhatWeDo/Publications/Documents/154/GuidanceOnThePrinciplesOfSafeDesign_2006_PDF.pdf.

Grandjean, E. 1988. *Fitting the task to the man: A textbook of occupational ergonomics* (4th Edition) (Taylor and Francis, London).

Hale, A., Kirwan, B. and Kjellén, U. 2007. "Safe by Design: where are we now?" *Safety Science*, 45, pp. 305–327.

Horberry, T., Larsson, T., Johnston, I. and Lambert, J. 2004. "Forklift safety, traffic engineering and Intelligent Transport Systems: a case study", *Applied Ergonomics*, 35 (6), pp. 575–581.

Horberry, T., Burgess-Limerick, R. and Steiner L.2010. *Human Factors for the Design, Operation and Maintenance of Mining Equipment* (CRC Press, USA).

Horberry, T. and Cooke, T. 2012. Safe and Inclusive Design of Equipment Used in the Minerals Industry. In P. Langdon et al. (ed) D*esigning Inclusive Systems: Designing Inclusion for Real-World Applications* (Springer-Verlag, UK), 23–32.

Horberry, T. 2012. "Better Integration of Human Factors Considerations within Safety in Design." *Theoretical Issues in Ergonomics Science*. DOI: 10.1080/1463922X.2012.727108.

Larsson, T. and Rechnitzer, G. 1994. "Forklift trucks- analysis of severe and fatal occupational injuries, critical incidents and priorities for prevention." *Safety Science* 17: 275–289.

Rasmussen, J. and Goodstein, L.P. 1985. *Decision support in supervisory control.* Risø National Laboratory, Denmark. RISØ-M-2525.

Stevens, A., Board, A., Allen, P. and Quimby, A., 1999. *A safety checklist for the assessment of in-vehicle information systems.* A user's manual (PA3536/99): (Transport Research Laboratory, UK).

Torma-Krajewski, J., Steiner, L.J. and Burgess-Limerick, R. 2009. *Ergonomics Processes: Implementation Guide and Tools for the Mining Industry.* Pittsburgh, PA: U.S. Department of Health and Human Services, Public Health Service, Centers for Disease Control and Prevention, National Institute for Occupational Safety and Health, DHHS (NIOSH) Publication No. 2009-107, Information Circular 9509, 2009 Feb, 1–149.

USER-CENTRED DESIGN OF VIRTUAL TRAINING FOR AUTOMOTIVE INDUSTRIES

Setia Hermawati & Glyn Lawson

Human Factors Research Group, Faculty of Engineering,
The University of Nottingham, UK

Operators on final assembly lines at automotive companies need to receive sufficient training to enable them to switch effortlessly in performing assembly operations on one model to the next. A virtual training system, "VISTRA", was designed to provide a supplementary approach for training to existing methods which rely on physical prototypes. A user-centred design approach was adopted to realise such system. Context of use and user requirements analysis were conducted; followed by providing design solutions through personas, scenarios, and storyboards. It was found that the combination of text-based scenarios and storyboards was beneficial for a team that consisted of multidiscipline and multinational members.

Introduction

In the automotive industry, operators in final assembly hold crucial roles as most operations have to be performed manually (Nof et al., 1997). In line with the increased competitiveness of the automotive market and customer demands for assemble-to-order products, several models of the same product share the capacity of one assembly line. This compels operators to be able to switch effortlessly in performing assembly operations for one model to the next. Therefore, when a new product and its variants are introduced, repeated exposure to the task training is crucial in developing operators' procedural skills (Gupta and Cohen, 2002).

Assembly operations are skill-based operations which require procedural skill i.e. an ability to execute action sequences to solve problems (Rittle-Johnson et al., 2001). Research has shown that procedural skill of a task can be developed from repeated exposure to the task (Gupta and Cohen, 2002). On the automotive assembly line, training is often performed on pre-series products (physical training). However, due to the high cost involved, pre-series products are limited in number and only developed for a small number of vehicle variants. Their usefulness in training is also limited further by wear from repeated assembly and disassembly. Unsurprisingly, these limitations have prompted a shift towards virtual training and resulted in the development of many virtual training systems to aid the acquisition of procedural skills related to assembly tasks. Unfortunately, most of these systems are aimed at training maintenance tasks in which knowledge of assembly/disassembly are part of the tasks (Peniche et al., 2011; Gutiérrez et al., 2010; Abate et al., 2009; Oliveira et al., 2007) and only few are focused on assembly tasks (e.g. Vizendo).

Our study aimed to develop a virtual training system that matched users' needs through adoption of a user-centred design process while simultaneously introducing the concept of serious gaming into the training. The system, VISTRA (Virtual Simulation and Training of Assembly and Service Processed in Digital Factories) was designed to support assembly training at final assembly lines for two automotive companies. In this paper we emphasise the use of scenarios and storyboarding as methods to produce design solutions. We begin with a brief overview of user-centred design. Next we describe the process of establishing context of use and user requirements and follow by describing the process used to produce design solutions. Finally, we discuss the findings and reflect on problems faced in the application of user-centred design in the development of virtual training.

User-centred design

User-centred design (UCD) is a common term, encompassing a philosophy and variety of methods, which refers to how end-users influence a design through their involvement in the design processes. UCD has been shown to have positive effects on various aspects i.e. the quality of final design; the speed of the design process; the match to the end users' needs or preferences; and end user satisfaction (Kujala, 2003). It has also been shown to contribute to the acceptance and success of products (Preece et al., 2002). ISO 9241-210:2010 (ISO, 2010) provided a human centred design framework which consisted of 5 phases: 1) plan the human centred design process; 2) understand and specify the context of use; 3) specify the user requirements; 4) produce design solutions to meet user requirements; 5) evaluate the designs against requirements, with phases 2 to 5 iterated where appropriate. As the VISTRA system is still in its initial development phase, this paper will only report the methodologies and results to date and is limited to the second, third and fourth phases of the framework.

Establishing the context of use

In the VISTRA study, one of the key elements of establishing the context of use is the identification of stakeholders. This was performed through discussion and liaison between human factors researchers and representatives of end users from the automotive industries. During the process, the end user representatives were encouraged to adopt broad definitions of stakeholders i.e. beyond end-users. This was then followed by a request to provide a general overview of the demographics and backgrounds for each identified group of stakeholders. Context of use analysis was continued with semi-structured interviews with identified stakeholders and field observation at final assembly lines. Collated data from the interviews and field observation were also used to specify the user requirements which are explained in detail in the next section.

The findings from establishing the context of use shows that the virtual training system involved four main user groups with diverse technology affinity, language skills, education as well as age. The four user groups identified were *trainee* (assembly operators), *trainer* (foreman), *training manager* (supervisors) and *data administrators* (engineers). It was also found that virtual training system usage could be differentiated into two: i) training for a new launch product; and ii) training for an existing product.

Specifying user requirements

Field observation and semi-structured interviews with identified stakeholders were used to specify the user requirements. A total of 45 participants were involved in the interviews which lasted approximately 30 minutes to one hour with between one and four participants at a time. During the interviews, participants were encouraged to talk freely. Involvement from the interviewer was limited to prompting participants to provide more details or expand on key issues. Due to commercial sensitivity and the resources available for analysis, responses from participants were mainly recorded through handwriting. Observations at the final assembly lines lasted between 45 minutes to 1 hour. Notes regarding final assembly tasks and the environment of assembly lines were taken. Short informal discussions with operators were also performed when required and possible, depending on the demands of their jobs.

The collated data were analysed with several methods: stakeholder analysis; time-line analysis; task analysis; link analysis; and thematic analysis. The results of the analyses were then used to produce user requirements and presented to the design team, which consisted of system development experts, end user representatives and human factor researchers. The representative end users then judged the importance (high, medium, low) of each identified user requirement while the system development experts judged the technical feasibility (high, medium and low) of each requirement. During the discussion, answers to questions such as "how important are the different requirements for the users?", "what are the consequences for users if we do not implement a specific requirement?" were continually assessed.

The findings from user requirements showed that both trainees and trainers user groups valued high cognitive and physical fidelity in the virtual training as well as ease of use. They also emphasised that a virtual training system should complement physical training instead of replacing it. They argued that virtual training could never replace all of the important elements of the physical training (e.g. "the feel of the part in your hands"). The study also found that there was a lack of support to share tacit knowledge of assembly operations among user groups which prompted the risk of losing useful knowledge to support faster completion of new product development projects, team performance, and innovation capabilities. This subsequently led to the identification of a virtual training system's potential use as a platform for knowledge sharing between user groups e.g. by allowing experienced operators to share their best practice and know-how not only with other operators but also with engineers who can then decide whether to exploit or rectify them. This additional use of a virtual training system was successfully identified solely due to the involvement of various stakeholders who provided a holistic overview of the current situation and how this could potentially be addressed by a virtual training system. Last but not least, as with other training systems, stakeholders also preferred that the cost and time to execute the virtual training system were kept to the minimum. Details of the user requirements results are published elsewhere.

Producing design solutions

The first step to producing the design solution was to detail the high level design of the VISTRA system (the left image in Figure 1), which was established in the beginning of the project. This involved creating relevant task for each user group while interacting with the system. As the human factors researchers are involved primarily with the development of the VISTRA knowledge sharing centre (VKSC), this paper will only report the results related to this subsystem.

A scenario-based design approach was used to produce a more detailed design of the subsystem. The main reasons for adopting this approach are twofold. First, scenario-based design is still more comprehensive than story-based design which was considered to have less breadth and representativeness (Vredenbrug et al., 2002). Second, scenario-based design can accommodate the communication of the design context to the design team without imposing difficulties on the members of the design team who do not have appropriate expertise in software engineering. Scenario-based design relies on concrete narrative descriptions and can be used to represent either the current situation or how a system will be used in the future. A scenario describes human activities or tasks in a story that allows exploration and discussion of context, needs and requirements from particular concrete situations (Caroll, 2000). Multiple scenarios are needed to reflect the different situations and views that occur. Scenarios have setting, actors, actors' goals and a plot; and can be represented further as prototypes through the use of storyboarding. In this study, a combination of text and storyboards was used to depict scenarios. Storyboards are sequences of snapshots, single visual images that capture a significant possible interaction. Several studies have successfully incorporated the use of storyboard

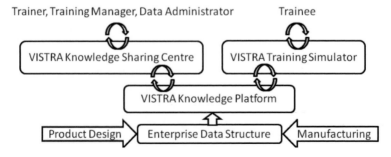

Figure 1.   High level design of the VISTRA system.

through photo images (Pedell, 2004), or videos (Mackay et al., 2000) to support scenario-based design.

As part of the scenario-based design, personas were created for each user group. The personas were created based on information from the context of use and user requirements specifications. A total of eighteen text-based scenarios were also produced and subsequently developed further into storyboards. Each scenario depicted how the virtual training system will be used in a specific circumstance. The number of scenarios created for each user group was determined by the variety of tasks in that user group. For example, eight scenarios were created for the trainer user groups while only two scenarios were created for the training manager user group. Figure 2 shows a snippet of one of two scenarios for the training manager user group. The scenarios and storyboards were presented to the design team to seek their feedback. The system development experts and end user representatives found the combination of text-based and pictorial storyboards easy to understand, irrespective of their background disciplines and nationalities. From the discussion, possible issues such as compliance to the organisation's policy on data security/confidentiality, practicalities of training review procedure and technical feasibility of scenarios were identified. After undergoing iterative refinement, the final versions of the scenarios and storyboards were then utilised to guide the development of detailed user interfaces. This was initiated by creating hand drawn user interfaces, which represent a sequence of user interaction with the system for each scenario. Principles of usability were continuously used as reference during this stage. Upon completion, a user interface flow was created and the user interfaces were recreated electronically to support the design legibility. Similar to the scenarios, the user interfaces were also refined iteratively. Figure 3 shows an example of user interface for user management.

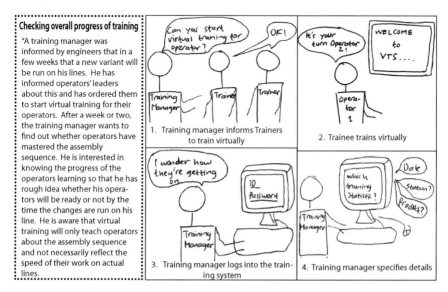

Figure 2.   An example of a scenario for training manager user group.

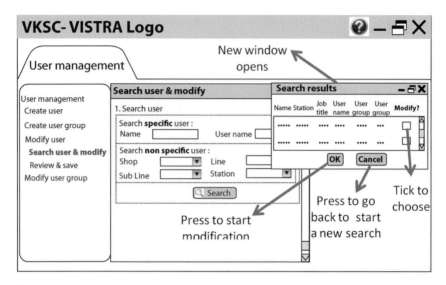

Figure 3.   Example of user interfaces for the data administrator user group.

Discussion

Within VISTRA context, this study has identified the additional need of a virtual training system in manufacturing i.e. as a platform for knowledge sharing between user groups. As it has been mentioned previously, the identification of a virtual training system's additional use was due to the involvement of various stakeholders who provided a holistic overview of current situation and shared their foresight towards a virtual training system. This was clearly made possible by involving the end user representatives early in the UCD as they were able to identify stakeholders with ease and accuracy due to familiarity to their organisations. A drawback of this study is the lack of input from potential end users during the refinement of the user requirements. There is the possibility that misjudgement occurred while weighing the importance of each requirement due to representative end users leaning more towards a managerial rather than end user role, conflict of interest especially in circumstances where representative end users have stake towards the issue, etc. To minimise this, researchers of this study played a valuable role as "independent auditors" during the refinement process to provide more objective perspectives of stakeholder positions and interests.

Overall, the combination of text and storyboards for scenarios has assisted collaboration of a multidisciplinary and multinational design team. This was clearly demonstrated by the ease of the design team in understanding the envisioned context of use and focus on design solutions, irrespective of their background disciplines and nationalities. This was likely due to the fact that the visual images are much easier to understand and imagine regardless of the observers' backgrounds (Roam, 2008), unlike exclusive text-based scenarios which are subjected to different interpretation and largely dependent on language proficiency. In terms of resources required, the storyboards were neither costly nor time consuming as they were simply created by hand drawing. However, it is acknowledged that hand drawing could be challenging for individuals who are not comfortable with it (Truong et al., 2006). Truong et al. (2006) also reported that initiating storyboards can be challenging, both for novice and expert designers, as it requires designers to tell a story in small concise parts and consequently force them to make early decisions on the possible interactions to be captured. These difficulties were not encountered during this study as the text-based scenario assisted the human factor researchers in identifying significant parts of the story that needs to be transformed into storyboards. However, it is acknowledged that adopting this approach means that extra effort was required. Last but not least, the approach described in this paper is likely applicable to other virtual training that is similar to VISTRA (i.e. with regards to context of use, training content, type of technology, etc.).

Conclusions

Through the involvement of both end users and their representatives, the potential to use a virtual training system to support knowledge sharing and different periods of usage during product life cycle was successfully identified. This study demonstrated that the use of a combination of text-based scenarios and storyboards has played a big role in supporting the design process for multidisciplinary and multinational team. This method is likely applicable to other virtual training that is similar to VISTRA.

Acknowledgements

The authors would like to acknowledge the contribution to this work from all members of the "VISTRA" project (ICT-285176), which is funded by the European Commission's 7th Framework programme.

References

Abate, A. F., Guida, M. G., Leoncini, P., Nappi, M., Ricciardi, S. 2009, A haptic based approach to virtual training for aerospace industry, *Journal of Visual Languages and Computing*, 20, 318–325.

Caroll, M. J. 2000, Five reasons for scenario-based design, *Interacting with Computers*, 13(1), 43–60.

Gupta, P., Cohen, N. J. 2002, Theoretical and computational analysis of skill learning, repetition priming, and procedural memory, *Psychological Review*, 109, 401–448.

Gutiérrez, T., Rodriguez, J., Vélaz, Y., Casado, A., Suescun, A., Sánchez, E. J. 2010, IMA-VR: a multimodal virtual training system for skills transfer in industrial maintenance and assembly tasks. In *Proceedings of 18th IEE International Symposium on Robot and Human Interactive Communication*, 428–433.

Haesen, M., Meskens, J., Luvten, K., and Coninx, K. 2009, Supporting multidisciplinary teams and ealy design stages using storyboards. In *Proceedings of the 13th International Conference on Human-Computer Interaction*, 616–623.

Hassenzahl, M., Wessler, R. 2000, Capturing design space from a user perspective: the repertory grid techniques revisited, *International Journal of Human-Computer Interaction*, 12(384), 441–459.

ISO, 2010. ISO 9241-210:2010 Ergonomics of human-system interaction – Part 210: Human-centred design for interactive systems.

Kujala, S. 2003, User involvement: a review of the benefits and challenges, *Behaviour and Information Technology*, 22(1), 1–17.

Mackay, W. E., Ratzer, A. V., and Janecek, P. 2000, Video artifacts for design: brigding the gap between abstraction and detail. In *Proceedings of the Conference on Designing Interactive Systems*.

Maclean, A., Young, R., Bellotti, V., and Moran, T. 1991, Design Space analysis: bridging from theory to practice via design rationale, *Technical report EPC-1991-128*.

Mylopoulos, J., Chung, L., Liao, S., Wang, H., Yu, W. 2001, Exploring alternatives during requirement analysis, *IEEE Software*, 18(1), 92–96.

Nof, S. Y., Wilhelm, W. E., Warnecke, H. J. 1997, *Industrial Assembly* (Chapman & Hall, London).

Oliveira, D. M., Cao, S. C., Hermida, X. F., Rodríguez, F. M. 2007, Virtual reality system for industrial training. In *Proceedings of 2007 IEEE International Symposium on Industrial Electronics*, 1715–1720.

Roam, D. 2008, *The back of the napkin: solving problems and selling ideas with pictures* (Portfolio, New York, NY).

Pedell, S. 2004, Picture scenarios: an extended scenario-based method for mobile appliance design. In *Proceedings of OzCHI 2004*.

Peniche, A., Diaz, C., Trefftz, H., Paramo, G. 2011, An Immersive Virtual Reality Training System for Mechanical Assembly. In *Proceedings of the 4th International Conference on Manufacturing Engineering, Quality and Production Systems*.

Preece, J., Rogers, Y and Sharp, H., 2002. Interaction Design: Beyond Human-Computer Interaction (John Wiley and Sons, New York).

Rittle-Johnson, B., Siegler, R., Alibali, M. W. 2001, Developing conceptual understanding and procedural skill in mathematics: an iterative process, *Journal of Educational Psychology*, 93(2), 346–362.

Truong, K. N., Hayes, G. R., and Abowd, G. D. 2006, Storyboarding: an empirical determination of best practices and effective guidelines. In *Proceedings of the 6th Conference on Designing Interactive System*, 12–21.

Vredenburg, K., Isensee, S., and Righi, C. 2002, *User-Centred Design: An Integrated Approach* (Prentice-Hall Inc, New Jersey).

HEALTHCARE

IDENTIFYING CAUSAL PATTERNS AND ERRORS IN ADVERSE CLINICAL INCIDENTS

Rebecca Mitchell, Ann Williamson & Brett Molesworth

School of Aviation, University of New South Wales, Australia

This study identified precursor (PE) and associated contributing factors (CFs) of clinical incidents in health care. A random sample of 498 clinical incidents in Australia were reviewed. Staff action was the most common type of PE identified. Correspondence analysis for all PEs that involved staff action by error type showed that rule-based errors were strongly related to performing medical or monitoring tasks and medication issues. Skill-based errors were strongly related to misdiagnoses. Factors relating to the organisation (66.9%) or the patient (53.2%) were the most commonly identified CFs for each incident. This study highlights the need for targeted approaches to tackling clinical incidents, based on an understanding of why they occur.

Introduction

There are various approaches that have been adopted to identify the precursor and contributing factors of adverse clinical incidents involving patients in health care. While human factors classifications systems that have been developed for health care are diverse in their structure, what is common across the literature is that human factors, particularly human error, plays a leading role in clinical incidents. Errors, defined as 'the failure of a planned action to proceed as planned' (US Institute of Medicine 2000), have been retrospectively analyzed in health care, but studies differ in the way that medical errors are classified. Many analyses used job-related descriptions of the nature of errors. For example, in a study of errors in radiology, the error classification included 'request for wrong patient', 'illegible request' or 'duplicate request' (Martin 2005). This type of approach is informative in providing direction in which task or job areas where errors are most likely to occur, but it is not descriptive in terms of the type of cognitive failure that explains why the particular error type occurred. The obvious advantage of cognitive classifications of error is that they provide insight into the nature of error itself which is helpful in understanding why it occurred.

Within New South Wales (NSW) Australia, clinical incidents that are categorized as serious have an Root Cause Analysis (RCA) investigation conducted by health care teams not involved in the incident. While RCAs can be useful to identify local issues, aggregated analysis of RCA findings could be useful to implement system-wide improvements (Wu et al. 2008; Nicolini et al. 2011). The aim of this research is to identify precursor and associated contributing factors to clinical incidents in a hospital setting using the Human Factors Classification Framework (HFCF) for patient safety.

Method

All clinical incidents in 2010 and 139 incidents in 2009 in NSW with a RCA investigation report were randomly reviewed (totaling 498 incidents). The RCA text-based reports were classified using a systematic coding framework.

Human factors classification framework for patient safety

The HFCF for patient safety was developed to identify information from narrative reports of clinical incidents (Mitchell et al. 2011). The HFCF for patient safety was adapted from an existing framework that identified the role of human factors in work-related fatalities (Williamson and Feyer 1990).

Precursor events (PEs) leading to the clinical incident were defined as discrete events which played a role in the occurrence of the incident and were linked in time to the incident. While time separating the events is variable and may range from seconds to days, PE1 is always closest to the incident and PE2 occurs prior to PE1 in the temporal sequence. The framework also incorporates any other factors that play a causal role. These are called contributing factors (CFs) and are defined as factors, circumstances, actions or conditions that *pre-existed* before the precursor event sequence began. CF's will have played a part in the origin or development of the incident or to increased risk of the incident occurring (World Health Organization 2009).

Each PE was coded into one of five categories then a number of subcategories. The five main categories included: (i) Equipment; (ii) Work environment; (iii) Staff action; (iv) Patient; and (v) Other factors. The framework allows classification of up to four PEs leading to the clinical incident. Contributing factors were classified into seven categories and a number of sub-categories. These included the same five categories as the PEs focusing on pre-existing conditions and two additional categories of: (i) Organizational; and (ii) Individual factors. Further detail regarding the role of error was also classified for staff action-related classifications in the PEs or CFs using Rasmussen's (Rasmussen 1982) skill, rule or knowledge-based error classifications, or a violation (Reason 1997).

Inter-rater reliability

Inter-rater reliability between four coders using the HFCF for patient safety is high, with average percent agreement between four coders for PE's at the first level of coding 97% and at 73% for error type (Mitchell et al. 2011).

Data analysis

Data were analyzed using SAS version 9.3 (SAS Institute 2012). Descriptive statistics were used to describe the types of PEs and CFs. Correspondence analysis was used to examine the relationships between the staff action PEs and error type. Correspondence analysis is an exploratory technique that represents categories as points on a two dimensional plot based on the chi-square distances between the categories (Clausen 1988). Categories that appear close together on the plot have a stronger association than categories that appear far apart.

Results

Of the 498 RCA investigative reports, 60% involved the death of a patient. The four most common incident types were procedures involving the wrong patient or wrong body part (22.9%), misdiagnoses or missed diagnoses (15.5%), in-hospital falls (8.8%), and inadequate treatment or care (7.0%).

Precursor events

Almost all (98.8%) clinical incidents had at least one PE identified, 67.1% had at least two PEs, 29.9% had three PEs and 11.6% had four PEs identified. Staff action was the most common type of event identified for all PEs (Table 1).

Table 1.   Type of precursor event (PE) involved in the clinical incident.

	PE1 (n = 492)		PE2 (n = 334)		PE3 (n = 149)		PE4 (n = 58)		Total
Precursor	n	%	n	%	n	%	n	%	n
Equipment	20	4.1	6	1.8	3	2.0	–	–	29
Work environ[1]	1	0.2	–	–	–	–	–	–	1
Staff action	399	81.1	292	87.4	120	80.5	35	60.3	846
CD issues[2]	81	16.5	68	20.4	32	21.5	15	25.9	196
MTF[3]	157	31.9	105	31.4	33	22.1	8	13.8	303
Monitoring	13	2.6	27	8.1	17	11.4	2	3.4	59
Delay	55	11.2	46	13.8	17	11.4	9	15.5	127
Misdiagnosis	73	14.8	34	10.2	12	8.1	1	1.7	120
Medication	17	3.5	9	2.7	8	5.4	–	–	34
Staff act nec[4]	3	0.6	3	0.9	1	0.7	–	–	7
Patient factors	50	10.2	16	4.8	5	3.4	5	8.6	76
Other factors	3	0.6	2	0.6	3	2.0	–	–	8
Not known	19	3.9	18	5.4	18	12.1	18	31.0	73
TOTAL[5]	492	100	334	100	149	100	58	100	1,033

[1]Work environment. [2]Communication or documentation issue. [3]Medical task failure. [4]Not elsewhere classified. [5]Six RCA investigations where expert review thought that not likely to have been preventable deaths.

Table 2.   Select precursor events for staff action factors by error type.

Staff action precursors	Total n	% of action
Medical task failure	**303**	
Skilled-based	94	31.0
Rule-based	173	57.1
Knowledge-based	3	1.0
Violation	1	0.3
Not known	2	0.7
Monitoring tasks	**59**	
Skilled-based	12	20.3
Rule-based	43	72.9
Knowledge-based	1	1.7
Not known	3	5.1
Misdiagnosis	**120**	
Skilled-based	103	85.8
Rule-based	8	6.7
Knowledge-based	8	6.7
Not known	1	0.8
Medication issue	**34**	
Skilled-based	6	17.6
Rule-based	21	61.8
Timing	3	8.8
Not known	4	11.8

Of the staff action-related PEs, medical task failures, monitoring tasks and medication issues were commonly identified as rule-based errors, while misdiagnoses were commonly identified as skill-based errors. Knowledge-based errors were relatively rare (Table 2). The correspondence analysis produced a two-dimensional map that showed that skill-based errors were strongly related to misdiagnoses (identified by their closeness in the figure). Rule-based errors were strongly related to both monitoring tasks and misdiagnoses as well as medication errors (Figure 1).

Contributing factors

Almost all (93.6%) clinical incidents had a least one CF identified. Just over two-thirds (64.9%) of incidents had two CFs identified, 31.5% had three CFs and 11.8% had four CFs identified. Organizational factors (66.9%), predominantly work practices, policies and procedures (49.8%), and patient factors (53.2%), predominantly pre-existing physical health (44.0%), were the most common factors identified that were identified as CFs that contributed to each incident (Table 3).

Discussion

The HFCF for patient safety provides a hierarchical classification system to identify multiple causation factors that are involved in the occurrence of clinical events. This

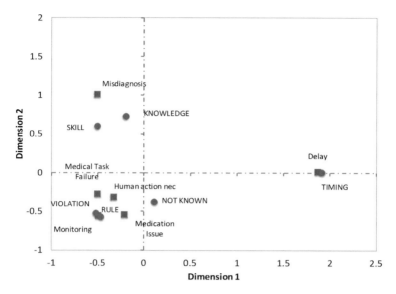

Figure 1. Results of correspondence analysis showing the relationship between precursor events that involved staff action by type of error.

research showed that staff action was the most common type of PE identified. This is not a surprising given the essential role of staff in the delivery of health care. Health professionals work in a complex environment that often involves changing and deteriorating patient conditions, and frequent interactions among multiple health care staff (Amalberti and Hourlier 2007); all circumstances that have the potential to result in an error. Despite their best endeavours, individuals are not perfect and some degree of error can occur in any workplace (Leape et al. 1991). In complex systems, such as health care, even relatively minor errors can have adverse consequences (Leape et al. 1991; Nyssen and Blavier 2006).

Examination of the role of error by type of staff action using correspondence analysis identified that error types differed by the type of action performed. There were only a relatively small proportion of knowledge-based errors, instead the most common types of errors identified were either rule- or skill-based errors. Medical task failures were strongly related to rule-based errors. Failures in monitoring tasks and medication issues were also most strongly related to rule-based errors. Misdiagnoses were strongly related to skill-based errors. The nature of the different types of errors indicate that different prevention strategies should be targeted for specific types of health care activities.

The CFs identified largely represented latent conditions that increase the likelihood of a PE occurring. The ability to identify these latent conditions provides additional information on the key characteristics that led to the clinical incident. Organizational factors, predominantly work practices, policies or guidelines, played a role

Table 3.   Type of contributing factors identified as involved in the event.

Contributing factors (CFs)	Contributed to each incident (n = 498)	
	n	% Cases
Equipment	**23**	**4.6**
Lack of equipment	7	1.4
Medical equipment failure – design	10	2.0
Medical equipment failure or breakage	5	1.0
Medical equipment not elsewhere classified	1	0.2
Work environment	**20**	**4.0**
Staff action factors	**176**	**35.3**
Communication or documentation issues	115	23.1
Medical task failure	23	4.6
Monitoring	10	2.0
Delay	14	2.8
Misdiagnosis	5	1.0
Medication issue	8	1.6
Staff action nec[1]	1	0.2
Patient	**265**	**53.2**
Physical health-pre-existing	219	44.0
Health state	21	4.2
Communication issues	16	3.2
Medication	2	0.4
Toxicology	2	0.4
Patient not elsewhere classified	5	1.0
Organizational factors	**333**	**66.9**
Work practices, policies or guidelines	248	49.8
Supervision	12	2.4
Organizational resources	28	5.6
Work pressure	45	9.0
Individual factors	**91**	**18.3**
Experience	36	7.2
Training	44	8.8
Fatigue	3	0.6
Stress	1	0.2
Individual factors not elsewhere classified	7	1.4
Other factors	**4**	**0.8**
Not known	**32**	**6.4**

in just over two-thirds of incidents. In most cases, these involved existing work practices, policies or guidelines not being followed.

The clinical incidents reported in this research represent the most serious clinical incidents, along with incidents that are required to be reported nationally in Australia as sentinel clinical events. In the majority of cases, these are out of the ordinary cases, where one thing goes wrong and, due to the rarity of the event and the alignment of different combinations of other factors, the clinical incident occurs. For the majority of clinical incidents, they are multi-causal in nature, often with multiple prime causes identified, making it difficult to target one single common approach for prevention. However, by using the HFCF for patient safety to establish the PE causal sequences and to identify the associated CFs by clinical incident type, common threads between incidents are able to be identified.

There are limitations of the current study including the ability to identify the causal sequence of events and associated CFs is only as good as the information available. Some RCA investigative reports lacked detail and identification of PEs and CFs were difficult to determine for these incidents. This occurred in 1.2% of cases for PEs (where no PE was identified) and 6.4% cases for CFs. It is possible that some clinical incidents may go unreported and thus be absent from this analysis. By grouping similar patterns of clinical incidents, specific detail regarding the incident is lost. However, grouping similar incident patterns enable establishment of the magnitude of the issue, can influence resource allocation and clinical policy.

Conclusion

This study was able to provide detailed information regarding the role of human factors, particularly human error, in clinical incidents in health care. The results confirm the complex, multi-causal nature of clinical incidents and the need to identify the sequence of events, examine the relationships between the events and CFs, and to identify the importance of each PE or CF to the causal event chain to be able to identify appropriate strategies for prevention.

Acknowledgements

This research was funded by an Australian Research Council linkage grant (LP0990057) and the NSW Clinical Excellence Commission and the NSW Ministry of Health. The views expressed in this paper are the views of the authors do not necessarily reflect the views or policies of the funding agencies.

References

Amalberti, R. and S. Hourlier (2007). Human error reduction strategies in health care. *Handbook of human factors and ergonomics in health care and patient safety*. Carayon, P. Mahwah, NJ, Lawrence Erlbaum Associates: 561–577.

Clausen, S. (1988). *Applied Correspondence Analysis: An Introduction*. California, Sage University.

Leape, L. L., T. A. Brennan, N. Laird, A. G. Lawthers, A. R. Lacolio, B. A. Barnes, L. Hebert, J. P. Newhouse, P. C. Weiler and H. Hiatt (1991). "The nature of adverse events in hospitalized patients. Results of the Harvard Medical Practice Study II." *New England Journal of Medicine* 324(6): 377–384.

Martin, C. (2005). "A survey of incidents in radiology and nuclear medicine in the West of Scotland." *The British Journal of Radiology* 78: 913–921.

Mitchell, R., A. Z. Chung, A. Williamson and B. Molesworth (2011). "Human factors and healthcare-related deaths: review of coronial findings using a human factors classification framework." *Ergonomics Australia- HFESA 2011 Conference Edition* 11: 18.

Nicolini, D., J. Waring and J. Mengis (2011). "Policy and practice in the use of root cause analysis to investigate clinical adverse events: Mind the gap." *Social Science and Medicine* 73: 217–225.

Nyssen, A. and A. Blavier (2006). "Error detection: a study in anaesthesia." *Ergonomics* 49(5–6): 517–525.

Rasmussen, J. (1982). "Human errors: a taxonomy for describing human malfunction in industrial installations." *Journal of Occupational Accidents* 4: 311–333.

Reason, J. (1997). Hazards, defences and losses – the 'Swiss cheese' model of defences. *Managing the risks of organizational accidents*. Reason, J. Aldershot, Ashgate Publishing Ltd: 1–20.

SAS Institute (2012). SAS: statistical software, version 9.3. Cary, North Carolina, SAS Institute.

US Institute of Medicine (2000). To Err is Human: Building a Safer Health System. Washington DC, National Academy Press.

Williamson, A. and A. Feyer (1990). "Behavioural epidemiology as a tool for accident research." *Journal of Occupational Accidents* 12: 207–222.

World Health Organization (2009). The conceptual framework for the International Classification for Patient Safety. Geneva, World Alliance for Patient Safety, WHO.

Wu, A., A. Lipshutz and P. Pronovost (2008). "Effectiveness and efficiency of root cause analysis in medicine." JAMA 299(6): 685–687.

BRIDGING THE RESEARCH PRACTICE GAP IN HEALTHCARE HUMAN FACTORS AND ERGONOMICS

Patrick E. Waterson & Janet Anderson

Loughborough School of Design, Loughborough University, UK
Florence Nightingale School of Nursing and Midwifery,
King's College London, UK

This short paper summarises past work which has looked at the research-practice gap within human factors and ergonomics (HFE). Some of this work has focused on Human Factors Integration (HFI), whilst other literature discusses the challenges involved in carrying out HFE research and influencing practice across a range of domains (e.g., defence). The paper serves as an introduction to a wider set of presentations and experience reports from other researchers and practitioners in healthcare HFE which will presented at the conference.

Introduction

This paper considers some of the lessons that can be drawn from previous work examining the difficulties involved in applying HFE skills and knowledge within industry and work contexts. One of the motivations for this is to examine what light this sheds on current attempts to apply HFE within healthcare. The paper is organised into two main sections: (1) applying HFE within industry – lessons learnt; and, (2) applying HFE within healthcare – current challenges.

Applying HFE within industry – some lessons from the past

Accounts of the problems involved in applying human factors within industry date back to some of the earliest work within human factors and ergonomics. Over the years a range of problems and barriers have been identified (see Waterson and Lemalu-Kolose, 2010, for a brief review). These include: the perception that human factors data and information was inaccessible relative to other formats (e.g., graphs, tables); skepticism regarding the value and cost-effectiveness of HFE; the relative weak position and low profile of the ergonomist within organisations; and, differences in terms of "mindsets" and values which exist between HFE practitioners and their colleagues (e.g., social science vs. engineering). More recently, attempts to formalize the process of integrating HFE have led to the development of human factors integration (HFI) and other initiatives. Some indication that HFI and the gap between research and practice remain important considerations can be drawn from the fact that it has been the focus of two recent IEHF lectures (Peter Buckle in 2010 and Barry Kirwan in 2011).

Applying HFE within healthcare – current challenges

There have been a number of recent calls to increase the application of HFE within healthcare and patient safety (e.g., Gurses et al., 2012). At the same time, there has been an acknowledgement that this poses a number of challenges. One of these might be termed the 'victims of our own success' problem. Many clinicians and healthcare workers have seized on HFE as a way of solving their problems whilst at the same time not understanding the scope of what HFE and what is realistically achievable. A related problem is that many of the 'solutions' that derive from HFE work are seen as applicable across a range of healthcare settings and systems. Whilst this may be true of some types of intervention (e.g., attempts to improve clinical handover), it is unlikely to apply to larger-scale interventions which are almost certainly shaped by other factors (e.g., national culture, healthcare policy). Much of the work for example, on large-scale healthcare information technologies (e.g., electronic patient records) might be said to fit into this category. An additional problem is that a number of HFE methods and techniques within healthcare are relatively immature and in need of further scientific development (e.g., measures of patient safety culture). It is hoped that these challenges and other issues for HFE and healthcare will make up some of the debates and discussions within the conference session at IEHF 2013.

References

Gurses, A.P., Ozok, A.A. and Provonost, P.J. (2012). Time to accelerate integration of human factors and ergonomics in patient safety. *BMJ:Quality and Safety, 21,* 347–351.

Waterson, P.E. and Lemalu-Kolose, S. (2010). Exploring the social and organisational aspects of human-factors integration: a framework and case study. *Safety Science, 48,* 482–490.

HEALTHCARE HUMAN RELIABILITY ANALYSIS – BY HEART

James Ward[1], Yi-Chun Teng[2], Tim Horberry[1] & P. John Clarkson[1]

[1] *Engineering Design Centre, University of Cambridge, UK*
[2] *Murray Edwards College, University of Cambridge, UK*

As part of the investigations into a surgical incident involving the accidental retention inside a patient's venous system of a guide wire for central venous catheterisation (CVC), the Human Error Assessment and Reduction Technique (HEART) was used to examine the potential for further occurrences. It was found to be time-efficient and to yield plausible probabilities of human error, although its use in healthcare has challenges, suggesting adaptation would be beneficial.

Introduction

CVC involves the insertion of a small tube – a catheter – into a vein. A guide wire is used to help insert the catheter, but occasionally it can be pushed too far into the vein, and subsequently retained within the patient's body, without immediate detection. Following an incident, we used a range of methods, including HEART, to examine the risks surrounding guide wire use. We report on our experiences.

Methods

HEART estimates the probability of human error in a task, by identifying a Generic Task Type (GTT), combined with an analysis of Error Producing Conditions (EPCs) (Williams, 1988). HEART was chosen as it is quick to use, is publicly available, is applicable across a range of domains and helps suggest areas for improvement. A hierarchical task analysis was developed to describe the catheter insertion procedure. Using HEART, three critical sub-tasks were analysed individually, by a team comprising a safety engineer, a human factors expert and a medical student. To expedite the process, we used an abbreviated form of HEART, comprising 12 EPCs (EUROCONTROL, 2004), rather than the original 38 (Williams, 1988). The analysis also helped identify a number of candidate solutions.

Results and discussion

The HEART took approximately two hours to complete, although much of this time was spent debating the applicability of each GTT and the relevance and – in particular the proportion – of each EPC to the sub-tasks. The nominal likelihood of failure was similar for each of the sub-tasks – approximately 0.01. This is around one order of magnitude greater than that measured through incident reporting over a six year period in another hospital. So it is plausible, but seems high intuitively.

Several points can be made about the applicability of HEART to healthcare: 1) we had difficulties interpreting and translating many of the descriptions – originally developed for use in the nuclear industry – to healthcare. A careful translation of these to the healthcare domain would be helpful. 2) For validation, a lack of accurate data (e.g. through incident reporting) means that comparing HEART results against reality is problematic. Audits of practice could go some way to helping address this. 3) Our experiences of healthcare have been that large variability exists in the context of use (e.g. between staff or EPCs in different locations), which may give rise to large differences in risk from one scenario to another. Care must be taken to specify precisely the situation to be analysed. 4) The HEART method was fairly easy to learn and apply time-efficiently. Both are essential, with such time pressure in healthcare, and a fix-it-quick culture meaning that lengthy analysis would be hard to justify to participants, and their peers and managers. 5) At present a HF expert – scarce in healthcare – would need to facilitate the HEART. 6) As a result of our analysis using HEART and other human factors methods (e.g. interviews) we proposed 14 possible solutions. In evaluations by CVC users, three stood out as particularly practical and effective, and will be tested further for implementation.

Conclusions

This study suggests that HEART may be useful in healthcare, although other methods should be used in addition. To facilitate its adoption in healthcare, minor adaptations should be made. We propose HEARTH – HEART for Healthcare.

References

EUROCONTROL, 2004. Review of techniques to support the EATMP safety assessment methodology. Volume I, (Eurocontrol, Brétigny-sur-Orge, France).

Williams, J. C. 1988. *A data-based method for assessing and reducing human error to improve operational performance*, 4th IEEE conference on Human factors in Nuclear Power plants, Monterey, CA.

DEVELOPING A SIMULATOR TO HELP JUNIOR DOCTORS DEAL WITH NIGHT SHIFTS

Michael Brown[1], Paul Syrysko[2], Sarah Sharples[1], Dominick Shaw[3], Ivan Le Jeune[3], Evridiki Fioratou[1] & John Blakey[4]

[1]*Human Factors Research Group, University of Nottingham, UK*
[2]*Propheris Limited, Nottingham, UK*
[3]*Respiratory Medicine, University of Nottingham, UK*
[4]*Research and Innovation, Nottingham University Hospitals NHS Trust, UK*

Outside of the Monday to Friday, 9 to 5 of hospital operation a skeleton staff of junior doctors, support workers, and nurses must provide safe and timely care in hospitals. Junior doctors currently receive little training to help them deal with the unique demands of 'Out of Hours' care. We aim to rectify this situation by exploiting a unique data resource to accurately simulate a night shift. This paper will explore the need to develop a simulator to increase junior doctor's skills for prioritizing their high and changing workload, making key decisions on contacting other services and specialists, planning complex routes around large and potentially unfamiliar hospitals, managing their personal needs for rest and sustenance, and handing over succinctly and accurately at the end of their shift.

Introduction

Only 25% of the hospital year falls between 9am and 5pm Monday to Friday. During the remaining 75%, known as 'Out of Hours' (OOH), a skeleton staff of junior doctors, support workers, and nurses must provide safe and timely care in hospital. These challenges are magnified by the increased stress and fatigue when working OOH. Failure to perform well may both delay patient care and impact on its quality with well-documented effects on error rate and adverse outcomes (Aylin et al, 2010).

Junior doctors under pressure

Over the past decade, individual junior doctors' hours have fallen by more than 35% to comply with the European Working Time Directive (Pickersgill, 2001). In parallel admissions to secondary care have risen by more than 15% and individual patients tend to have more complex medical problems and treatments (Barnett et al, 2012). These changes have placed an increased pressure on staff, with higher work intensity during shifts (Royal College of Physicians, 2012).

In teaching hospitals, there are added complications during the August changeover period, as the problems associated with OOH care are amplified at the start of the academic year when newly qualified junior doctors begin their first OOH shifts.

Internal reports and academic publications (Teahon, 2009; Vaughan et al, 2011) have highlighted concerns over the preparedness of junior doctors and nurse coordinators for this increasingly demanding aspect of their job. Indeed, there is persisting national concern over a potential increase in mortality during the yearly junior doctor changeover (Jen et al, 2009; Young et al, 2011). There is also growing evidence that this high demand role leads to emotional burnout and low quality of life (West et al, 2011).

Towards a solution

Previous work has shown that targeted digital innovation can help with the OOH situation in general. A wireless OOH working solution deployed in two large teaching hospitals in the UK has been shown to have significant advantages including limitation of information loss, fewer inappropriate cardiac arrest calls, shorter lengths of stay and a reduction in reported incidents relating to handover of information or OOH response from a 22% chance of experiencing an incident in a single day without the system to 7% with the system (Blakey et al, 2012). In addition to these objective improvements, staff preferred the system.

By examining the data from this wireless taskflow system we have begun to gather quantitative data on intermediate processes leading up to adverse events in OOH care when new Junior Doctors work their first shifts. Crucially, we now have data on the denominator for adverse events, enabling the findings of investigations into adverse events to be put into an appropriate context. Examples of quantitative data suggesting risk in the system include (Blakey, Fearn & Shaw, manuscript under review):

- More potentially dangerous omission-type errors if drug cards are rewritten OOH
- A greater proportion of OOH tasks are classed as urgent suggesting failure to recognise deterioration in daytime hours
- Longer time to complete complex tasks for particularly unwell patients as new doctors are uncertain how to order certain tests, complete forms, where guidelines are located, or when to contact seniors for advice
- Evidence of non-urgent tasks being done more quickly as new doctors fail to prioritize or are doing work in batches to simplify route planning.

These findings are consistent with previous qualitative research in this domain (Sheehan, Wilkinson & Bowie, 2012).

Recent innovative information technology developments such as this wireless taskflow system permit for the first time an accurate description of work undertaken OOH (Blakey et al, 2012) allowing us to accurately model the variations in night shift behaviour and relating the outcomes of simulator based interactions with real world patient outcomes. Using these data we are working with a commercial partner (Propheris ltd) to create a night shift simulator using video game technology and sensibilities to engage and challenge new staff.

Having explored data collected from the wireless task flow system and reviewed relevant literature we have identified the following skills as paramount for efficient operation of junior doctors during OOH shifts:

- Prioritizing high and changing workloads
- Decision making about contacting other services and specialists
- Route planning
- Managing personal factors: stress, fatigue and sustenance
- Succinct and accurate handover at the beginning and end of a shift

Thus, we will develop a simulator with the aim of training junior doctors in these skills before and during their first weeks of working OOH by exploiting both high detail real world logs of night shift activity and high accuracy mapping.

The simulator will involve users prioritising tasks and guiding an avatar through a detailed recreation of the hospital that they work in. They will experience an accelerated shift when attempting to undertake tasks in an efficient manner to prevent patient deterioration, complaints from patients or other staff, or fatigue that impairs task completion accuracy. Following each simulator session users will receive tailored constructive feedback on their performance and links to dedicated on-line resources to aim further skill development.

Simulators and games in medical education

The use of simulation and games in medical education is well-established. Teaching demanding tasks using simulation delivers better results than traditional methods (Sroka et al, 2010), and the pattern of errors in simulated tasks is similar to that seen in actual practice (Chopra et al, 1994). The use of such training methods is looked on highly favourably by students and junior doctors (Kron et al, 2010). Educational games have proven efficacious for patients (Kato et al, 2008), nursing staff (Aebersold et al, 2012), and clinicians (Kanthan & Senger, 2011) in a variety of scenarios. Users report greater personal learning and engagement, reduced stress and a desire to compare their performance against that of their peers.

Interactivity, repetition, and feedback have been shown to be the key features of successful electronic learning tools (Cook et al, 2010). Game-like software should be engaging, encourage repetition to improve scores and advance, and provide specific and tailored feedback to users. In the case of our simulation, these educational messages will be agreed by a panel of experienced clinicians, and draw on programs with proven benefits, such as those developed by McCue and Sachs (1991).

A simulator or a game?

While it is not the main thrust of this article, it is important to mention the difference between the terms Game and Simulation. Traditionally games can be seen as systems of rules whose primary purpose is entertainment, whereas simulations focus on accurate representation of a real-world system (Nichols, 1988). However, concepts such as Simulation Games and Serious Gaming (Abt, 1970) mean that the line between the two is far from clear. For the purposes of our research we have labelled the proposed technology a simulator in an attempt to emphasise the safety critical nature of our developments and the accuracy of the workflow data and mapping that informs it, but acknowledge and aim to build on the existing work around serious games and game development in general.

Human factors consideration

Designing for stress

There are a number of ethical challenges associated with a simulator that is designed to represent a stressful situation. Traditionally within experimental design the inclusion of a task that may induce stress can be perceived as a 'risk' (Milligan, 2012) that require a number of considerations, including preparation of the participant for the anticipated experience of stress, ability to withdraw from the task at any time, and management, monitoring and de-briefing of the experience of stress during task completion. However, the experience being simulated in our proposed application is by its nature stressful, and one of the important challenges of training a junior doctor is to ensure that they are able to cope with the stressful tasks that they encounter, and manage their behaviour effectively. Whilst we are used to designing tasks that do not have stress 'risks' – in this case it is important that the user is prepared for the stress, and it is preferable that they experience a stressful or negative outcome in the simulated context than in the real hospital setting.

The representation of negative outcomes also needs to be carefully considered. The users need to be completely aware of when a patient has not survived, or needs further emergency treatment. The simulation offers the opportunity to allow the user to trace data to understand the reasons for a negative outcome – whether it was inevitable, or if a different management of treatment could have led to a more positive outcome.

Complexity

Within the domain of virtual reality design there has been much debate about the importance of photorealism and the need to simulate to a high level of detail and complexity in a range of contexts (Eastgate, D'Cruz & Wilson, 1997). This debate also applies there – if we consider that the element of the task being trained is primarily of rule and knowledge based behaviour (Rasmussen 1983), then it is important that these elements of the task are those that are represented in detail (Eastgate, Nichols & D'Cruz, 1997). These elements do not necessarily demand high levels of photorealism, for example. Thus our focus is on situational realism rather than sensory realism. Determining which aspects of working a night shift must be replicated in order to maintain the important elements will first be explored via stakeholder meetings and requirements gathering and later through iterative testing and re-design of prototypes.

Ubiquitous navigation

Hospitals are generally large, complex and often multi-building environments which new doctors must quickly learn to navigate effectively. Previous research have shown success using virtual environments to teach both indoor (Ruddle, Payne & Jones, 1997) and outdoor (Brunye, Gardony, Mahoney & Taylor, 2012) navigation in new environments, but little work exists bridging the two. The differences that have been identified between strategies for navigating in outdoor and indoor environments (Nurmi et al, 2011) suggests that optimising for both within a single system may not be a trivial undertaking.

Measuring performance during simulation

Our literature review suggests that a key indicator of a doctor who is coping with their task load is that they are acting proactively rather than reactively, an issue analogous to the air traffic control domain (Reason, 2000; Uang, Rakas & Bolic, 2008). There are a number of different ways in which 'proactive behaviour' could be inferred – firstly by identifying behavioural markers or visibly observable behaviour, where, for example, an increased amount of passive monitoring of a system could suggest that a user is effectively managing their workload (Sharples, Millen, Golightly and Balfe, 2011). A second way could be by asking participants to mimic the task of handover and analyse the output in comparison with both other participants' performance and real world data. A further approach would be to score outcomes such as number of patients deteriorating, number of jobs pending, or distance walked, but unless the scoring system is carefully defined, it is potentially possible to promote inappropriate strategies that lead to apparently positive outcomes.

Discussion

Having identified many of the difficulties facing junior doctors during their first OOH shifts we now aim to develop a simulator to help train these doctors in personal, task and location management.

Measuring impact

An important issue to considering training simulators is how to determine the effectiveness of the system and in turn justify implementing it on a large scale. This problem is exaggerated in the complex, dynamic and generally difficult to predict world of health care. As the ultimate goal of such interventions, patient outcome in terms of morbidity, length of stay and satisfaction would be an obvious choice, but the wide range of uncontrollable factors that influence these parameters means more sensitive intermediate phenotypes will be measured to ensure sample size and test duration are realistic. We will use a wide range of tools to examine the effect of the simulator on both hospital staff and the patients they treat and other systems in operation.

Plan of work

We have extracted detailed data on workload and geography for the simulation, extensively photographed the location, and are improving the algorithms and graphics. The next steps are to complete the prototype development, and test the application in partnership with stakeholders in the training of junior doctors. This work will begin with focus group style stakeholder meetings in order to validate and expand upon the human factors issues highlighted above. The results of these meetings will feed into the development of a prototype system that will be developed iteratively with user testing at each stage of implementation.. By comparing actual performance through the taskflow system and indoor positioning technologies we also aim to robustly assess the relation of the simulation to actual practice and its impact on junior doctors. In the medium term developments will permit simulation of potentially challenging and costly medical scenarios, such as ward moves or critically high admission rates.

Acknowledgements

We would like to acknowledge the financial support from Propheris Limited, the Special Trustees for Nottingham University Hospital, the East Midlands Health Innovation and Education Cluster, and the National Institute for Health Research, without which this project would not be possible. This work was carried out in collaboration with Horizon Digital Economy Research, through the support of RCUK grant EP/G065802/1 and the MATCH project thought the support of EPSRC grant *EP*/F063822/1.

References

Abt, C. 1970, *Serious Games*. University Press of America,

Aebersold M, Tschannen D, Bathish M. 2012, Innovative simulation strategies in education. *Nurs Res Pract*. 765212.

Aylin P, Yunus A, Bottle A, Majeed A, Bell D. 2010, Weekend mortality for emergency admissions. A large, multicentre study. Qual Saf Health Care, 19(3), 213–217.

Barnett, K., Mercer, S., Norbury, M., Watt, G., Wyke, S., Githrie, B. 2012, Epidemiological of multimorbidity and implications for health care, research and medical education: a cross-sectional study. *The Lancet*, 380 (9836), 37–43.

Blakey, J. D., Guy, D., Simpson, C., Fearn, A., Cannaby, S., Wilson, P., Shaw, D. 2012, Multimodal observational assessment of quality and productivity benefits from the implementation of wireless technology for out of hours working. *BMJ Open*, 2(2), e000701.

Blakey, J., Fearn, A., Shaw, D. Under Review, What Drives the "August Effect"? A Quantitative Study of the Effect of Junior Doctor Changeover On Out Of Hours Work. *Journal of the Royal Society of Medicine*.

Brunye, T., Gardony, A., Mahoney, C., Taylow, H. 2012, Going to town: Visualized perspectives and navigation through virtual environments. *Computers in Human Behaviour*, 28 (1), 257–266.

Chopra, V., Gesink, B. J., de Jong, J., Bovill, J. G., Spierdijk, J., Brand, R. 1994, Does training on an anaesthesia simulator lead to improvement in performance? *Br J Anaesth*, Sep; 73(3):293–7.

Cook, D. A., Levinson, A. J., Garside, S., Dupras, D. M., Erwin, P. J., Montori, V. M. 201, Instructional design variations in internet-based learning for health professions education: a systematic review and meta-analysis. *Acad Med.* May; 85(5):909–22.

Eastgate, R., D'Cruz, M., Wilson, J. 1997, A Strategy for Interactivity within Virtual Environment Applications: A Virtual ATM Case Study. *Proceedings of Virtual Reality WorldWide*, Santa Clara, California.

Eastgate, R., Nichols, S & D'Cruz, M. 1997, Application of human performance theory to virtual environment development. In, D. Harris (ed), *Engineering Psychology & Cognitive Ergonomics, Volume 2 – Job Design and Product Design*, pp. 467–475. Ashgate: Aldershot.

Jen, M. H., Bottle, A., Majeed, A., Bell, D., Aylin, P. 2009, Early in-hospital mortality following trainee doctors' first day at work. *PLoS One*, 4(9):e7103.

Kanthan, R., Senger, J. L. 2011, The impact of specially designed digital games-based learning in undergraduate pathology and medical education. *Arch Pathol Lab Med*, Jan; 135(1):135–42.

Kato, P. M., Cole, S. W., Bradlyn, A. S., Pollock, B. H. 2008, A video game improves behavioral outcomes in adolescents and young adults with cancer: a randomized trial. *Pediatrics*, Aug; 122(2):e305–17.

Kron, F. W., Gjerde, C. L., Sen, A., Fetters, M. D. 2010, Medical student attitudes toward video games and related new media technologies in medical education. BMC Med Educ, 10 (50).

McCue, J. D., Sachs, C. L. 1991, A stress management workshop improves residents' coping skills. *Arch Intern Med.*, Nov; 151(11):2273–7.

McGaghie, W. C., Issenberg, S. B., Cohen, E. R., Barsuk, J. H., Wayne, D. B. 2011, Does simulation-based medical education with deliberate practice yield better results than traditional clinical education? A meta-analytic comparative review of the evidence. *Acad Med.*, Jun; 86(6):706–11.

Nichols, S. 1998. Games Psychology. *Games Monthly.* October, 1998.

Nurmi, P., Salovaara, A., Bhattacharya, S., Pulkkinen, T., Kahl, G. 2011, Influence of Landmark-Based Navigation Instructions onUser Attention in Indoor Smart Spaces. *Proceedings of the 16th international conference on Intelligent user interfaces*, 33–42.

Milligan, C. 2012, *Social Science Research Ethics.* Available at: http://www.lancs. ac.uk/researchethics/index.htmlaccessed on the 26th of September, 2012.

Pickersgill, T. 2001, The European working time directive for doctors in training. *BMJ.*, 323(7324).

Rasmussen, J. 1983, Skills, rules, and knowledge: Signals, signs, and symbols, and other distinctions in human performance models. *IEEE Transactions on Systems, Man, and Cybernetics*, SMC-13(3): 257–266.

Reason, J. 2000, Human Error: models and management, *British Medical Journal.* 320, pp. 768–770.

Royal College of Physicians.2012, *Hospitals on the edge? The time for action.* Available from http://www.rcplondon.ac.uk/projects/hospitals-edge-time-action accessed on the 28th of September 2012.

Ruddle, R., Payne, S., Jones, D. 1997, Navigating buildings in "desk-top" virtual environments: Experimental investigations using extended navigational experience. *Journal of Experimental Psychology*, 3(2), 143–159.

Sharples, S., Millen, L., Golightly, D., Balfe, N. 2011, The Impact of Automation in Rail Signaling Operations. *Proceedings of the Institution of Mechanical Engineers, Part F: Journal of Rail and Rapid Transit*, 225, 179–191.

Sheehan, D., Wilkinson, T., Bowie, E. 2012, Becoming a practitioner: Workplace learning during the junior doctor's first year. *Medial Teacher*, 1–10, early online edition, available from http://informahealthcare.com/doi/pdf/10.3109/0142159X.2012.717184

Sroka, G., Feldman, L. S., Vassiliou, M.C., Keneva, P. A., Fayez, R., Fried, G. M. 2010, Fundamentals of laparoscopic surgery simulator training to proficiency improves laparoscopic performance in the operating room-a randomized controlled trial. *The American Journal of Surgery*, 199, 115–120.

Teahon K. 2009, *Report on a Study of May 2009.* University Hospitals Nottingham, NHS Trust.

Uang, G., Rakas, J., Bolic, T. 2008, Proactive, reactive and interactive risk assessment and management of URET implementation in air route traffic control centres. *International Conference on Research in Air Transportation (ICRAT)*, Fairfac VA,

Vaughan L, McAlister G, Bell D. 2011, August is always a nightmare: results of the Royal College of Physicians of Edinburgh and Society of Acute Medicine August transition survey. *Clin Med.*, 11(4):322–6.

West CP, Shanafelt TD, Kolars JC. 2011, Quality of life, burnout, educational debt, and medical knowledge among internal medicine residents. *JAMA.*, 7; 306(9), 952–60.

Young JQ, Ranji SR, Wachter RM, Lee CM, Niehaus B, Auerbach AD. 2011, "July effect": impact of the academic year-end changeover on patient outcomes: a systematic review. *Ann Intern Med.*, 155(5), 309–15.

CERTAIN MEDICAL DEVICES REQUIRE MORE ATTENTION ON DESIGN AND USABILITY

Abdusselam Selami Cifter & Ilgim Eroglu

Industrial Product Design Department, Mimar Sinan Fine Arts University, Istanbul-Turkey

The medical devices market is continuously growing in Turkey; however, 85% of the market depends on import trade. Although there are Turkish brands competing in the market, it is argued that there is a prejudice against these products. It was hypothesised that this prejudice originate from the design and usability shortcomings of 'made in Turkey' branded medical devices. This paper summarises and discusses the results of an interview study with 12 medical device retailers located in Istanbul, in an effort to understand their point of view regarding the general image of 'made in Turkey' branded medical devices in terms of their design and usability.

Introduction

The medical devices market is one of the fast developing markets in Turkey. Due to its stable growth, it is forecast that the market value will reach to 3.12 billion US dollars by 2015 (TPIPA, 2010). However, 85% of the market depends on import trade which is mainly dominated by far eastern countries due to price advantage (Kocak, 2008). Some of the important countries Turkey imports medical devices from are the United States, Germany, UK, Japan, France, Holland and China (Kocak, 2008).

According to a report published by Republic of Turkey Prime Ministry Investment Support and Promotion Agency, most of the 'made in Turkey' (MiT) branded medical devices are low-tech products, while the high-tech systems or materials to be used in these devices are mainly imported (TPIPA, 2010). Some of the medical devices produced in Turkey are, e.g. operating tables, dentistry units and equipment, hospital beds, gynaecological tables, orthopaedic prosthesis, patient headwalls, sterilizers and etc (TOBB, 2009). As the medical devices market is considered as one of the most important sectors in Turkey, new investments in R&D are supported by the Turkish Government. Recently, the medical devices market is designated as one of the 12 sectors which are supported by "Large-Scale Investment Incentives Scheme" (TOBB, 2009; Eren, 2009). Due to these developments in the sector, the requirement for industrial designers working in this field has also increased.

One of the shortcomings of the sector is that, as suggested by The Union of Chambers and Commodity Exchanges of Turkey (Kocak, 2008), there is a prejudice against MiT branded medical devices in terms of their reliability. However, all the medical devices manufactured in Turkey require a CE mark in order to be sold in the Turkish market. This means that these devices go through the same regulatory procedure with medical devices produced in European Union (EU).

Regulatory requirements for medical devices in Turkey

Based on the "Free Movement of Goods" principle, a medical device which fulfils the harmonised requirements of the European Commission (EC), can be freely sold all through the EU market (EC, 2012). Since Turkey is a candidate country for the EU, there is a harmonisation process of Turkish legislation to the EU (Sener & Ozdemir, 2011). This harmonisation also covers the healthcare sector including medical devices. Therefore, a medical device produced in Turkey needs to fulfil the same regulatory requirements (Sener & Ozdemir, 2011). The three European Commission's Directives relevant to medical devices are integrated into Turkish legislation. These EU Directives are:

- The Council Directive on Medical Devices Directive, 93/42/EEC
- Council Directive on In-Vitro Diagnostic Medical Devices, 98/79/EC
- Council Directive on Active Implantable Medical Devices, 90/385/EEC

The situation is also the same for the European Standards. For example, in order meet the Quality System requirements; a Turkish medical device manufacturer requires proving compliance with EN ISO 13485 (Medical Devices – Quality Management Systems – Requirements for Regulatory Purposes). Some of these Standards are translated into the Turkish language as well.

Is the problem based on design?

Medical devices can be stated as a reasonably new field of work for industrial designers in Turkey. However, to date, their position and contributions has not been clearly understood in this field. On the March 14th 2012, a meeting was organised by the Industrial Designers Society of Turkey (ETMK) regarding the design of medical devices in Turkey. In the meeting, the designers and producers met and discussed the issues relevant to the sector. According to the meeting report, there is a communication problem between designers and manufacturers. However it was also emphasised that more designers are required and should be encouraged to contribute in this sector.

Due to the fact that MiT branded medical devices require assuring the level of quality expected to fulfil the harmonised regulatory requirements of the EC, it was hypothesised that the prejudice against MiT branded medical devices might be arising from design and usability related shortcomings. In a literature review, no previous research was identified regarding the design or usability aspect of MiT branded medical devices. Therefore an interview study was carried out with 12 medical device retailers located in Istanbul.

Study method

Medical device retailers are selected as the sample group of this study, because:

- They are in communication with both manufacturers and end-users
- They sell both import trade and MiT branded medical devices
- They sell a diverse range of medical devices
- End-users frequently get in communication with their retailers if they are not satisfied with the device they bought.

In total 12 medical device retailers took part in the study. During the sampling, an effort was made to include a variety of medical device retailers selling different types of medical devices, including products for professional users and lay users. The product range included: hearing aids, orthesis and prosthesis, general hospital fixtures and equipment, dentistry equipments, basic laboratory equipment, assistive devices/rehabilitation products. Therefore purposive sampling was used as the sampling method (Robson, 2011).

Interviews are a particularly good and flexible method when the intention of the researcher is to gain in depth understanding of the subject area (Robson, 2011). Semi-structured interviews are used as the main study method. The objectives were to:

- Understand the current image of MiT branded medical devices regarding their designs.
- Investigate the current image of MiT branded medical devices in terms of their usability.
- Make a comparison of MiT branded medical devices with import trade products in terms of their design and usability aspects.

When preparing the interview questions, the aim was to keep the interview session to less than 30 minutes. In order to facilitate answering and decrease time taken in certain questions, five-point Likert scale questions were also used (some of the interviewees neutrality for these questions were also considered; therefore a neutral response was designated as '3'). For the analysis of the data, SPSS Statistical Analysis software was utilised. On the other hand, the participants were encouraged to give verbal expressions regarding the reasons behind their answers. In total, 7 questions were prepared. All the interviews were conducted at the retailers own store. During the interviews note-taking technique was utilised.

Results

The results are presented in accordance with the three main objectives of the research.

Table 1.   General image and reliability perception of MiT branded medical devices (N = 12).

	Very Bad	Bad	Mod.	Good	Very Good	Median
Current Image of MiT Branded Medical Devices	8.3%	25%	16.7%	50%	0%	**3.5**
Reliability perception of MiT branded Medical Devices	8.3%	8.3%	8.3%	75%	0%	**4**

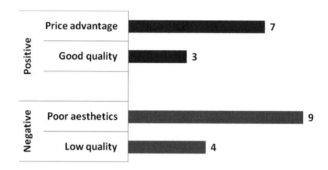

Figure 1.   Number of participants reporting positive and negative aspects of MiT branded medical devices (N = 12).

The current image of MiT branded medical devices

Firstly, the participants were asked to rate the current 'general image' and the 'reliability perception' of MiT branded medical devices by using two likert scale questions (five-point: '1' very bad; '5' very good). The results over '3' were accepted as positive responses. The percentages for each option and the median values are presented in Table 1.

As can be seen from the table, the current image of MiT branded medical devices has a median score of 3.5; fifty percent of the participants rated 3 or less. On the other hand, the 'reliability perception' received a better median score of 4 where 75% of the participants rated as 'good'. None of the participants rated MiT devices as 'very good' for either question. The participants were asked about the reasons for their answers as well. The positive and negative comments were coded and grouped under two categories. The main reasons are presented in Figure 1.

The most advantageous feature of MiT branded medical devices is the price aspect. This was mentioned by seven interviewees. On the other hand, most of the interviewees argued that in terms of how they look, MiT branded medical devices are generally poor; even though end-users value aesthetics. They mentioned that more importance should be given to the devices' aesthetics.

Table 2. Results of the Likert scale questions regarding usability (N = 12).

	1	2	3	4	5	Median
End-users' attention on usability features	0%	8.3%	0%	25%	66.7%	5
Perceived image of MiT branded medical devices in terms of their usability features	0%	0%	16.7%	83.3%	0%	4
Turkish producers' attention level to usability	8.3%	58.3%	8.3%	0%	25%	2

One interesting result identified a confliction between the negative and positive comments in terms of the quality of MiT branded medical devices. In this paper, quality is referred to the feeling given by the material used, robustness and smoothness of the design. It was found that the confliction was based on the quality perspective of the interviewees; when the interviewees compared MiT branded medical devices with poor quality import trade products, they argued that MiT branded medical devices are of good quality. This suggests that although MiT branded medical devices are perceived to have a degree of quality, improvements are still necessary. Two participants also expressed that the quality image has been changed positively over the last five years, which also supports this idea.

Usability of MiT branded medical devices

In order to understand the usability aspect of MiT branded medical devices, the interviewees were asked to respond to three likert scale questions (five-point) and then they clarified the reasons behind their responses. The likert scale questions were as follows:

- Regarding your experiences, to what degree do end-users pay attention to usability features of the products they buy? ('1' never, '5' always)
- When compared with import trade products, what is the perceived image of MiT branded medical devices in terms of their usability features? ('1' very bad, '5' very good)
- To what degree do you think Turkish medical device manufacturers pay attention on usability features of the products they produce? ('1' never, '5' always)

The percentages for each option and the median values are presented in Table 2.

The results suggested that according to the retailers, end-users pay a significant amount of attention to the usability features of the products they buy (91.7% of the participants rated 4 or 5). On the other hand, even though the perceived image of MiT branded medical devices received a positive score (most of the participants rated as 4), it was quite interesting to see that the retailers do not think that Turkish medical device manufacturers pay sufficient attention to usability when developing products, where 66.6% of the participants rated as 2 or 1 to this question.

During the interviews, generally the interviewees answered questions regarding their own experiences and product range. Some of the interviewees indicated that most of the medical devices manufactured in Turkey are low-tech products. Therefore they considered these products as easy to use, since they are simple devices anyway. On the other hand, three interviewees expressed that one of the main advantage of MiT branded medical devices is the 'language' factor. This is particularly valid for devices having a visual display, where having software integrated in Turkish language can be a preference for end-users.

In terms of Turkish manufacturers' attention to usability, half of the interviewees mentioned that for the sake of cost cutting, usability is often neglected. Three participants also mentioned that, rather than investing in usability testing, some of the Turkish manufacturers develop their products on the basis of other successful import trade products in the market.

Comparing of MiT branded medical devices with import trade medical devices

The interviewees were asked to indicate their opinions about why MiT branded medical devices are preferred or not preferred above import trade medical device. The results are presented in Figure 2.

The results suggest that MiT branded medical devices are mainly preferred because they are cheaper and provide a better technical service since the manufacturers are located in Turkey. Interestingly, two interviewees also expressed that some end-users prefer MiT branded medical devices just because they are made in Turkey.

On the other hand, the aesthetics of the devices and the quality feeling they give were the most mentioned shortcomings of MiT branded medical devices. Three

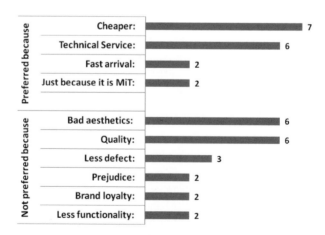

Figure 2.   Number of participants stating why MiT branded medical devices are preferred or not preferred (N = 12).

interviewees also mentioned that end-users often think that these devices are more defect-prone. This consideration may originate from the inadequate quality feeling as well.

Some interviewees expressed that brand loyalty is an important factor in healthcare sector which significantly influences end-users' purchase decision. Due to the fact that many Turkish brands are new to the market, they are less preferred just because end-users simply prefer the brands they trust and know.

Discussion and conclusion

In recent years, medical devices have become an important market segment in Turkey. However the market is mainly dominated by import trade products. It is also argued that there is a prejudice against MiT branded medical devices (Kocak, 2008). Due to the fact that all MiT branded medical devices go through the same regulatory procedure with the EU, these devices need to prove they have the level of quality expected before going in the market. Therefore it was considered that design and usability related shortcomings may have an effect on this prejudice, because as suggested by Wiklund & Wilcox (2005), the marketplace demands medical devices that not only satisfy functional requirements but also user needs and preferences. "Devices that are easy to use, as well as appealing to view and touch will engender greater user satisfaction" (Wiklund & Weinger, 2011).

In order to understand the general image of MiT branded medical devices in the market and their usability aspect, an interview study was conducted with 12 medical device retailers located in Istanbul. The main findings are summarised below:

- MiT branded medical device producers tend to use price as an advantage in the market. However the overall image of MiT branded medical devices requires more improvements.
- From the retailers' point of view, Turkish manufacturers do not pay sufficient attention to usability when developing products. However usability is generally regarded as an important factor by end-users.
- More attention should be paid to the aesthetics of MiT branded medical devices. In addition the quality feeling should be increased by giving more consideration to their overall design.

All these findings point out the necessity of more user research and integration of usability in Turkish medical devices sector. Adoption of the EU Standard EN 62366:2008 (Medical devices – Application of usability engineering to medical devices) can be of help to manufacturers to systematically apply usability procedures in product design and development process. In addition, in conjunction with EN ISO 14971 (Medical devices – Application of risk management to medical devices), these two standards allow manufacturers to decrease possible risks of the device and help them to optimise its usability. However, at the moment these standards are not available in Turkish language.

This research only presents the current situation of MiT branded medical devices from a limited perspective. Therefore in the following research the users and manufacturers of medical devices will be interviewed separately in an effort to understand their point of views.

References

BSI. 2003, BS EN ISO 13485:2003 Medical Devices – Quality Management System – Requirements for Regulatory Purposes. British Standards Institute

BSI 2008, BS EN 62366:2008 – Medical Devices – Application of Usability Engineering to Medical Devices. British Standards Institute

BSI. 2012, BS EN ISO 14971:2012 Medical Devices – Application of Risk Management to Medical Devices. British Standards Institute

EC. 1998, The In Vitro Diagnostic Medical Devices Directive (98/79/EC). Official Journal of the European Communities

EEC. 1990, The Active Implantable Medical Devices Directive (90/385/EEC). Official Journal of the European Communities

EEC. 1993, The Medical Devices Directive (93/42/EEC). Official Journal of the European Communities

Eren, I. 2009, Public Support and Incentives for Industry in Turkey. In: UCTEA Chamber of Mechanical Engineers (ed.) Proceedings of the Third National Medical Devices Production Industry Conference and Exhibition, (Guven Offset, Samsun), 27–31

Kocak, A. 2008, The Report of Medical Devices Sector, The Union of Chambers and Commodity Exchanges of Turkey, Ankara

Robson, C. 2002, Real World Research: A Resource for Social Scientists and Practitioner, (Blackwell Publishing)

Sener, O. & Ozdemir, A. 2011, New Approaches in Medical Device Regulations, In: UCTEA Chamber of Mechanical Engineers (ed.) Proceedings of the Fourth National Medical Devices Production Industry Conference and Exhibition, (Guven Offset, Samsun), 3–8

TPIPA, 2010, The Report of Healthcare Sector in Turkey, Republic of Turkey Prime Ministry Investment Support and Promotion Agency

TOBB. 2009, Medical Sector in Turkey, The Union of Chambers and Commodity Exchanges of Turkey, (Yorum Press, Ankara)

Wiklund, M, & Weinger, M, B. 2011, General Principles, In: M. B. Weinger, M. E. Wiklund & D. J. Gardner-Bonneau. (eds.) Handbook of Human Factors in Medical Device Design, (CRC Press, Boca Raton), 1–22

Wiklund, M, & Wilcox, S. 2005, Designing Usability into Medical Products, (CRC Press, Florida)

EC. 2012, Single Market for Goods – Information Package, URL: http://ec.europa.eu/enterprise/policies/single-market-goods/documents/index_en.htm   [Last accessed: 30-09-2012]

AN ERGONOMICS APPROACH TO SPECIFY A SOLO-RESPONDER PARAMEDIC BAG SYSTEM

Mike Fray, Daniel Joyce, Andrew Adams & Sue Hignett

Loughborough Design School, Loughborough University, UK

Three different carrying systems were evaluated using a range of ergonomics methods. Four focus groups were completed with paramedics to identify design issues. A clinical scenario based user trial explored usability and postural requirements, and a laboratory trial compared the physiological demands. This multi-faceted investigation found that all 3 carrying systems had positive features that need to be considered for the final design specification. An optimal carrying system would combine elements of both the Green rucksack and CURE systems to design an effective clinical and efficient physically carrying system.

Introduction

Previous research has explored the potential of standardising portable ambulance equipment (Hignett et al, 2010; 2011; 2012) for the role of the Emergency Care Practitioner (ECP) in pre-hospital (urgent) care (Department of Health, 2005). To achieve this, a new technology support system (CURE: Community Urgent Response Environment) with portable and mobile (vehicle) components was designed to aid the delivery of pre-hospital care (assess, diagnose, treat and discharge). A limitation of the previous research was the difficulty some participants reported in differentiating between the requirements for emergency and urgent care responses (Hignett et al, 2012). This project evaluated 3 carrying systems for use by paramedics as the first on scene to an emergency call in a solo response. In addition to clinical efficiency the effort of transporting equipment (manual handling) has raised concerns about musculoskeletal risks.

Bag systems

Current carrying systems used in the NHS ambulance services include the Green Rucksack system and a Blue Shoulder Bag system. This project compared the use of these 2 systems with a new design (CURE) which offers a modular carrying system with flexibility to provide for all paramedic and ECP duties. In some systems items were carried separately e.g. Lifepak defibrillator and oxygen (Table 1).

The CURE bag system has 3 pods carried in a single wearable rucksack with the ECG/defibrillator device (LifePak 1000) carried separately. The Green Rucksack was available with different layouts and size, in this project a 3 bag system was used. The Blue Shoulder bag had shoulder and rucksack fittings but was used as both a 2 and 3 component system.

Table 1.   Bag systems.

System	Items
CURE	1. Black backpack with three pods
	2. LifePak 1000
Green	1. Green Rucksack
	2. Oxygen and ventilation bag
	3. LifePak 1000
Blue	1. Blue Shoulder Bag
	2. Oxygen barrel bag
	3. LifePak 1000

Aim

This project aimed to compare 3 equipment carrying systems to produce a carrying system specification that could be used to purchase replacement solo responder bags. This project was confirmed to be an extension of CURE NHS ethical approval REC ref. 10/H0406/81.

Focus groups

Four focus groups were completed with 19 participants (16 operational staff) from 12 ambulance stations including paramedics, fleet and procurement. Participants provided background information about working experience and history of musculoskeletal problems in the last 12 months using the modified Nordic Pain Questionnaire (Dickinson et al, 1992).

Participants

The 16 operational staff were trained as solo responders (but not as ECPs), with a range of experience in the ambulance service (4–32 years, mean 13.5 years, median 9.5 years) and experience as a solo responder (from 1–17 years, mean 5.5 years, median 4 years). Thirteen staff reported musculoskeletal problems in the last 12 months, 4 reported that it had prevented them from doing their normal work and 5 reported trouble in the previous 7 days. The ache, pain or discomfort could be felt in single and multiple areas of the body (Figure 1).

Data collection

As participants were familiar with the Green Rucksack and Blue Shoulder Bag, they were given an introduction to the CURE system at the start of each focus group. The key points of the introduction were:

1. Single carrying backpack with 3 pods for solo responder use.
2. Design concepts included in CURE (small footprint, standardised equipment and layout improved storage and transportation).
3. Range of pods available (assessment, wound respiratory, trauma etc.).

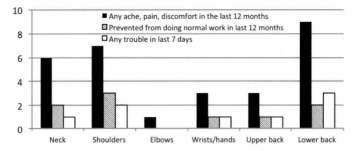

Figure 1. Musculoskeletal problems.

Each bag system was discussed separately with a varied order of presentation to minimise bias. After reviewing each system participants completed a semi-structured evaluation proforma to explore design features, clinical functionality and ideal bag preferences.

Results

The data for each bag system were analysed individually and then combined for comparison using NVivo9. Overall the Green Rucksack was viewed most positively for carrying, and equally as positive as the CURE system for usability in clinical practice. The Blue Shoulder bag had very few positive comments except minimising the number of bags to be carried. The content from all the discussion showed that four higher level codes emerged; carrying, exterior bag design, preparing and using the work station, usability in clinical practice

Participants were asked whether carrying or clinical functionality should be the highest priority for design. Both aspects were considered important with different perspectives (and rationales) about priorities:

- *'Getting equipment to patient safely. Ergonomics is key!! Cost of injuries to staff/Trust is key factor'*
- *'(1) carrying: the help needs to arrive speedily and in good condition and be able to provide aid/treatment (2) patient'*
- *'Design should be with patient in mind, so when paramedic arrives they can be treated as quickly as possible, the bag should be easy access, with no clutter, whilst maintaining essential items'*
- *'Most important: design for the patient. By this I mean I need a bag where cannulation equipment, intubation etc. are readily accessible, but is tidy, easy to find'.*

User trials

Protocol

Two clinical scenarios were developed with patient actors to give a wide range of use of the bag systems. Each paramedic used all bag systems (n = 3) and repeated

Table 2.   Trial and Scenario order.

Subject	Trial 1	Trial 2	Trial 3	Trial 4
1	A Blue	B Green	A CURE	B Blue
2	B CURE	A Blue	B Green	A Green
3	A Green	B CURE	A Blue	B CURE
4	B Blue	A Green	B CURE	A Blue
5	A CURE	A Green	B Blue	B Green
6	B Green	B Blue	A CURE	A CURE
7	A CURE	A Green	B Blue	B Green

(Key: A/B = Clinical Scenario; Blue/Green/CURE = Bag systems)

each scenario (n = 2) with two bag systems. Seven participants completed the trial giving 28 datasets. Random allocation of the trials was not possible so a structured order was created (Table 2). The scenarios were run in real time with the guidance to treat the activity with the same importance as a real patient call. The scenario route included lifting the bag system from a car boot, going through doors, ascending stairs, going along corridors and working in small (confined) treatment spaces. The physical trial ceased after the patient had been treated and the evaluation was then completed.

Data collection

The scenarios were video recorded for analysis. Two analyses were completed, a qualitative analysis of user opinion and a physical evaluation.

a) User Opinion (Adams, 2012)
   • Questionnaires were completed at the end of each clinical scenario to evaluate carrying, preparation of work area, and clinical usability
   • A semi-structured interview was carried out after data collection (4 trials) to explore preferences and design issues for all bag systems
   • A final focus group was carried out after the completion of all the user trials with three research champions to discuss the emerging issues.
b) Physical Evaluation (Joyce, 2012)
   • Postural and biomechanical evaluations of video data were completed after the trial. Key postures were selected and analysed with Rapid Entire Body Assessment tool (Hignett and McAtamney, 2000).
   • Physiological evaluations of carrying the 3 bag systems were completed with a group of volunteer participants (n = 6) who carried each of the bag systems over a representative route. Heart rate, expired air and a short questionnaire were collected for analysis.

Results

Table 3 summarises the responses from the questionnaires. The evaluation of carrying (questions 1–6) found the Green Rucksack was preferred. Even though the

Table 3.   User Trials Questionnaire Results.

Item	Most Preferred		Least Preferred
Carrying	Green	CURE	Blue
Work Area	Green	CURE	Blue
Usability	Green	CURE	Blue
Exterior Design	CURE	Green	Blue

CURE pods contained hard sterile work surfaces the Green Rucksack system was preferred for assessing and treating patients (questions 10–14). General usability of the bag systems, including access to contents and standardisation, recorded similar rating for Green and CURE systems (questions 15–18). Finally the questionnaire recorded views about exterior design (durability, future flexibility) when the CURE system was preferred (questions 19–22). The responses to questions 7–9 were incorporated into the physical evaluation.

Semi-structured interviews were used to compare the bag systems after the 4 trials. The topics explored included carrying, operational use, equipment, modularisation and familiarity. The analysis used 3 stages for open, axial and selective coding (Robson, 2002). Ten categories were created to compare the bag systems:

1. Access to Bag (opening and reaching equipment).
2. Bag Type for carrying.
3. Carrying (comfort and effort).
4. Durability.
5. Individual Features including the fold-out table and cannula board.
6. Item Identification (finding and identifying items).
7. Modularisation (layout and compartments).
8. Number of components in the bag system.
9. Protection (to prevent content damage).
10. Sterility.

The Green Rucksack performed very well in the categories of carrying, item identification and access into bag but badly in durability, number of components, protection, and sterility. The CURE bag performed well in most categories, particularly for modularisation, durability, protection, sterility and item identification but the weight and size was criticized for carrying effort and comfort. The Blue Shoulder bag performed poorly overall except for number of components.

The physical evaluation collected data from the 7 user trial participants for the 3 carrying systems. Musculoskeletal discomfort was rated highest for the Blue Shoulder bag with CURE second. The rate of perceived exertion (RPE) for carrying (Borg, 1998) was highest for the CURE system and second for the Blue bag. Given the relationship between weight and RPE and discomfort it is notable to see a different order in these criteria for the Blue and CURE systems.

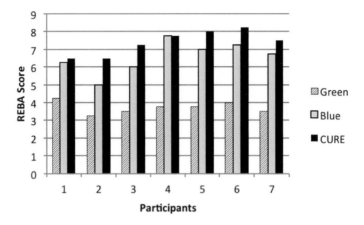

Figure 2.   Posture scores for user trial (REBA).

Four postures were analysed with REBA: lifting the bag from the vehicle, opening a door, walking upstairs, and placing the bag on the ground. The REBA scores are shown in Figure 2 (n = 144). The Green Rucksack consistently recorded low scores whereas CURE, due to its size and weight, had the poorest postures. None of the 21 postures analysed reached the highest level of risk in the REBA scale (>11). Twelve of the postures analysed for the Green Rucksack recorded the lowest level of risk (=1) where no changes are required.

The physiological evaluation created 3 datasets:

- The subjective assessments of the pick-up, walking, incline walk and placement. The Green Rucksack gave the best results.
- There was no major effect in the average heart rate readings across the 3 bag systems. For some participants the discomfort and uneven loading had more effect on heart rate than the weight of the CURE system.
- The extra weight in the CURE system had some effect but not across all participants (Figure 2). The lack of difference in VO_2 showed that the carrying mechanism for the CURE system compensated for weight (CURE 19.65 kg, Blue 11.27 kg, Green 11.65 kg).

The physical evaluation found that the CURE system had a higher risk of musculoskeletal injury (REBA score), with the Green and Blue systems having low and medium risks respectively. There were no physiological performance differences between the carrying systems.

Discussion

The data recorded in this multi-faceted investigation showed that all 3 carrying systems had positive attributes that need to be considered in the final design.

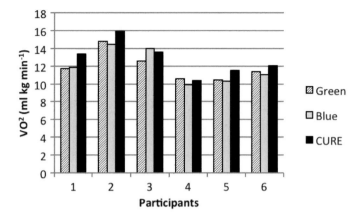

Figure 3.   Oxygen Consumption.

The Green Rucksack was the superior carrying system in the physical evaluation; it was unlikely to cause a musculoskeletal injury. However it did not contain all the required equipment in one bag. In comparison the CURE system held all the equipment in one bag but was heavy despite the improved design of the shoulder and waist fixings, and handles to support a mid-range power grip. Although the Blue Shoulder bag was mostly poorest performing system it supported modularisation and the preferred option of a single bag.

The relationship between the physical ergonomics of carrying equipment to an emergency call and the design of the clinical treatment delivery system created an interesting dichotomy in this evaluation (Hignett et al, 2012). The perceived need to carry all treatment options to an emergency results in large volume and high weight in the portable system.

It is suggested that the optimal design for a solo responder carrying system will need to balance clinical effectiveness and physical efficiency. The following criteria were recommended for inclusion in the product specification for the solo responder paramedic bag system:

- Single bag to carry all required equipment.
- Reduction of size of equipment (e.g. oxygen cylinder and defibrillator).
- Regularly review the list of essential contents.
- Comfortable padded rucksack design with shoulder and waist straps.
- Visibility of the products to be optimized through the use of the pods and packs design e.g. clear fronted packs, colour coding and labelling.
- Contents of paramedic bag to be standardised with specific task pods functionally grouped to aid selection during emergency procedures.
- Protection of equipment/consumables provided by robust physical design of individual sections and bag exterior (straps, zips etc.).
- Maximum weight of the fully packed bag system should not exceed 15 kgs.

References

Adams, A. (2012). *A Qualitative Comparison of the CURE Modular Treatment Pods against Existing Equipment Carrying Systems Found in Pre-Hospital Care.* Unpublished B.Sc. Dissertation. Loughborough Design School, Loughborough University.

Borg, G. (1998). *Perceived Exertion and Pain Scales.* Human Kinetics.com, USA.

Department of Health (2005). *Taking Healthcare to the Patient. Transforming NHS Ambulance Services.* The Stationary Office, London.

Dickinson, C., Campion, K., Foster, A., Newman, S., O'Rourke, A., Thomas, P. (1991). Questionnaire development: an examination of the Nordic Musculoskeletal Questionnaire. *Applied Ergonomics,* 23, 3: 197–201.

Hignett, S., Fray, M., Benger, J., Jones, A., Coates, D., Rumsey, J., Mansfield, N. (2012). CURE (Community Urgent Response Environment) Portable Work Stations. *Journal of Paramedic Practice,* 4, 6, 352–358.

Hignett, S., Jones, A., Benger, J. (2011). Portable Treatment Technologies for Urgent Care, *Emergency Medicine Journal.* 28, 192–196

Hignett, S., Jones, A., Benger, J. (2010). Modular treatment units for pre-hospital care *Journal of Paramedic Practice,* 2, 3, 116–122

Hignett, S., McAtamney, L. (2000). Rapid Entire Body Assessment (REBA). *Applied Ergonomics.* 31: 201–205

Joyce, D. (2012). *An Evaluation of the Effects that Different Carrying Systems have on Paramedics in terms of Physical and Physiological Strain.* Unpublished B.Sc. Dissertation. Loughborough Design School, Loughborough University.

Kluth, K., Strasser, H. (2005). Ergonomics in the rescue service-Ergonomic evaluation of ambulance cots. *International Journal of Industrial Ergonomics,* 36, 3, 247–256.

Robson, C. (2002). *Real World Research* (2nd ed.), Blackwell, London.

DEVELOPING A HUMAN FACTORS CURRICULUM FOR FRONTLINE STAFF TRAINING IN THE NHS

L. Morgan[1], S. Pickering[3], S. New[2], P. McCulloch[1], R. Kwon[1] & E. Robertson[1]

[1]Nuffield Dept. Surgical Sciences, University of Oxford, UK
[2]Saïd Business School, University of Oxford, UK
[3]Warwick Medical School, University of Warwick, UK

The benefits of Human Factors and Ergonomics principles are recognised by the healthcare industry, however the process and the content that is delivered under this premise is not standardised. This paper discusses the development of a training program that aims to address this.

Introduction

Since the well recognised report by the Department for Health (2000) the realization of the requirement for human factors in healthcare has been increasing. The use of the aviation crew resource management (CRM) model of teaching human factors is common, with research programs investigating the benefits to such approaches (Catchpole et al., 2010). Others have suggested that human factors principles and concepts are included into the core medical curriculum (Glavin & Maran, 2003). In addition there has been the growth in lean process improvement approaches such as The Productive Series (NHS Institute, 2006). Whilst all of these approaches have their benefits, it was our aim to attempt to combine the highlights from each to establish if a core curriculum covering the basis of patient safety improvement could be developed. We also sought to investigate novel ways in which the content could be delivered, learning from others in the HF field (Shorrock, 2012). The course development is discussed herein.

Course development & curricula

Due to the limited availability of the hospital teams for training, constraints were naturally placed on the length of any program developed. With this in mind we sought to develop modular based content where delivery could be flexible. Members of the research team (LM, SP, ER, RK) developed an outline plan of principles of HF relevant to the healthcare domain. The content of other patient safety based training programmes for the NHS were evaluated.

A 3 tier approach was identified with a differing curriculum for each. The aim was to allow every individual within the hospital department to be exposed to a level of training, but that this would be stratified across the needs of the dept. with members from each section of the multidisciplinary teams attending each layer of training.

This was preceded by a series of drop-in sessions where staff were introduced to the research program that this training forms part of, and invited to contribute suggestions for areas of focus for improvement after the training sessions.

Tier 1: Extended training – this training is provided to interested volunteers sought from across the organisation. This training is spread across 2 full day sessions, and bespoke follow-up support. The training covers the principles of systems ergonomics, with case studies from other industries, analysis of hospital processes, principles of data collection, analysis and project improvement. The use of human factors design principles in solutions, error trapping and building resilience.

Tier 2: All-team training – this training is provided to all team members, and is a half day covering of principles of patient safety and human factors in healthcare. The content focusses more on non-technical skills, communication types and mitigating error.

Tier 3: Summary sessions – 30–60 min. These sessions are aimed to give attendees a summary of the research project and the aims of the improvement projects, the principles of systems ergonomics and patient safety are introduced, but not discussed at length. It is intended that these sessions will encourage individuals to attend a tier 2 or tier 1 session.

Discussion

The aim of the course development was to ascertain if human factors principles, wider than those classically covered by the CRM-type content, could be combined with lean principles, to the greater benefit of the hospital. Of course the key question of any educational program is does it change anything? Our continuing program research surrounding this training development will seek to evaluate this through process measures and clinical outcomes, and will be published in turn. We are 70% through delivery of the course (*when writing*) and so far the course has been well received by staff from the department. Contents of the course, issues with development and delivery will be discussed in the associated verbal presentation.

References

Catchpole, K. R., Dale, T.J., Hirst, D.G., Smith, J.P., and Giddings, T. 2010, A multicenter trial of aviation-style training for surgical teams. *J Patient Safety*, 6, 180–6.

Department of Health. 2000, Organisation with a memory (The Stationery Office, London).

Glavin, R. J., and Maran, N. J. 2003, Integrating human factors into the medical curriculum, *Medical Education*, 37, 59–64.

Shorrock, S.T., 2012. Discussion cards for understanding and improving safety culture. *Contemporary Ergonomics and Human Factors 2012,* (Taylor and Francis, London) 321–328.

GREEN ERGONOMICS

LEAN, CLEAN AND GREEN: A CASE STUDY IN A PHARMACEUTICAL CLEANING DEPARTMENT

Margaret Hanson[1] & Michel Vangeel[2]

[1]*WorksOut, Edinburgh, UK*
[2]*Janssen (Johnson & Johnson), Geel, Belgium*

Interventions that reduce an organisation's ergonomics risks can also reduce their environmental impact. This case study demonstrates how the major redesign of a work process and area tackled ergonomics, environmental, industrial hygiene and quality concerns. Initiatives undertaken in the central cleaning department of a pharmaceutical company included a new layout of the area, and new work processes and equipment to facilitate cleaning and handling items. This resulted in significant improvements to the ergonomics risks, productivity and quality, and a reduction in solvent use and the requirement for PPE. This win-win intervention allowed ergonomics, environmental, industrial hygiene, productivity and quality concerns all to be addressed.

Introduction

The Central Cleaning Department at Janssen in Geel, Belgium (a Johnson & Johnson company), cleans all the mobile tools and equipment at the chemical production and development site to the required industry standards. Annually over 36,500 items are cleaned, varying from simple pieces (e.g. containers) to complex equipment (e.g. pumps).

In recognition of health, safety and environmental risks present in the process, in 2006 a cross-functional team planned an upgrade of the whole cleaning process. The aim was to implement ergonomic improvements, to reduce exposure to noise, chemical agents and active pharmaceutical ingredients, reduce the environmental impact, comply with more stringent quality requirements and improve the workflow and cycle time.

Interventions

Building upgrade and layout

The work flow within the building was redesigned to be more lean and logical. Activities involving solvent use were consolidated into specific work areas which helped facilitate solvent extraction and minimize operator exposure.

Introduction of a drying room

This eliminated the need to dry equipment manually with compressed air, which had posed significant manual handling and noise related risks.

Ultrasonic installation for cleaning small tools

The manual cleaning of small tools with solvents was replaced with a custom-made automatic ultrasonic installation. This significantly reduced the ergonomics risks and operator time required, and replaced solvents with detergents.

Hose cleaning installation

Manually cleaning heavy, inflexible hoses with spraying pistols and a 1.5 m high attachment point was time consuming, posed ergonomics and slips and trips risks, and required full PPE. The new, custom-made installation was less labour-intensive, and reduced exposure to solvents, noise and manual handling risks.

Installation of lifting devices in the staging and assembly areas

Custom-made lifting devices were installed which allow items to be disassembled, assembled and inspected by operators in good working postures.

Benefits

- **Ergonomics risks:** All ergonomic high risks reduced to moderate or low.
- **Industrial hygiene risks:** Reduction in respirator use by 67% and elimination of the need for hearing protection.
- **Environmental risks:** Solvent use reduced by 73% and reduced detergent concentration, thus reducing the demand on water/air treatment processes.
- **Productivity:** Automation reduced the required operator hours by 55%.
- **Quality:** there was a 40-fold increase in quality.
- **Return on investment (ROI):** The capital investment of €1,587,000, led to annual savings €507,800 (2008 prices); ROI is estimated as 3.125 years.

Conclusion

This case study demonstrates the multi-dimensional benefits of systematically addressing workplace risks. It has shown that tackling ergonomics risks can be integrated into a programme that also addresses environmental and productivity concerns leading to win-win results: higher quality, improved efficiency, reduced personal exposure, and reduced ergonomics and environmental risks. The new work arrangements have been well received by both management and workforce.

DESIGN PRINCIPLES FOR
GREEN ERGONOMICS

**Andrew Thatcher[1], Gabriel Garcia-Acosta[2,3] &
Karen Lange Morales[2,4]**

[1]*Psychology Department, University of the Witwatersrand,
South Africa*
[2]*School of Industrial Design,
Universidad Nacional de Colombia, Colombia*
[3]*Centre de Disseny d'Equips Industrials,
Universitat Politècnica de Catalunya, Spain*
[4]*Institute of Ergonomics, Technical University of Darmstadt, Germany*

This paper provides an outline of the green ergonomics approach and the relationship between green ergonomics and human factors for sustainable development. Green ergonomics is based on ergo-ecology, the broader multidisciplinary field. Four design principles for green ergonomics based on ecological and ergonomics science are proposed and introduced. The principles explained are: (1) evaluation, design and innovation for eco-efficiency, eco-effectiveness, and eco-productivity; (2) evaluation, design, and innovation consistent with ecological resilience; (3) evaluation, design and innovation for indigenous/vernacular solutions; and (4) acknowledge how natural systems value "design".

Introduction

Given the large and growing number of ecological crises, and humanitarian and economic catastrophes it is important to understand how ergonomics might contribute to the long-term survival of all life on our planet. The area of ergonomics that is currently gaining attention in this regard is *sustainable development and human factors* (Steimle and Zink; 2006). Steimle and Zink (2006) drew on a combination of Brundlandt's (1987) definition of sustainable development, Elkington's (1998) notion of the "triple bottom line" (TBL), and Docherty et al.'s (2002) concept of sustainable work systems as the theoretical background. The TBL approach encourages people to think in terms of a balance between three types of capital (economic, social, and natural capital) in order to attain sustainability. Sustainable work systems refer to work that meets the physical, physiological and psychological limits of human functioning while still allowing sufficient rejuvenation opportunities (i.e. recreation and rest) to recover. More recently, *ergoecology* (Garcia-Acosta et al., iFirst) has been proposed as a multidisciplinary field for developing more comprehensive approaches to intervene in socio-technical systems (the built environment) in relation to ecosystems (the natural environment). The term *green ergonomics* (Thatcher, iFirst) has been suggested to explore the synergies between human work

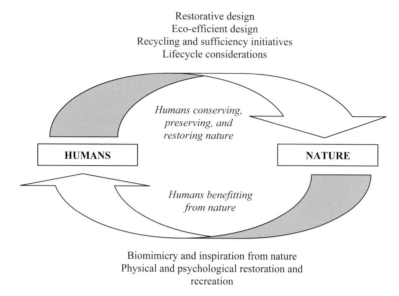

Restorative design
Eco-efficient design
Recycling and sufficiency initiatives
Lifecycle considerations

*Humans conserving,
preserving, and
restoring nature*

HUMANS

NATURE

*Humans benefitting
from nature*

Biomimicry and inspiration from nature
Physical and psychological restoration and
recreation

Figure 1.   Bi-directional relationships for green ergonomics

systems and the natural environment. Green ergonomics is a component of sustainable development and human factors within the multidisciplinary framework of ergoecology.

Green ergonomics defined

Green ergonomics is concerned with ensuring human and natural system wellbeing through understanding the bi-directional relationships between natural systems and human systems (as shown in Figure 1). Natural systems provide a range of ecological services (Daily, 1997) that provide sustenance and resources that enable human wellbeing and health whereas humans require a conservation ethic to ensure the preservation and restoration of natural environments. The ecological crises alluded to in the introduction produce environmental conditions that are not conducive to human health and wellbeing and have been linked to a range of human health and social upheaval impacts (Pimentel et al., 2007).

Green ergonomics plays a role on the one side of the relationship in the conservation and preservation of natural systems and, more actively, the restoration of natural systems. These activities are meant to ensure that ecological services continue to provide environments conducive to human wellbeing and health. On the other side of the relationship, natural systems provide a range of services that can by harvested by humans for a range of human benefits. From the perspective of ergonomics, such benefits might include the design of work-rest cycles, biomimetic or sustainable bionic designs, and creativity benefits. See Thatcher (iFirst) and Hanson (in press) for more detailed examples.

Contributions are achieved through efforts towards eco-efficiency, eco-effectiveness, socio-efficiency, socio-effectiveness, ecological equity, and sufficiency. While efficiency means creating and producing products or services that use fewer resources, eco-efficiency includes the reduction of negative environmental impacts. Eco-effectiveness refers to products or services that specifically enhance natural systems. Socio-efficiency describes the relationship between financial capital and ensuring the wellbeing of people and social systems. Socio-effectiveness refers to the positive impacts that can be made for social systems. Sufficiency means using resources at or below the rate of regeneration. Ecological equity involves balancing social and cultural demands with natural systems for long-term stability. While Steimle and Zink (2006) call for a balance between the three dimensions, most traditional ergonomics' interventions have arguably focused on economic and social aspects and have largely neglected the impact on nature. Even in the work of "human factors for sustainable development", Thatcher (2012) noted that the majority of early studies favoured the economic-social arm of the triad. In some respects green ergonomics attempts to address some of this imbalance.

Green ergonomics, ergoecology, and ecological science

Because of the synergies with natural systems, green ergonomics requires an understanding of theory from ergonomics and ecological science. Ecology, like ergonomics, is a multidisciplinary science which involves understanding the interactions between biological organisms and their environment within various systems of interdependence as well as understanding patterns of the geographical diversity of organisms (Krebs, 2008; Schmitz, 2007). From our anthropocentric viewpoint, humans are clearly the most important biological organisms, but we are very obviously not the only such organisms and all other organisms are also important. Within ecological theory, an ecosystem is "the lowest level in the ecological hierarchy that is complete – that is, has all the components necessary for function and survival over the long term" (Odum, 1997; p. 43). Ecosystem models (e.g. Odum, 1997) are highly similar to open systems' models in ergonomics with environmental inputs, system transformations, and outputs back into the environment. Garcia-Acosta et al. (2012) take these ideas further and propose a multidisciplinary field called ergoecology. Ergoecology looks specifically at the relationships between sociotechnical systems and ecosystems, concentrating on aspects such as energy and other resource exchanges. Ergoecology is concerned with the human, the task, and the immediate technological environment as well as the transformations within the system (e.g. the resulting waste products and energy conversions) as well as the transformations of the ecosystem. In this sense, ergoecology is embedded in an anthropocentric approach but is deeply compatible with an ecospheric approach. As a broad framework for the principles of ergoecology Garcia-Acosta et al. (iFirst) introduce the need for (1) a systemic approach; (2) focused on sustainability; that is (3) anthropocentric. These provide a guiding framework for considering the specific principles in the following section.

Green ergonomics principles

In this section, the broad preliminary principles of green ergonomics are suggested. These principles have been drawn from the ecological science literature (Krebs, 2008; Odum, 1997) as well as related ecological "fields" such as ecological engineering (Bergen et al., 2001), ecological design (Todd and Todd, 1994), and sustainable education (Sterling, 2001).

Principle 1: Evaluation, design and innovation for eco-efficiency, eco-effectiveness, and eco-productivity

In ergonomics, efficiency and effectiveness refer specifically to the work system. In green ergonomics, eco-efficiency and eco-effectiveness also refer more generally to an understanding of how the energy flows, nutrient cycling, and resource paths extend beyond the work system to the ecosystem.

(a) Green ergonomics, recognises human energy efficiency as an important goal, but invites us to extend efficiency to include an understanding of the life-cycle flow of energy (e.g. embodied energy) and other nutrients/raw materials (e.g. water, carbon, sulphur, nitrogen, etc.). Garcia-Acosta et al. (2012) refer to exergy (energy required by the system to maintain the system and that becomes embodied in the system's outputs) and anergy (residual energy left over; i.e. waste). Energy eco-efficiency is achieved when maximal use is made of available exergy and minimal anergy results. Garcia-Acosta et al. (2012) applied the same notions to energy embodied in raw materials and to knowledge frameworks.

(b) In ergonomics effectiveness refers to the ability of the system to achieve a particular task or goal. Eco-effectiveness extends the definition to incorporate the broader ecosystem. Task or goal achievement cannot be considered eco-effective if the energy flows consistently impact ecosystems without allowing for reasonable replenishment. Eco-effectiveness means examining the capacity of the system to transform energy, materials, resources, and information without waste or harm to other systems. Eco-effectiveness involves the appropriate use of non-renewable materials and energy, giving a rational use to them beyond the short-term economic interest of exploitation.

(c) Garcia-Acosta et al. (iFirst) extend eco-effectiveness through eco-productivity. In ergonomics, productivity refers to the output of a given product or service within a given time frame. Eco-productivity in ergoecology considers the balance of outputs (i.e. products or services) with inputs (i.e. available natural resources) over the long term. Eco-productivity is achieved when the system remains in equilibrium over sufficiently long periods of time. As an example, the cradle-to-cradle philosophy (McDonough and Braungart, 2002), suggests that products be designed either as biological nutrients (that will rapidly and efficiently return to the biological cycle) or as technical nutrients (that are almost infinitely and efficiently recycled or even upcycled). Focusing on these aspects requires taking a life-cycle approach (Guinée, 2002). The potential for exergy and anergy production is the sum of these impacts at each stage of the life cycle. This would involve eco-efficiency, eco-effectiveness, and eco-productivity considerations at the design, production, utilisation (including maintenance), recycling, and disposal stages (Wenzel et al., 1997). To be truly ecological though, one would also need to adhere to the sufficiency principle. In nature each system only uses what it requires, no more. Where the sufficiency principle is abandoned, ecosystems generally collapse. Anthropocentric systems will also have to re-learn the sufficiency principle if they are to be sustainable.

Principle 2: Evaluation, design, and innovation consistent with ecological resilience

Sociotechnical systems have important bi-directional relationships with ecosystems at the micro- and at the macro-level. Ecological science has a long history of trying to understand the complex systemic relationships within ecosystems. Important ecological principles to consider would include resilience, diversity, adaptability, and interdependence. Beyond the usual definition of resilience in engineering terms, in green ergonomics the term refers specifically to ecological resilience. Ecological resilience refers to "how large a disturbance an ecosystem can absorb before it changes its structure and function" (Bergen et al., 2001; p. 205); i.e. how much turbulence the system can withstand before becoming chaotic. Fiksel (2004) defines ecologically resilient systems as diverse, adaptable, (eco-)efficient, and cohesive.

(a) Diversity refers to the incorporation of cultural, social, economic, and biological diversity (Sterling, 2001) as well as technological diversity (Garcia et al, iFirst). In ergonomics, diversity has been used to describe "requisite variety" (Imada, 2008; p. 135) but the concept is usually applied to macroergonomics and aspects such as task variety. In green ergonomics diversity means allowing for variety within the work system. In this way, each component of the work system will demand different resources from the ecosystem. Diversity refers to an understanding of the multiplicity of approaches, technologies, and inter-related systems. This could refer to product design, task design, people diversity, technological diversity, and environmental diversity, but more likely a combination of all these factors.

(b) Adaptable systems are those that demonstrate flexibility to change. Bergen et al. (2001) refer to systems that take "advantage of evolutionary opportunities" (p. 206). There are two different types of adaptable systems relevant to this discussion. The first type concerns technology that is adaptable to environmental conditions. This type of adaptability is also useful when considering anthropotechnology and the transfer of technology between contexts (see Principle 3), with the inclusion of the natural environment as part of the context. The second type concerns the adaptability of the larger ecosystem itself. The ecosystem should contain sufficient diversity, interdependence, and resilience to be able to adapt to larger perturbations in the eco-sphere. For green ergonomics, adaptable systems should be able to respond cooperatively with the other ecosystems.

(c) Interdependent and cohesive systems are those that display a high degree of interdependence with society, the economy, and the natural environment. Cohesive systems are those systems that encourage frequent (and possibly redundant) linkages with other systems.

Principle 3: Evaluation, design, and innovation for indigenous/vernacular solutions

Wisner (1997) has long advocated the term anthropotechnology to describe the constraints and issues related to technology transfer. The anthropotechnological approach encourages us to examine the anthropological implications of technology transfer. Green ergonomics suggests that an understanding of the local ecosystem system is also required. For example, transferring green building technology from a cold and wet climate to a warm and dry climate is unlikely to lead to a successful transfer. Indigenous/vernacular solutions encourage the ergonomist to understand all of the local conditions (including the social, political, cultural, and natural environment) and to facilitate localised solutions. Indigenous/vernacular solutions are best represented by the participatory ergonomics approach. Kogi's (2008) participatory approach in developing countries are illustrative of the value of these interventions. There are two aspects to indigenous solutions that are important for green ergonomics. First, the use of local knowledge, expertise, and raw materials means that these resources do not have to be imported (e.g. the carbon footprint is reduced). Second, indigenous design implies that the people who design the systems also live with these systems. Orr (2002) argued that people were more likely to develop sustainable solutions if they also had to live with the consequences of those system designs. While participatory ergonomics has been shown to have economic and social benefits (Kogi, 2008), pro-nature benefits have not yet been demonstrated. Green ergonomics could be used as a vehicle to emphasise solutions that take the *sense of place* (or *genius loci*) into account.

Principle 4: Acknowledge how natural systems value "design"

In the final principle, green ergonomics encourages the adoption of lessons that nature has learnt as it adapts and innovates. In practical terms, this principle means embracing uncertainty, taking a precautionary stance (Orr, 2002), and acknowledging the rights and needs of future generations.

(a) In some respects, uncertainty emerges from Wisner's (1997) anthropotechnology in that it acknowledges that what may work in one context, may not be transferable to other contexts. Uncertainty is also inherent in cohesive natural systems. In all complex, cohesive systems there are multiple relationships that are unknown and that behave in unpredictable and chaotic ways. The green ergonomics approach embraces uncertainty and accounts for unpredictability and complexity in the design. Embracing uncertainty means taking a precautionary approach to design.

(b) Since much of the relationship between sociotechnical systems and natural systems is still very much unknown (particularly the complex interrelationships between our human functioning, technological artifacts, and the natural environment), green ergonomics suggests a precautionary approach. The precautionary approach invites us to acknowledge the uncertainty and take a discretionary stance in design. This is what Bergen et al. (2001; p. 208) refer to as "safe-fail solutions" – approaches that attempt to minimise unintended consequences.

(c) Current ergonomics approaches work on a time scale that is ecologically short. Time frames in ergonomics usually span seconds to minutes or hours (for a proportion of work-based activities), and in some instances for months or a few years (for example, with repetitive strain injuries and musculoskeletal disorders). Human factors for sustainable development acknowledges the anthropometric focus but encourages the ergonomist to think beyond a single human lifespan to future generations and longer term sustainability. In line with this movement, green ergonomics approaches encourage consideration of intergenerational consequences and interactions that usually span years, decades, and centuries.

Conclusions and future work

Sustainability and sustainable development are themes that concern the human factors and ergonomics community. In line with this movement, ergoecology proposes a multidiscipline for systematically studying the relationship between human beings and the environment. Green ergonomics offers an approach for developing ergonomics interventions with a pro-nature focus. While the principles of green ergonomics are outlined it is also obvious that further work is required to define the specific terms and the principles. Some of the terms used in this paper are already in use within ergonomics science, but may have separate general ergonomic meanings that are different from those intended under the specific area of green ergonomics (e.g. resilience is common parlance in many ergonomics studies).

References

Bergen, S.D., Bolton, S.M. and Fridley, J.L. 2001. Design principles for ecological engineering. *Ecological Engineering, 18(2)*, 201–210.
Brundlandt, G.H., 1987. *Our Common Future. Report of the World Commission on Environment and Development* (Oxford: Oxford University Press).
Daily, G.C. 1999. Developing a scientific basis for managing Earth's life support systems. *Conservation Ecology, 3(2)*, 14, [Online] Available from: URL: http://www.consecol.org/vol3/iss2/art14/ [Accessed 22 September 2011].
Docherty, P., Forslin, J. and Shani, A.B. Eds. 2002. *Creating sustainable work systems: emerging perspectives and practice* (London: Routledge).
Elkington, J. 1998. *Cannibals with forks: the Triple Bottom Line of 21st Century business* (Oxford: Capstone).
Fiksel, J. 2003. Designing resilient, sustainable systems. *Environmental Science and Technology, 37(23)*, 5330–5339.

Garcia-Acosta, G., Saravia Pinilla, M.H. and Riba i Romeva, C. 2012. Ergoecology: evolution and challenges. *Work: A Journal of Prevention, Assessment and Rehabilitation, 41(1)*, 2133–2140.

Garcia-Acosta, G., Saravia Pinilla, M.H., Romero Larrahondo, P.A. and Lange Morales, K. iFirst. Ergoecology: fundamentals of a new multidisciplinary field. *Theoretical Issues in Ergonomics Science*, DOI:10.1080/1463922X.2012.678909.

Guinée, J.B. ed. 2002. *Handbook on life cycle assessment: operational guide to the ISO standards* (Dordrecht: Kluwer Academic Publishers).

Hanson, M. in press. Green ergonomics. Challenges and opportunities. *Ergonomics*, in press.

Imada, A.S. 2008. Achieving sustainability through macroergonomic change management and participation. In: K.J. Zink, ed. *Corporate sustainability as a challenge for comprehensive management* (Heidelberg: Physica Verlag) 129–138).

Kogi, K. 2008. Participation as precondition for sustainable success: effective workplace improvement procedures in small-scale sectors in developing countries. In: K.J. Zink, ed. *Corporate sustainability as a challenge for comprehensive management* (Heidelberg: Physica Verlag) 183–198.

Krebs, C. 2008. *The ecological world view*. Collingwood: CSIRO Publishing.

McDonough, W. and Braungart, M. 2002. *Cradle to cradle: remaking the way we make things* (New York: North Point Press).

Odum, E.P. 1997. *Ecology: a bridge between science and society*. Sunderland, MA: Sinauer Associates.

Orr, D.W. 2002. *The nature of design: ecology, culture, and human intention* (New York: Oxford University Press).

Pimentel, D., Cooperstein, S., Randell, H., Filiberto, D., Sorrentino, S., Kaye, B., Nicklin, C., Yagi, J., Brian, J., O'Hern, J., Habas, A. and Weinstein, C. 2007. Ecology of increasing diseases: population growth and environmental degradation. *Human Ecology, 35(6)*, 653–668.

Schmitz, O.J. 2007. *Ecology and ecosystem conservation* (Washington, DC: Island Press).

Steimle, U. and Zink, K.J. 2006. Sustainable development and human factors. In: W. Karwowski, W., ed., *International encyclopedia of ergonomics and human factors (2nd ed.)* (London: Taylor & Francis) 2258–2263.

Sterling, S. 2001. *Sustainable education: re-visioning learning and change (Schumacher briefing No. 6)* (Bristol: Schumacher, UK).

Thatcher, A. 2012. Early variability in the conceptualisation of "sustainable development and human factors". *Work: A Journal of Prevention, Assessment and Rehabilitation, 41(1)*, 2892–3899.

Thatcher, A. iFirst. Green ergonomics: definition and scope. *Ergonomics*, DOI: 10.1080/00140139.2012.741798.

Wenzel, H., Hauschild, M. and Alting, L. 1997. *Environmental assessment of products (Vol. 1)*. London: Chapman & Hall.

Wisner, A. 1997. *Anthropotechnologie: vers un monde industriel pluricentrique.* (Toulouse, Octarès Éditions).

RE-INVENTING THE TOILET: CAPTURING USER NEEDS

Diane Gyi, Ruth E. Sims & Elaine-Yolande Gosling

Loughborough Design School, Loughborough University, UK

It is estimated that 2.5 billion people worldwide lack access to sanitation. A multidisciplinary team from Loughborough University received funding from the Bill and Melinda Gates Foundation to conduct research into re-inventing the toilet. The challenge was to make the toilet environmentally friendly, user-friendly, accessible, and socially acceptable. Users are recognised as important stakeholders in design where the relationships with such products are formed. The research in this paper engages with experts, primary users and secondary users to facilitate more thoughtful design of a new re-invented toilet experience.

Introduction

In 2011, the Bill and Melinda Gates Foundation challenged universities to re-invent the toilet and process human waste in a way that is clean, safe and affordable, but without the need for piped water, electricity or a sewer system. The solution should be viable for both wealthy countries as well as the developing world. A team from Loughborough University was one of eight universities selected for this challenge and received second prize in this prestigious international competition in August 2012 for their new technology involving carbonisation of human waste (Loughborough University, 2012). The team involved experts in design, ergonomics, materials science, chemical engineering, and water and waste management systems. The research presented in this paper has engaged with experts, primary users (including those with diverse needs such as older people), and secondary users (such as cleaners). The aims were to investigate how the new re-invented process will potentially affect sanitation behaviours in users and identify the design challenges for a user-centred re-invented toilet.

Method and findings

A four phased approach was used for the research (Table 1). Design challenges and a 'wish list' for the new re-invented toilet were identified, for example:

- Waste disposal behaviours. The system involves a carbonisation process. Putting non-biodegradable items down the toilet is potentially a problem if the 'char' is to be put in the soil. A design challenge is how to discourage the latter.

Table 1.   Programme of work.

Phase	Research Activities
1 Literature review	Best practice user-centred design requirements collated from the literature e.g. age, gender, culture.
2 Workshop	Defined the process for the reinvented toilet. Documented the potential effects on user behaviour.
3 In-depth interviews	In-depth interviews conducted to understand the context of implementing new technologies in developing countries.
4 User trials	User engagement trials with primary users, using a walkthrough approach. Purposive sampling strategy.

- The flushing liquid. A brown 'coffee coloured' liquid is used in the flushing process. Although users won't see this, there is a slight caramel smell. This will need further exploration e.g. an alternative smell or good ventilation.
- The toilet lid. The new re-invented toilet requires a sealed system to flush the waste with less liquid; the mechanism of achieving this is important – squat or sit-type.
- Cleaning and maintenance. Design for secondary users and caring for the system is paramount e.g. checking the pressure.

Conclusion

User-focussed design challenges have been captured for the new re-invented toilet. In addition it is important to ensure that the design is as intuitive, robust, sustainable and as inclusive as possible for communities: understanding the make-up of users and local sanitation behaviours will be paramount for successful design. The next stages will involve iterative exploration of solutions to the design challenges.

Reference

Loughborough University 2012, "*Loughborough University wins Reinvent the Toilet award from Bill & Melinda Gates Foundation.*" Retrieved 20th October 2012, from: http://www.youtube.com/watch?v=979iKiaplww.

SUSTAINABILITY AND USABILITY OF PUBLIC BATHROOM TAPS

**Aneesa Alli, Musa Maluleke, Sarika Bhana,
Talia Solomon, Yashar Klipp & Andrew Thatcher**

*School of Human and Community Development,
University of Witwatersrand, Johannesburg, South Africa*

This study assessed how the design of three common public bathroom taps was able to limit water wastage while allowing the user to engage easily with the taps. A mixed method research design was implemented including think-aloud procedures, questionnaires, and an objective water collection measure. The results of this study indicated that the simple, traditional tap had the highest score on both usability and sustainability criteria. These findings suggest that usability and sustainability can be thought of as mutually inclusive and complementary components.

Introduction

It is important to understand how the design of the tap influences the usability and the sustainability of this appliance. Research suggests that the usability of a product can support sustainability (Anjos, Matias & Gontijo, 2012). Integrating the needs of the users into the design process will facilitate knowledge of users' habits and mental models that assist or hinder the adoption of sustainable designs by users. This study utilised cognitive engineering methods to determine if the design of three public bathroom taps sufficiently enable the user to utilise the tap effectively, whilst minimising water wastage.

This study examined the three taps below in isolation from the systems to which they belong. These included: (a) a traditional type tap (tap A), (b) a mixer tap (tap B), and (c) an automated tap (tap C) as seen in the figure below.

Usability

Usability is defined by the International Organisation of Standardisation (ISO-9241-11, 1998) as "the extent to which a product can be used by specified users to achieve specified goals with effectiveness, efficiency and satisfaction in a specified context of use". The specified users of each tap range from young children to the elderly, and might include the mentally or physically handicapped. The primary task for users is to wash their hands. The specified context refers to the users, the tasks and the environments of use (Bevan, 1995).

Figure 1.   The three taps utilised in the study.

Based on the ISO-9241-11 (1998) definition, this study measured usability in terms of ease of use, efficiency and satisfaction of use. The ease of use of a piece of technology comprises three main elements: (a) anatomical characteristics, (b) physiological characteristics and, (c) psychological characteristics (Fisher, Katz, Miller & Thatcher, 2003). Anatomical and physiological characteristics inform how easy the tap is to use physically, whereas the psychological properties of the tap inform the human user's ability to make sense of how to go about using the tap.

Sustainability

Sustainability is defined as a general worldview in which people should fulfil their needs in a way that does not compromise the ability of future generations to meet their own needs (Docherty, Kira & Shani, 2009). The concept of sustainability places emphasis on the satisfaction of needs, not wants. The developmental state of technology and social organisations has a powerful impact on the environment. Thus, through technological and social advances, ecological sustainability can be achieved (Docherty et al., 2009). It is possible to reduce ecological damage and prevent further risks through the use of technological innovations (Docherty et al., 2009) such as designing taps that enforce sustainable water usage. While engineering advances allow greater efficiency of product operation, the user's decisions and habits have a significant effect on the resources used by the product. Therefore there is a need to look at how the design influences the behaviour of the user. By enforcing sustainable forms of behaviour through new designs, devices can automatically direct the actions of the human users towards outcomes that are eco-friendly (Jelsma, 2003). Sustainability was measured in terms of the mean water usage of each user during hand washing.

Research questions

1. Which tap has the highest usability score?
2. Which tap uses the least water?

Methods

Usability evaluation methodology

The usability of three types of public bathroom taps was assessed using Usability Evaluation Methodologies (UEMs). Through UEMs, the users' interface with the system, the functionality of the system and the ease of use of the system were analysed. The Think-Aloud UEM technique was employed and has been recognised for its effectiveness in identification of usability problems (Dix, Finalay, Abowd & Beale, 2004). Through this technique, researchers gained an understanding of the mental models and cognitive processes of the users while using the taps. A quantitative questionnaire with a Tap Usability Scale (TUS) was distributed at the University by means of convenience sampling. Additional open-ended questions were included in the questionnaire to attain a more in-depth understanding of the users' perceptions of the taps. Furthermore, individual water usage was quantitatively measured. This was done during the think-aloud sessions and during a later water collection phase. Basins were plugged up during hand-washing, and the water inside was put into a measuring beaker. A syringe was used to ensure that all of the water was measured.

Research sample

The main study employed a non-probability convenience sample of 178 individuals (prior to exclusions) from the University of Witwatersrand. This included 90 questionnaires administered (30 per tap). The final sample (after exclusions) was 171 individuals (57 per tap). The sample consisted of 46.78% females and 52.63% males. 105 participants (90 questionnaires issued without water collection and 15 questionnaires issued with the think-aloud) completed the questionnaire, ages ranged from 17 to 42 with a mean of 20 and a standard deviation of 3.7 years. Racially, 23.3% were White, 52.4% were Black, 2.9% were Asian, 17.5% were Indian, and 3.9% were Coloured.

Pilot tests

A non-probability convenience sample of 29 individuals from the University of Witwatersrand was utilised for a pilot study to test the TUS Scale that was administered electronically.

Think-aloud procedure

The think-aloud procedure was completed by 15 individuals randomly selected from the 178 individuals in the main study; 5 participants per tap. Two individuals were removed from this sample as they demonstrated a hygiene pathology in that they used an excessive amount of water to wash their hands.

Water collection

For the 15 individuals that completed the think aloud procedure water measurements were taken as an objective measure of sustainability. Further water collections took place (an additional 23 for tap A, 24 for tap B, and 26 for tap C) with a total of 86 water collections (including those that were taken during the think-aloud procedure excluding the two with a hygiene pathology). Five individuals were randomly excluded from the sample so that the groups were all of equal size. The final sample size utilised for water collection was 81, 27 per tap.

Measuring battery

The TUS is an 11 item scale developed by the researchers to measure the usability principles of the tap, that is, efficiency of use, ease of use, and pleasant to use. The TUS scale is scored on a 5-point likert scale ranging from strongly disagree to strongly agree with a higher score indicating high usability. See table 1 for all the items.

Procedure

After the electronic pilot study, the psychometric properties of the scale were analysed using SPSS. Results of the analysis indicated that the scale had an acceptable reliability. Exploratory factor analysis indicated the three-factor structure of the scale, ease of use, efficiency of use, and user satisfaction. Items 1, 2, 4 and 5 had salient loadings on the efficiency of use criteria; items 3, 6, 7, 10, and 11 loaded on ease of use; whilst items 8 and 9 loaded on user satisfaction. For the

Table 1.   Descriptive data for each item in the TUS.

Item	N	Mean	Std. Deviation	Tap A Mean	Tap B Mean	Tap C Mean
1. I was able to adjust the tap to obtain the desired water flow.	105	3.45	1.315	3.97	3.71	2.73
2. I was able to adjust the tap to obtain the desired water temperature.	105	2.61	1.326	2.91	3.15	1.73
3. I knew exactly how to turn this tap on.	105	4.15	1.142	4.53	3.91	4.06
4. I was able to prevent unnecessary water wastage while using the tap.	105	2.98	1.301	3.50	3.41	2.06
5. I knew exactly how to turn this tap off.	105	3.84	1.385	4.68	4.18	2.70
6. I had to think about how I was going to use the tap before I used it (Reverse Scored).	105	3.60	1.465	4.37	3.03	3.48
7. When I saw the tap, I automatically knew how to use it.	105	3.97	1.259	4.56	3.71	3.64
8. I think the tap is visually attractive.	105	3.15	1.262	2.03	3.71	3.76
9. I felt satisfied after using the tap.	105	3.60	0.957	3.32	3.65	3.60
10. I thought the tap was easy to use.	105	3.93	1.059	4.53	3.53	3.94
11. I imagine that most people would learn to use this tap very quickly.	105	4.11	1.041	4.38	4.03	3.94

main study, people passing the buildings with the taps were randomly approached, and asked to wash their hands and then fill in the questionnaire. During the think-aloud procedure, users were instructed to verbalise their thoughts, actions and any potential difficulties they may encounter whilst washing their hands as per a standard script of instructions. Thereafter, participants were asked to complete the TUS questionnaire.

Data analysis

The TUS results were analysed using a one-Way ANOVA as all parametric assumptions were met. A second ANOVA was performed on the water collection data of 81 participants. Groups were compared to see which of the taps was the most environmentally sustainable in terms of its use of water.

Results

Results from the thematic content analysis are summarised in Table 2. The ANOVA for the TUS indicates that results were significant with an $F_{(2,1)} = 11.203$; $p < 0.01$.

Table 2.   Themes emerging from the open-ended questions and think-aloud protocols.

Themes from Tap A: Positive Aspects	Themes from Tap A: Negative Aspects
Universal and simple design	No control over the water temperature
Using the knobs makes it easy to open and close tap	Unhygienic to touch the knobs
Hot and cold indicators on each tap	Difficult to open if it is closed too tightly
Easy to adjust the water flow	Does not always close properly which
Familiarity with using the tap	results in dripping and water wastage

Themes from Tap B: Positive Aspects	Themes from Tap B: Negative Aspects
Aesthetically pleasing	No indicators for hot and cold (they were
Lever handle allowed for ease of use	present but difficult for users to see)
Lever made it easy to adjust water temperature	No indication of which direction to turn for water
Lever made it easy to adjust water pressure	Temperature
Lever did not require strength to use; was not stiff	No indication of which direction to turn for water pressure
Able to use tap without touching it much: increases hygiene	Did not know how to open the tap
One outlet for hot and cold water made it easier to control	Unfamiliar with tap
	Unhygienic due to touching the lever

Themes from Tap C: Positive Aspects	Themes from Tap C: Negative Aspects
Push button made it easy to turn on	Cannot adjust temperature or pressure
Tap switches off automatically therefore prevents water wastage	Difficult to press
	Time limit prevents user control
Hygienic: do not need to touch the tap after washing hands to switch tap off	Tap may switch off while still using it
	Did not know if required to switch tap off

Table 3. Mean scores and standard deviations for the TUS and water collection data.

	TUS Score					Water Collection				
Tap	N	Mean	SD	Min	Max	N	Mean	SD	Min	Max
A	35	42.56	6.619	15	53	27	23.807*	6.981*	12.65*	42.43*
B	35	40.03	5.645	28	52	27	29.729*	9.380*	12.65*	56.12*
C	35	35.63	6.318	23	49	27	30.704*	7.329*	20*	52.92*

*The water collection data presented represents the normalised data after square root transformations

In order to determine which taps were significantly different from each other, Fisher's LSD post-hoc tests were conducted. Results reflect that taps A and C as well as taps B and C were significantly different from one another. Taps A and B were not statistically significant from each other; this could be explained by the similarity in certain aspects of the taps; they both offer control over temperature and pressure and the user has freedom to switch the taps on and off at will. Table 3 displays the mean scores for each tap.

The water collection data was initially skewed to the right and square root transformations were performed to normalise the data as recommended by DeCoster (2001). Thereafter, a one-way ANOVA for the water usage data revealed that there was a significant difference between taps as $F_{(2,78)} = 5.925$; $p < 0.01$. A Fisher's LSD post hoc test was performed and revealed that tap A was significantly different from taps B and C whilst taps B and C were not significantly different from one another at $\alpha = 0.05$. The non-transformed water collection data showed that tap A displayed the least variability in water usage with a range of 350 ml–1480 ml as compared to taps B (425 ml–3150 ml) and C (900 ml–2800 ml). Here Tap A clearly demonstrates its sustainability and these results were further emphasised on expansion of the study. Tap A had a range of 160 ml–1800 ml, as compared to taps B (160 ml–3150 ml) and C (400 ml–2800 ml). Thus, tap A shows the potential for sustainability in terms of amount of water. Therefore, tap A used the least amount of water in our sample and is therefore the most sustainable.

Discussion

To answer research question 1, tap A was found to be the most usable. The user had control over pressure and found the tap practical, simple, universal and conventionally designed. In addition, the majority of participants indicated that they are most familiar with tap A that explains the constant reinforcement of mental models relating to that tap. With tap B, users had the greatest amount of control over temperature and pressure. It was also the most aesthetically pleasing, relatively hygienic whilst requiring minimal physical effort to use. In contrast, the lack of individual control over temperature and pressure may have contributed to tap C having the lowest usability score. In addition, such automated taps are relatively new tap designs for public bathrooms, thereby reducing the familiarity of the tap for participants. There are also a variety of automated taps available which could make it difficult for users to form mental models of automated taps. Furthermore, because tap C operated for a limited time, some of the respondents reported washing their hands quicker than they would have liked to. This explains why tap C rated lower on usability and user satisfaction.

Tap C offers a Design with Intent or behaviour-steering solution which enforces sustainability by steering consumer behaviour to use less water (Lockton et al., 2008). The system design does not rely on behavioural changes of the user when operating the tap. Despite these advantages, results revealed that tap A was the most sustainable; this answers the second research question. Tap A offers the user a higher degree of control than tap C, although the knobs for tap A were more difficult to open and the effort required to achieve the desired temperature meant having to open two taps instead of one. Thus, tap A had the highest scores for both usability and sustainability.

Limitations of this study

Sustainability was perhaps operationalised too specifically as the study only looked at sustainability in terms of the amount of water used; the study did not look at electricity usage with regard to the three taps. The use of self-report questionnaires also has potential drawbacks as participants may have responded in a socially desirable manner (Snell & McDaniel, 1998). Another limitation was the manner in which usability was defined and then measured which discriminated against the automated tap (tap C) which was built to eliminate control, a key feature of usability. Therefore, should usability be defined differently, this would have different implications for the type of measurement of usability and potentially the results. Furthermore, the usability definition examined user satisfaction; however it only loaded on two factors in the factor analysis. Therefore it is questionable whether user satisfaction is a separate usability factor.

The use of taps is routine; questioning the cognitive processes of individuals' tap usage was difficult as users tend to wash their hands in an automatic manner. This process may have led users to wash their hands for longer during the think aloud procedure as their talking aloud may have distracted the users by increasing their mental load during a routine task. Observation may have caused users to modify their actions and the process of verbalising their thoughts may have distracted users from performing tasks in their usual manner (Preece, 1993). Alternatively, users may have been engrossed in the task to the extent that they forgot to verbalise their actions.

Conclusion

In conclusion, usability and sustainability can be allies. Tap A scored highest in terms of both usability and sustainability and no trade off had to take place. Thus, the design of a tap can and should incorporate sustainability and usability criteria. Individuals differ with regard to their water saving behaviours, and thus it is important that sustainability and usability are built into the design of taps. Such designs will optimise user experience with taps and simultaneously minimise water wastage.

References

Anjos, T. P., Matias, M., & Gontijo, L. A. 2012. The usability of aproduct can be an ally of sustainability. *Work, 41* , 2117–2121.

Bevan, N. (1995, July). Usability is quality of use. *Proceedings of the 6th International Conference on Human Computer Interaction,* Yokohama.

DeCoster, J. 2001. *Transforming and restructuring data.* Retrieved 15 September 2012, from http://www.stat-help.com/notes.html

Dix, A.J., Finalay, J.E., Abowd, G.D. & Beale, R. 2004. *Human-computer interaction. 3rd Ed.* Harlow Essex, Pearson Education Limited.

Docherty, P., Kira, M. & Shani, A. B (eds.) 2009. *Creating sustainable work systems: developing social sustainability.* London: Routledge.

Fisher, J., Katz, L., Miller, K., & Thatcher, A. (2003). *South Africa at Work: Applying Psychology to Organisations.* Johannesburg: Witwatersrand University Press.

ISO. 1998. Ergonomic requirements for office work with visual display terminals (VDTs) – Part 11: Guidance on usability. *International Organization, 28*(2). International Organization for Standardisation Genève, Switzerland. doi:10.1038/sj.mp.4001776

Jelsma, J. 2003. Innovating for sustainability: involving users, politics and technology. *Innovation, 16,* 103–116. University, Delft Univesity of Technology.

Lockton, D., Harrison, D.& Stanton, N.A. 2008. Making the user more efficient: design for sustainable behaviour. *International Journal of Sustainable Engineering*, 1(1), 3–8.

Preece, J. 1993. *A guide to usability: human factors in computing.* Wokingham: Addison-Wesley.

Snell, A. F., & McDaniel, M. A. 1998.Faking: getting data to answer the right questions. Paper presented at the *13th Annual Conference for the Society for Industrial and Organisational Psychology*, Dallas, TX.

SPSS Inc. 2009. *SPSS for Windows*. Chicago: SPSS Inc.

ACCESSIBILITY

MAKING SELF-SERVICE ACCESSIBLE: TALKING AUTOMATIC TELLER MACHINES (ATMS)

P.N. Day[1], E. Chandler[2], M. Carlisle[1] & C. Rohan[1]

[1] *Consumer Experience, NCR Financial Solutions Group Ltd, Dundee*
[2] *Innovation Unit, RNIB, Peterborough*

In this paper key features are presented that help to make a self-service machine accessible. The importance of not only offering these features but also raising awareness within the user community is discussed.

Introduction

Self-service machines, like kiosks and automated teller machines (ATMs), allow consumers to conveniently access financial services, check in for a flight, or scan shopping. They need to be usable by everyone, without prior training or experience, including aging populations and those with disabilities.

Key design features

ATMs have evolved over time and there have been significant improvements on the accessibility based on research and technological improvements (e.g. greater processing power, powerful text to speech engines). These improvements in capability have enabled the development of a feature set that help to make ATMs more accessible. Some of these key features are described below to illustrate how a self-service product can better accommodate consumers with different needs and capabilities.

Private audio lead through

This consists of two parts, a 3.5 mm audio socket with integral volume button and tactile indicators located near to the card reader, and software which vocalises the menus and options, often by use of an integrated text-to-speech engine. The tactile indicator and shape of the socket enable a blind user to locate where to put the audio jack, along with consistently positioning the audio socket to the right of the machine.

The user inserts their earphones into the audio socket, which is funnelled to aid in correctly inserting the jack. The user is then typically presented with a choice (on screen and through the audio) as to whether they would like the screen blanked or not, thus enabling privacy without penalising those who would rather use a combination of visual and auditory feedback.

Audio lead through is used to supplement the screen instructions for those with sight problems and has also been used in some emerging markets to support those with lower literacy levels (Gill, 1998). Historical note: NCR first introduced audio on ATMs in mid 1980's (with a cassette tape solution); private audio was introduced in 1995.

The growth in popularity of mobile phones has meant that more people who are blind or partially sighted have become used to using earphones and carrying them with them (Huddy, 2011). In the self-service context, offering public audio is not suitable for privacy reasons, offering headphones permanently attached to the device is not preferable for hygiene reasons but also prone to vandalism, and the use of telephone-style handset forces the user to always have one hand taken up with the handset (along with the aforementioned hygiene and vandalism problems). For this reason the provision of a headphone socket for audio feedback has become a *de facto* standard in the self-service industry.

Keyboard mapping

This software adaptation facility enables the user to select from the displayed menus using the numeric keypad, which when coupled with audio provides benefit to blind and partially sighted people. It also has benefits for others in reducing the need for extended reach from the keypad to the touch screen or function keys around the display.

Tactile markings on keypad

The majority of ATMs and point of sale terminals have tactile markers on keys, such as the "5", "Enter", "Clear" and "Cancel" to aid orientation and identification. The use of tactile symbols is recommended over Braille as they are more inclusive. Historical note: NCR played an important role in recognising, developing and promoting their adoption internationally via standards bodies.

Restricted height and depth of the physical interface

The physical layout of an ATM is tightly defined by the accessibility standards (e.g. ADA 2010, CAE 2002). This is to ensure that interface elements can be used with limited reach and hand and eye movements. A bespoke set of guidelines created from the most restrictive requirements has been created to make sure that NCR ATMs meet the needs of customers. In addition human modelling tools like SAMMIE can be used to assess the accommodation of a design (Summerskill *et al.*, 2010) which when combined with expert reviews from the context of specific capabilities (e.g. reach when seated in a wheelchair, or grip with severely reduced manual dexterity) can help to ensure that the solution is usable for those with reduced mobility or dexterity.

Display clarity

Glare, reflections, and markings like fingerprints on screen can cause significant problems with using a self-service device like an ATM. Many of these issues can be alleviated by offering bright, high-contrast displays designed to be legible in bright sunlight and the use of anti-reflective coatings or anti-glare etchings to reduce the impact of reflections (Day *et al.*, 2010). It is also important to provide appropriate colour and tonal contrast within the on-screen content (Silver *et al.*, 1995).

Evaluation of solutions

Providing features like those described above is potentially beneficial, but only if they have been validated to be usable and so the importance of user testing should not be underestimated. This can be helpful for example when considering a new feature on a machine or an entirely new product (Day *et al.*, 2012, Johnson & Coventry, 2001). When considering accessibility solutions it is also useful to ensure the testing includes people with particular capabilities (Coventry *et al.,* 2002, Day *et al.*, 2011).

Importance of these features

If an ATM does not have the features previously described, a blind user would have no way of using the device. The dominant use case by sighted people is to walk up to the device, insert a card and enter a personal identification number (PIN), read the instructions on screen, identify and select the required option, then complete the transaction. The provision of audio, tactile markings and features and use of the keypad to select options enable blind consumers to use the machine without assistance. Similarly the provision of high-contrast screens, lighting around interface slots, clear sans-serif fonts on legends and tactile features enables people with partial sight to use the machine.

When considering the needs of people in wheelchairs one of the major challenges is reaching to the elements on the interface, with the height that the ATM has been installed at being a critical factor in how usable it is to seated users. This height requirement therefore needs to be considered when designing an ATM to ensure it accommodates wheelchair users.

Campaigns and awareness

Although there has been significant work to make ATMs accessible as previously described, this has not transferred to the high street. In the USA 1 in 4 ATMs are accessible to blind and partially sighted people, whereas in the UK it is 1 in 1000 (Huddy, 2011). As a result RNIB initiated a campaign to improve the provision of ATM accessibility which focused on partnering with the banking industry rather than just using legislation. This combination of providing accessible solutions, partnerships between industry and special interest groups and raising awareness offers an effective approach to the area of accessibility.

References

ADA. 2010. http://ada.gov.

CAE. 2002. *Access to ATMs: UK design guidelines.* Centre for Accessible Environments, London. ISBN 0 903976 33 1.

Coventry, L., G.I. Johnson, A. De Angeli. 2002, Achieving accessibility through personalisation. In: *People and Computers XVI – Memorable yet invisible, Human Computer Interaction (HCI) 2002 Proceedings, Vol. 2.* (Springer Verlag, London) 66–70.

Day, P.N., J. Colville, C. Rohan. 2010, An Evaluation of Sunlight Viewable Displays. In *People and Computers XXIV Proceedings of HCI 2010.*

Day, P.N., M. Carlisle, C. Riley, C. Rohan, P. Gregor, I. Ricketts. 2011, The use of biometric fingerprint technology with the over 65s: a case study. In: *Contemporary Ergonomics and Human Factors 2011.* (CRC Press) 379–386.

Day, P.N., I. Ricketts, M. Carlisle, P. Gregor, C. Rohan. 2012, An Evaluation of Bulk Cheque and Cash Deposit Usability. In: *Contemporary Ergonomics and Human Factors 2012,* (Taylor and Francis, London), 47–54.

Gill, J. 1998, *Access Prohibited? Information for Designers of Public Access Terminals,* London, RNIB.

Huddy, H. 2011. *Make money talk – why banks should provide accessible and talking ATMs to assist blind & partially sighted customers.* London, RNIB.

Johnson, G.I., L. Coventry. 2001, You talking to me? Exploring voice in self-service user interfaces. *International Journal of Human-Computer Interaction* 13(2): 161–186.

Silver, J. H., Gill, J. M. & Wolffsohn, J. S. W. 1995, Text display preferences on self-service terminals by visually disabled people. *Optometry Today,* 35(2): 24–27.

Summerskill, S.J., R. Marshall, K. Case, D.E. Gyi, R.E. Sims, P.N. Day, C. Rohan, S. Birnie. 2010, Validation of the HADRIAN System using an ATM evaluation case study. In *International Journal of Human Factors Modelling and Simulation (IJHFMS), Special Issue on Application of Digital Human Modelling Tools in User Centred Design Processes,* 1(4): 420–432.

BLIND PEOPLE AND APPS ON MOBILE PHONES AND TABLETS – CHALLENGES AND POSSIBILITIES

Sabine Croxford & Cathy Rundle

Innovation Unit, RNIB (Royal National Institute of Blind People), Peterborough, UK

Whereas in the past blind and partially sighted people might have been able to use a mobile phone to make a phone call and possibly send a text message, nowadays, with new text to speech (TTS) developments like VoiceOver on Apple and TalkBack on Android, Smart Phones are getting more and more accessible, opening up a world of possibilities with apps that could make life a whole lot easier. Unfortunately, having an accessible Smartphone doesn't automatically mean that all apps on that phone are accessible and usable as well. This paper discusses usability and accessibility issues that blind and partially sighted people face when using apps.

Introduction

Access to information is something most people take for granted. However, if you can't see what's on the screen this suddenly becomes a lot more difficult, if not impossible. Using text to speech and different gestures on a touch screen can potentially make your Smart Phone and its apps accessible and usable even if you can't see the screen. The way an app is developed and presented makes the difference between an accessible and usable app, one that's accessible but unusable and one that you just can't use at all.

Using Text to Speech (TTS)

For a blind person there is one main way of accessing the information. This is through TTS, where a synthetic voice gives you information on what is on screen. If properly coded, all items on the screen (including headings, buttons, text etc.) will be read aloud so that the user can identify where they are within the app and what to do next. When using a touch screen with TTS then the use of different types of gestures is needed so that the user can work out what they are touching before activating it. For example, when touching an icon, the TTS tells the user what it is and a double-tap anywhere on the screen will activate it. Other gestures (for example with VoiceOver on Apple) include a swipe to move onto the next or previous item (such as words, headers, links, images etc.), skipping over items in between. In this way the user can move through a list of items without having to see where they are on the screen, and without having to read through all the details (e.g. a long list of destinations in alphabetical order could be navigated quicker by going directly to the required letter of the alphabet as long as these are identified correctly as headings).

In general it can be said that designing for people with a disability can benefit everybody, and although TTS might not be for everybody, there might be times where it could be useful. Think about reading books or listening to GPS information when walking.

What are the problems?

Overview: One issue is that it will take longer to get an overview of the location on the page and the options available when using speech. A sighted person can scan a page in a few seconds, a speech user hears the information in a linear way, starting at the top and working down.

Structure: A well-structured page with good use of headings will help give a quicker overview so the user will know what's on the page and what to expect. It is key to use headings properly though and not just for visual effect (don't have the main section as a heading 3 and less important text under a heading 1). This is particularly important for people scanning a page by heading levels.

Labelling: The correct labelling of headings, buttons and edit boxes is key as well. It is not very useful to hear "button 33454" followed by "Station chooser" followed by "Selected Button 99554" etc. It would be more useful to hear: "Cancel, button" followed by "Station chooser, heading" followed by "Selected, Stations A–Z button" etc. This way people know what page they are on as this is indicated by 'Heading', they know where they are on the page as this is indicated by the word 'selected', and they know what actions they can carry out as these are indicated by 'Button' (with a sensible name instead of a number).

Efficiency: It also takes longer to listen to something being read out than it does to read it visually and it's harder to ignore unnecessary words. It's therefore important to know the key information at the start of a sentence (or start of a heading or button etc.). Unnecessary words should be left out or be placed at the end of a sentence so a blind user can skip to the next item. A good example is a list of buttons. If there is a list of buttons then the user needs to know that these are buttons as opposed to general text for information. However, especially for people who are more familiar with the app, it is more useful to hear: "Find a branch near you, button", "Log in, button", "View balance, button", "Make payments, button", rather than starting each item with the word 'button'. If the user wants to quickly skip to the button they need they can do that as they will hear the useful information first and can skip the information at the end till they need it, e.g. "Find a branch", "Log in", "View balance", Make payments". Users really familiar with the app will probably only need: "Find", "Log", "View", "Make", and then they know they are on the button to 'Make a payment'.

Visually inherent information: Another important issue is that the speech will automatically only tell the user what is in focus at that point and not what else is there. For most things this is OK, you need to know what the button does before you select it. However, in other cases the sighted user will be looking at other information that is important to their understanding but is displayed visually elsewhere on screen and not within the focus. This could be as simple as that they have opened a list of options on item 4 of 12 and therefore have 3 items above the focus. If the user can't see this and is not told they might assume they are on item 1 and then miss the other possibilities. Alternatively if the user is looking at a programme guide then the location of the programme in the grid gives the sighted person additional information (channel and date/time it is on). If the voice simply speaks the programme title then the blind user doesn't know what time it starts, when it finishes and what channel it is on.

Aesthetics: It is also crucial to consider whether the visual designs are used for aesthetic effect only, or if they also have a function. If they are only there for aesthetic reasons it's unlikely that they need to be voiced. If they are there for clarity to aid understanding, then it is important to make sure that this is also provided in the speech output. This could be the case when using 'greyed out' text to indicate something is not available or 'red' to indicate an error message where it is important for the speech user to know that an option is unavailable so the use of the words "unavailable option" or "greyed out" will help. In the second case the error message needs to contain the word 'error' at the start to visually reinforce the red colouring and to provide this information to people who cannot see the colour. Placing the word 'error' at the start lets the speech user know that this is an important message as soon as they hear this word. Placing 'error' at the end is likely to mean many people will miss it.

Pop-Up windows: Within an app it is vital that the focus is clear. Often pop-up windows are used to display information and it is vital to inform the screen reader that this has been done and to transfer the focus to the new window. Otherwise, the speech could read something different from what is on the screen and this will not be clear to people who are unable to see it.

Programming/coding: The TTS will often include some information automatically. For example, Voice Over on Apple products will tell the user if something is a button as long as it is correctly coded. However, it is still important to test the finished product with speech to make sure that the information provided makes sense in context and is not unduly repetitive.

Self-voicing app: When an app is self-voicing it is key to make sure that this also works with TTS (this is generally an issue) or that the self-voicing facility can be turned off and the app can be used with just TTS. It is also important to keep in mind that gestures, when using the phone or tablet with TTS, might be different to gestures that need to be used with the self-voicing app and if they are different this is a big usability issue, making the app difficult or unusable for people who are unable to see the screen. In addition, if it is difficult to turn the speech on or off (for example if it is necessary to go through the phone settings option) then the user has to return to the home screen and find the self-voicing app without the aid of speech (and screen if you are totally blind).

Other considerations: When using an app with TTS it is important to also consider privacy. Users might want to turn off the screen so other people are unable to read information or might want to use headphones so other people do not hear the TTS. Interference from the environment is also an issue. Especially in noisy locations, for example in a train station with lots of audible announcements, it will be more difficult to hear the TTS. Headphones and easy to access volume control will then be important. Headphones can also be used in public or open settings to ensure that other people around are not adversely affected.

The use of beeps rather than, or in addition to speech, is useful as well, although not everybody is able to distinguish between different beeps, or remember what the different beeps mean. The number of different beeps for different functions should be limited, but if designed with usability in mind then this can help users.

Testing (technical, expert and user testing)

Lastly, but most importantly, try out the app yourself without looking at the screen or better still watch someone else who is not familiar with the app, have a go. You'll probably learn a whole lot about what makes your app difficult or, hopefully, easy to use with speech.

SIMULATING VISION LOSS: WHAT LEVELS OF IMPAIRMENT ARE ACTUALLY REPRESENTED?

Joy Goodman-Deane, Sam Waller, Alice-Catherine Collins & P. John Clarkson

Engineering Design Centre, Engineering Department, University of Cambridge, UK

Capability loss simulators give designers a brief experience of some of the functional effects of capability loss. They are an effective method of helping people to understand the impact of capability loss on product use. However, it is also important that designers know what levels of loss are being simulated and how they relate to the user population. The study in this paper tested the Cambridge Simulation Glasses with 25 participants to determine the effect of different numbers of glasses on a person's visual acuity. This data is also related to the glasses' use in usability assessment. A procedure is described for determining the number of simulator glasses with which the visual detail on a product is just visible. This paper then explains how to calculate the proportion of the UK population who would be unable to distinguish that detail.

Introduction

As the population of the developed world ages, there is a growing awareness of the need for more accessible and inclusive designs. Products and services need to be usable by people with a wider range of capabilities and characteristics. To do this, designers need better information about people's capabilities and capability loss and their impact on usability. However, written methods of presenting such information can often seem dry and difficult to relate to product use (Nickpour and Dong, 2009).

An alternative is the use of capability loss simulators. These give designers and other stakeholders a brief experience of some of the functional effects of capability loss for themselves (Nicolle and Maguire, 2003; Cardoso and Clarkson, 2006). A functional effect is the effect of a medical or other condition on a person's physical or sensory capabilities, such as the ability to see fine detail. Thus simulation can be achieved through wearing equipment that restricts one's motion or reduces one's sensory capability. This ranges from whole body suits such as the Third-Age Suit (Hitchcock et al., 2001) to individual sets of gloves or glasses (e.g. Goodman-Deane et al., 2008). Some authors also advocate using simple techniques to reduce capability in a more approximate manner, such as taping buttons onto knuckles or smearing glasses with petroleum jelly (Nicolle and Maguire, 2003). Alternatively, simulation software can be used to show how things would appear or sound to someone with a sensory impairment (e.g. Goodman-Deane et al., 2008).

Simulators can encourage greater empathy with users with capability loss, and provide a more personal understanding of impairment. They can also be worn while using products and prototypes to provide some insight into the effect of capability loss on product use and to identify usability problems.

Simulation does have limitations, as it provides a constrained experience of capability loss. It does not fully convey the frustration, social consequences or coping strategies involved in living with an impairment day-to-day. It is also usually not possible (or not ethical) to simulate the pain and other symptoms associated with an impairment. As a result, simulation is not intended to be used on its own. Rather, it should be used in combination with user involvement and expert appraisal methods. It can supplement these by helping a designer to internalize information obtained through other methods. It can also provide initial usability feedback to help correct some of the major issues before designs are taken to users.

Calibrating simulators

Another limitation of simulation is that it is often difficult for designers to tell what levels of impairment simulations correspond to, and thus whether they affect many of the target user group or just a few. This is particularly problematic for the rough-and-ready simulation techniques, but can also affect pre-produced simulators. For example, the Third-Age Suit reduced visual acuity, increased glare sensitivity and added a yellow tint, to simulate some of the effects of aging on vision. However, it did not specify what level of these effects were used (Hitchcock et al., 2001).

Other simulators do cover a range of impairment. For example, software simulators often allow levels of vision or hearing loss to be manipulated through interface controls (e.g. eclipse, 2012). However, they lack the immediacy and immersiveness of wearable simulators or their ability to be used while examining products directly. Wearable simulators can also cover a range of capability loss. In vision simulation, sets often contain multiple pairs of glasses that mimic different levels of vision loss (e.g. Zimmerman Low Vision Simulation Kit, 2012). However, these often focus on higher levels of impairment, with simulated vision typically starting at about 20/60 or worse (affecting less than 0.8% of the population). They thus do not help designers to understand the lower levels of impairment that can cause many people problems with mainstream designs. Furthermore, the basis for stating that a certain pair of glasses simulates a certain level of impairment is also often unclear.

This study addresses this issue by calibrating the level of visual acuity loss simulated by a set of simulator glasses covering both lower and higher levels of vision impairment (Engineering Design Centre, 2011). Waller et al. (2008) calibrated an earlier version of these glasses using a self-report vision capability scale. Their purpose was to demonstrate how such data could be presented and used, rather than to give an accurate calibration. The current study takes this further, using a standard

Figure 1.   Cambridge simulation glasses.

vision test with 25 participants to give more reliable data, and comparing the results with data on visual acuity in the wider population.

The Cambridge simulation glasses

The study used the Cambridge Simulation Glasses, which restrict the ability to see fine detail and perceive contrast differences (Engineering Design Centre, 2011). The glasses are made from a thin, lightweight material so they can be layered to simulate greater levels of impairment (Figure 1). Gloves are also available that restrict the functional ability of the hands. The two can be used in combination to help designers understand the impact of a range and combination of impairments.

The glasses and gloves were developed from a previous toolkit by Cardoso and Clarkson (2006), which was adapted by Waller et al. (2008). The glasses in this paper were further developed to make them cheaper to manufacture and to make it easier to wear multiple pairs on top of each other. The glasses examined in this study were the "Level 2" glasses from the kit, which contain two sheets of filter material. Level 1 pairs, containing a single sheet, were not available at the time of the study. Furthermore, using Level 2 glasses meant that the range of the glasses could be examined while keeping the length of the study manageable.

Calibration of the simulators

Method

Participants' visual acuity (VA) was measured using the Landolt C chart from Test Chart 2000, with 8 possible orientations of the C. This eye test is a standardised and reliable method of measuring visual acuity. It was chosen because it is faster to run than EDTRS letter charts, yet produces a reliable measure. Speed was an important consideration as each participant's eyesight was measured multiple times.

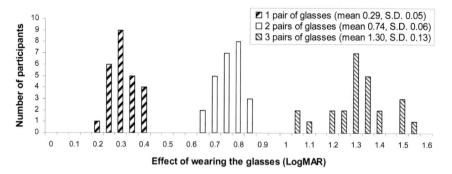

Figure 2.   The effect of wearing the simulator glasses. Scale points on the x-axis correspond to 0.05 LogMAR bins (0–0.05, 0.05–0.1 LogMAR, etc.)

Participants wore their usual vision setup (e.g. glasses or contact lenses) for all the tests. VA was first measured with three pairs of simulator glasses (plus the usual vision setup), then with two pairs, a single pair and finally no pairs of simulator glasses. This last was defined as the participant's base visual acuity (base VA).

Sample

Twenty-seven staff and students of the authors' research centre took part in the study. This recruitment strategy was used because the study depends primarily upon participants' levels of eyesight, rather than other participant factors. Two of the participants were removed from the analysis as their base VA was very low and would skew the sample. It was 0.3 LogMAR (Snellen 20/40) or worse, which is generally considered to impact the ability to drive. The remaining 25 participants had base VA varying from 0.22 to −0.12 LogMAR (mean 0.07, S.D. 0.10).

Results and analysis

The results are summarised in Figure 2, which shows the effect of wearing different numbers of simulator glasses. The effect is calculated as the difference between a person's VA when wearing the glasses and their base VA. The three peaks on the graph correspond to the results with one, two and three pairs of simulator glasses.

The effect of wearing one or two pairs of glasses is not correlated with the participants' base VA ($r = 0.05$, $p > 0.05$; $r = -0.12$, $p > 0.05$). However, there is a strong correlation between base VA and the effect of wearing three pairs of glasses ($r = -0.54$, $p < 0.01$). The better a participant's visual ability (lower LogMAR), the larger the effect of the glasses. The regression line is modelled by: *effect with 3 pairs = 1.34 − 0.67 ∗ (base VA)*. However, it should be remembered that this only accounts for about 29% of the variance in the results.

The actual VA experienced when wearing one or two pairs of glasses can be estimated by adding the wearer's base VA to the mean effect of those glasses. For three

Table 1.   Visual acuity when wearing different numbers of simulator glasses. Snellen 20/X figures have been rounded to the nearest whole number.

Base VA		VA with 1 pair of glasses		VA with 2 pairs of glasses		VA with 3 pairs of glasses	
LogMAR	20/X	LogMAR	20/X	LogMAR	20/X	LogMAR	20/X
x		$x+0.29$		$x+0.74$		$x+1.34$ $-0.67*x$	
0.2	20/32	0.49	20/62	0.94	20/17	41.41	20/514
0.1	20/25	0.39	20/49	0.84	20/13	81.37	20/469
0	20/20	0.29	20/39	0.74	20/11	01.34	20/438
−0.1	20/16	0.19	20/31	0.64	20/87	1.31	20/408

pairs of glasses, using a single mean value is not appropriate. Instead, the effect for a particular base VA can be calculated using the regression equation. The resultant visual acuities for various base VAs are summarised in Table 1.

Using the simulators for usability assessments

The simulator glasses can be used to provide initial feedback on a product's usability and the levels of visual demand it places on a user. To do this, we propose the following "simulator assessment procedure". Designers first test their own eyesight, while wearing their usual glasses or contact lenses (if appropriate). This is, by necessity, a rough test as it is self-administered. Ideally, the designers would use the Landolt C test as in this paper, although a letter chart is also possible. The designers then put on three or more pairs of simulator glasses (on top of their normal glasses). With these on, they examine the product. For example, they may try to read some text on the product and distinguish some markings. If they cannot do this with all the glasses on, they remove one pair and try again. The "simulator demand level" is the number of glasses with which the feature is just visible.

For example, imagine a designer with 0 LogMAR visual acuity (20/20 vision). If he/she can distinguish the controls on a product with one pair of simulator glasses, but not two, then the simulator demand level is one pair of glasses. This corresponds to the controls being discernable by users with a VA of 0.29 LogMAR but not by those with 0.74 LogMAR (Table 1). Note that, in practice, users may employ other strategies to help them use products, so this evaluation method should not be used in isolation. Nevertheless, it gives an initial indication of a product's visual inclusivity.

Population figures

It can be helpful to relate these levels of visual acuity to population figures, to help designers understand how many people this would actually affect. Survey data can

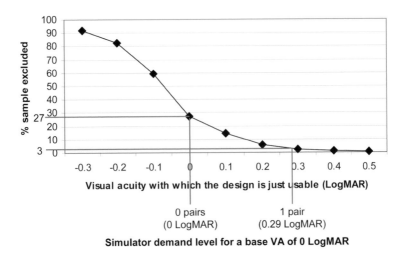

Figure 3.   The percentage of the sample (from a 362 person survey) that would be excluded by a design that is just usable at different visual acuities. This graph was modified from one in (Goodman-Deane et al., in progress).

be used to calculate the proportion of a population that would not have the level of visual acuity required by a product feature and thus would be excluded from its use (Keates and Clarkson, 2003).

The graph in Figure 3 was constructed using this method with data from a survey of 362 participants in the UK (Clarkson et al., 2012; Tenneti et al., in progress). It thus gives the proportion of the sample that would be excluded rather than the proportion of the whole UK population. However, the survey was postcode sampled and weighted for age and gender and thus can give a good indication of exclusion on a wider scale. This data source was used because it covers a wide range of vision and other capability measures meaning that exclusion for other impairments and combinations of impairments can also be calculated (Keates and Clarkson, 2003; Goodman-Deane et al., 2011). In addition, the survey was a pilot and it is hoped that funding will be obtained for a full UK representative survey which will give more comprehensive results.

The survey tested visual acuity using LogMAR EDTRS letter charts. The results from these are generally comparable with the results from the Landolt C charts (Kuo et al., 2011), although some studies have found a small difference between the measures of between 0.01 and 0.1 LogMAR (e.g. Wesemann, 2002).

Some points are marked on the graph in Figure 3. They correspond to different values of the "simulator demand level" for a designer with a base VA of 0 LogMAR. For example, if that designer can just distinguish the markings on a product when wearing one pair of simulator glasses, then the graph indicates that about 3% of the UK adult population would be unable to see those markings.

The exclusion with different base VAs can be determined by examining Table 1. This table gives the visual acuity corresponding to different numbers of glasses (and hence to different simulator demand levels) with various base VAs. This visual acuity can then be matched to the x-axis in Figure 3, and the percentage exclusion can be read off the graph.

Note that the exclusion for two and three pairs of glasses are off the x-axis scale on the graph. They thus correspond to less than 0.7% exclusion. This level of exclusion is sufficiently low that it may be of little interest to designers designing for a general population. However, they should still consider that about 180,000 people in the UK are registered blind with 1.3 LogMAR or worse, roughly corresponding to three pairs of glasses (Action for Blind People, 2012). Using two and three pairs of glasses can still be a good check on a design, and is particularly valuable if designing with an older population or visually impaired population in mind.

The graph in Figure 3 raises another issue. The points for zero and one pairs of glasses are spaced quite far apart. It may be useful to have glasses with smaller increments to provide more detail on visual inclusivity, and help designers to understand the effects of smaller changes in their designs. As mentioned above, this paper used the Level 2 glasses from the simulation kit. Since the study, Level 1 glasses have also been produced (Engineering Design Centre, 2011). These contain one rather than two sheets of filter material. Thus they should be approximately half the strength of the ones used in this study. However, their precise effect on visual acuity has not yet been measured.

Conclusions and further work

The effect of the Cambridge Simulation Glasses on visual acuity was tested with 25 participants. This paper describes a person's visual acuity when wearing different numbers of these glasses, for different starting levels of visual acuity. The glasses can be used to examine the visibility of product features, and the results used to calculate the proportion of users who would be unable to distinguish these features.

The study indicates the need for a finer level of simulation. Half-strength simulator glasses have been produced but further work is needed to calibrate them. Further work is also needed to use the glasses in practice to evaluate more products, and compare the results with usability findings from user trials and other methods.

References

Action for Blind People. 2012, "Facts and figures about issues around sight loss". Retrieved 12th Sep 2012 from http://www.actionforblindpeople.org.uk/about-us/media-centre/facts-and-figures-about-issues-around-sight-loss/

Cardoso, C. and Clarkson, P. J. 2006, Impairing designers: using calibrated physical restrainers to empathise with users. In *2nd International Conference for Universal Design*, International Association for Universal Design, Kyoto

Clarkson, P. J., Huppert, F. A., Tenneti, R., Waller, S., Goodman-Deane, J., Langdon, P., Myerson, J. and Nicolle, C. 2012, Towards Better Design, 2010 [computer file]. Colchester, Essex: UK Data Archive [distributor]. SN: 6997, http://dx.doi.org/10.5255/UKDA-SN-6997-1

eclipse, 2012. ACTF aDesigner. Retrieved 17th Sep 2012 from http://www.eclipse.org/actf/downloads/tools/aDesigner/

Engineering Design Centre, 2011, "Inclusive design tools". In *Inclusive design toolkit*. Retrieved 12th Sep 2012, from http://www.inclusivedesigntoolkit.com/betterdesign2/inclusivetools/ inclusivedesigntools.html

Keates, S. and Clarkson, J. 2003, *Countering design exclusion: An introduction to inclusive design.* (Springer, London)

Goodman-Deane, J., Waller, S. and Clarkson, J. 2008, Simulating Impairment. In *(re)Actor3/HCI 2008*. Liverpool

Goodman-Deane, J., Waller, S., Williams, E., Langdon, P., and Clarkson, P.J. 2011, Estimating Exclusion: A Tool to Help Designers. In *Proceedings of Include 2011*. (Helen Hamlyn Centre, London)

Goodman-Deane, J., Waller, S., Latham, K., Price, H., Tenneti, R., Clarkson, J. In progress, Vision measures and implications for design. Submitted.

Kuo, H.-K., Kuo, M.-T., Tiong, I.-S., Wu, P.-C., Chen, Y.-J., Chen, C.-H. 2011, Visual acuity as measured with Landolt C chart and Early Treatment of Diabetic Retinopathy Study (ETDRS) chart. *Graffes Archive for Clinical and Experimental Ophthalmology* 249(4): 601–605

Nickpour, F. and Dong, H. 2009, Anthropometrics without Numbers! An Investigation of Designers' Use and Preference of People Data. In *Proceedings of Include 2009*. (Helen Hamlyn Centre, London)

Nicolle, C. A. and Maguire, M. 2003, Empathic modelling in teaching design for all. In *Universal access in HCI, HCI International 2003*, Vol. 4, pp. 143–147

Tenneti, R., Langdon, P., Waller, S., Goodman-Deane, J., Ruggeri, K., Clarkson, P.J., and Huppert, F.A., in progress, Design and delivery of a national pilot survey of capabilities. Submitted to *Applied Ergonomics*

Waller, S. D., Langdon, P. M., Cardoso, C. and Clarkson, P. J. 2008, Calibrating capability loss simulators to population data. In *Contemporary Ergonomics 2008*, pp. 291–296

Wesemann, W. 2002, Visual acuity measured via the Freiburg Visual Acuity Test (FVT), Bailey Lovie Chart and Landolt Ring Chart. *Klinische Monatsblatter fur Augenheilkunde* 219(9): 660–667

Zimmerman Low Vision Simulation Kit, 2012. Retrieved 17th Sep 2012 from http://www.lowvisionsimulationkit.com

EVALUATING THE UNIVERSAL NAVIGATOR WITH BLIND AND PARTIALLY SIGHTED CONSUMERS

P.N. Day[1], M. Carlisle[1], E. Chandler[2] & G. Ferguson[3]

[1] Consumer Experience, NCR Financial Solutions Group Ltd, Dundee
[2] Innovation Unit, RNIB, Peterborough
[3] UX Consultant, South Falfield, Leven

Touch screens are increasing in usage in both consumer electronics and self-service devices, but pose a number of accessibility challenges. An alternative device for navigating on-screen content is presented, along with evaluations of this device with 48 people who were blind or partially sighted.

Introduction

This paper follows on from a prior publication (Day et al., 2012) that described the initial concept of an alternate input device called the Universal Navigator (uNav), and also described initial user testing with sighted consumers. In this paper we present further studies evaluating the use of this device by consumers who were blind or partially sighted as part of collaborations between NCR and RNIB.

The universal navigator is a physical input device which can be attached to a touchscreen such as that used on a self-service kiosk, and enables users who cannot see or reach the screen to activate the onscreen options. This is of increasing importance as there is a trend to use touch screens in self-service products (Penn et al., 2004, Digital Trends, 2011), along with ageing populations in many countries (UKONS, 2010, Zaidi, 2008). It has the potential to provide significant accessibility advantages to those with low visual acuity, reduced upper body mobility, and also low manual dexterity as it enables consumers to navigate any onscreen option using a tactually discernible, physically compact device with associated private audio feedback (i.e. through earphones rather than speakers). It consists of four direction keys laid out in a diamond shape, arranged around a central select button (Figure 1). The left and right buttons cycle through each on-screen option in turn, with the up/down buttons jumping to the next block of on-screen options (e.g. next row of an on-screen keyboard). An audio socket and volume button are also provided as auditory feedback is essential for people who are blind or partially sighted. It is a concept device and is currently not in development.

Background

The RNIB strategy for 2009–2014 is focused on ending the isolation associated with sight loss (RNIB, 2009). As part of the strategy the Innovation Unit within RNIB

Figure 1.   The uNav & sample screens from transaction.

Figure 2.   Participants using vertical & horizontal uNav and touchscreen.

has been tasked with improving the accessibility of touchscreen self service kiosks. The aim of this is to improve the independence of blind and partially sighted people to make journeys and control and access money when they encounter a touchscreen terminal (such as a ticket machine). In order to achieve this goal, RNIB collaborated with NCR on a solution that had been devised by them to understand whether or not it would match user needs and could successfully be used by people who are blind or partially sighted.

Method

The evaluation was a repeated measures design, with participants completing the check-in task twice using the uNav in the vertical and horizontal orientations, and those that felt able to use a conventional touch screen completed an additional task with this touchscreen (Figure 2). There were 48 participants (27 male, 21 female), with good representation of age (from 25–79). 22 were blind without any useful residual vision (i.e. could perceive light and dark only), 15 were blind with some useful residual vision and the remaining 10 were partially sighted. This evaluation used a modified version of an NCR travel check-in application (Figure 1) previously used (Day *et al.*, 2012), but with the addition of high-quality speech output. This task (checking in for a flight) was chosen as it included a number of complex options including using an on-screen alphanumeric keyboard and selecting a seat on the aeroplane.

Participants were asked background questions (age, gender, level of visual acuity, upper body mobility and manual dexterity) including self-efficacy questions pertaining to their attitude to technology. These questions were taken from Meuter et al., (2003), but only 6 questions used rather than the full 9 from Meuter. Those chosen were the highest 6 predictors of technology anxiety identified in this paper (confident learning skills, difficulty understanding, apprehensive using technology, worry about damaging technology, understand jargon, avoid technology because it is unfamiliar).

Procedure

The check-in task required each participant to enter an alphanumeric confirmation number, modify a seat selection, modify the number of bags and special items they needed before confirming the booking. Ordering of conditions was randomised to reduce the effect of ordering, no practice was given, and the entire evaluation took 20–30 minutes to complete. Performance metrics were: task completion times, number of button clicks or touches per screen, and number of times a participant navigated to a screen (with the latter two giving an indication of error). Subjective measures were also collected with participants scoring each device on a series of ratings and overall ranking for each condition (vertical/horizontal device, or touch screen). The ratings used were 5 point Likert scales with explicitly labelled points for (very easy/very difficult to use, very comfortable/uncomfortable) and were asked at each stage of the transaction.

Results

In order to analyse the results quickly measures such as times and error measurements (number of clicks and page counts) were not included as they were recorded in separate log files in a form that requires post-processing before use. Fuller analysis that includes these performance metrics will be presented in a later publication.

Overall the uNav was very well received by participants. When asked whether the device was acceptable for use, the majority (72%) stated that it was acceptable with minor changes, and the remainder (28%) that it was acceptable as is.

Comparison between touchscreen and uNav

16 out of 48 participants completed the task with the touchscreen. Among those who completed all three conditions (vertical & horizontal uNav and touchscreen), the ratings for each stage of the transaction (confirmation number, seat selection, add bags, confirm) show preference for the touchscreen, although the differences between touchscreen and the uNav in either orientation were not significant.

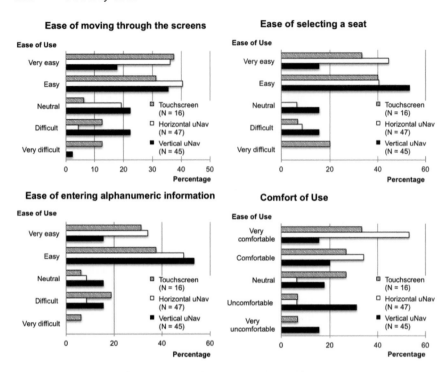

Figure 3.   Ratings for all 3 conditions.

Differences between horizontal & vertical uNav

The vast majority could use uNav to complete the task in both orientations, with only 2 participants failing to complete the check-in task. These two participants both found the check-in task itself confusing, but were able to use the device to navigate around the options. Respondents reported a clear preference for the close to horizontal orientation (94% of participants) which was also supported by individual ratings throughout the transaction.

When considering just the conditions where the uNav was used in the vertical and horizontal orientations, a Wilcoxon signed rank test revealed significant differences between ratings for ease of moving through screens ($S = 158.5, p < 0.001$), ease of selecting a seat ($S = 135, p < 0.001$), ease of entering alphanumeric information ($S = 85.5, p = 0.0002$), comfort of use ($S = 280, p < 0.001$) and when considering the median of all the ease of use questions ($S = 143, p < 0.001$).

Correlations with other factors

Upper body mobility. There were significant negative correlations between level of upper body mobility and a number of the ease of use ratings for the vertical uNav with participants tending to rate it harder to use as level of upper body impairment increased. Significant results from a Spearman's rank correlation coefficient were ease of use for moving through the screens ($\rho = -0.39$, $p < .005$), selecting a seat ($\rho = -0.36$, $p < .05$), median of all ease of use questions ($\rho = -0.37$, $p < .05$) and also ratings of comfort of use ($\rho = -0.40$, $p < .01$).

Height. There were significant negative correlations between participant height and ease of use ratings for the touchscreen with participants tending to rate it easier as height increased. Significant results were ease of use for moving through the screens ($\rho = -0.58$, $p < .01$), selecting a seat ($\rho = -0.44$, $p < .05$), entering alphanumeric information ($\rho = -0.57$, $p < .01$) and median ease of use ($\rho = -0.58$, $p < .01$).

Sight. There were significant correlations between sight and a number of responses. In general confidence with technology decreases with worsening sight, with significant results for "I feel apprehensive about using technology" ($\rho = -0.33$, $p < .05$), "When using technology, I worry that I might damage it or make mistakes that I cannot correct" ($\rho = -0.33$, $p < .05$) and "How often do you use self-checkout where available at a retail store" ($\rho = -0.49$, $p < .0005$). The raised features on the uNav were rated as more useful by those with worsening sight ($\rho = 0.29$, $p < .05$).

Comfort and perceived ease of use. Across a number of activities increased rating of comfort correlated with greater ease of use. This was significant for ratings of ease of use for entering alphanumeric information using vertical uNav ($\rho = 0.36$, $p < .05$), seat selection on horizontal uNav ($\rho = 0.31$, $p < .05$) and touchscreen ($\rho = 0.64$, $p < .005$), and for navigating between screens on touchscreen ($\rho = 0.66$, $p < .005$).

Age. Increasing age correlated with increasing confidence in using technology. Significant results were found for feeling apprehensive about technology ($\rho = 0.38$, $p < .01$), avoiding technology because it is unfamiliar ($\rho = 0.34$, $p < .05$), difficulty understanding most technology ($\rho = 0.42$, $p < .005$), worry about damaging or making mistakes when using technology ($\rho = 0.38$, $p < .01$). These results were counter to expectations. A significant negative correlation was found between age and understanding technology related jargon ($\rho = -0.36$, $p < .05$).

Age correlates negatively with ATM use ($\rho = -0.46$, $p < .005$) and a single ease of use question for navigating between screens for horizontal uNav ($\rho = -0.31$, $p < .05$) and positively for the comfort of Touchscreen use ($\rho = 0.66$, $p < .005$).

Voice Output. There were significant correlations between ratings of voice output and ease of use measures with the more helpful the user found the voice output, the easier they found the devices to use, particularly the touchscreen (Table 1). This is not related to sight as the voice output rating and sight rating do not have a significant correlation ($\rho = -0.21$, $p > .1$).

Discussion

All participants said the uNav was acceptable as was or with minor changes. This was extremely encouraging as the transaction chosen was extremely challenging

Table 1.   Correlations between voice output and ease of use.

Rating	Rating	ρ	P
Voice output rating	Ease of use for selecting a seat (vertical uNav)?	0.479	0.0009
Voice output rating	Ease of use for navigating between screens (vertical uNav)?	0.3719	0.0119
Voice output rating	Ease of use for entering alphanumeric information (horizontal uNav)?	0.344	0.0179
Voice output rating	Ease of use for navigating between screens (horizontal uNav)?	0.4671	0.0009
Voice output rating	Ease of use for selecting a seat (touchscreen)?	0.5964	0.0055
Voice output rating	Ease of use for navigating between screens (touchscreen)?	0.6085	0.0044
Voice output rating	Ease of use for navigating between screens (uNav)?	0.3516	0.0154

(including complex flows, on screen alphanumeric and a spatial task like selecting a seat along with the challenges inherent in using auditory feedback such as boredom if too much information is given, or confusion if not enough is given at the correct time) and the screen flows had not been enhanced for use specifically with the uNav. This was a deliberate decision to validate how well a device like this could be retro-fitted to existing self-service technology with minimal impact to the existing infrastructure. A common question from participants at the end was *"how soon will this be available"*?

Those who were sighted preferred using the touch screen over uNav, even those with limited partial sight, although the differences between ratings between the uNav and touchscreen were not statistically significant. Several commented that they would like the option of uNav in the future when their sight may deteriorate further. However, of those participants with partial sight there was also a desire to have a more accessible touch screen in addition to offering a device like the uNav.

When considering horizontal and vertical orientations there was a clear preference to the horizontal layout. However, virtually all participants managed to complete the task with the vertical uNav so either layout could be adopted. The horizontal layout should be used by default based on preference results and also based on the negative correlation between upper body mobility and ease of use for the vertical. In addition, there may be a slight detrimental impact on perceived usability if the uNav is installed vertically as seen in the difference in ease of use ratings between the two orientations. It should be noted that the majority of participants were standing for this evaluation, which could have an impact on this preference; further research is indicated to validate if this finding holds for participants in wheelchairs as well. As taller participants tended to rate the touchscreen easier to use than shorter participants, it may be that touchscreen usability could be rated worse for some wheelchair users, thus further supporting the addition of a device like the uNav.

The decline in confidence with technology usage is a worrying trend, as it may be that consumers who are blind or partially sighted due to this lack of confidence do not feel able to use new technological solutions. However, the positive feedback received on the features of the uNav suggest that the use of tactile markers, lighting, texture and other physical attributes offer benefit to this community and so should be considered in the future development of such a solution. The correlation between age and an increase in some measures of confidence in technology was an encouraging one, and suggests that ageing populations are not always resistant to using technology, particularly if it accommodates their needs and capabilities.

As part of this evaluation the participants offered a number of unsolicited comments. In general the device was seen to be a form of empowerment, with participants appreciating the opportunity to be enabled to use new self-service channels. Direct quotes from participants include:

> *"I would love to be able to go and do whatever everyone else does without having to ask for help, like walk into Euston station and buy my own ticket"* (blind with some residual vision, aged 35–44).

> *"This is brilliant, I love it"* (blind with no residual vision, aged 25–34).

> *"I've done user trials before, and usually have to say the technology is not good enough for me, but I love this, and want it released as soon as possible"* (blind with no residual vision, aged 25–34).

Conclusion

The device evaluated here showed benefit for consumers who were blind or partially sighted with participants being able to complete a travel check-in kiosk without prior training or experience of such an application. It also affirms RNIB's strategy to improve the independence and confidence of people with sight loss, and the benefits of having self-service machines that are accessible. The results demonstrate a preference for having the device in an orientation close to horizontal, and also demonstrate the importance of offering those with low visual acuity the choice to use touch screen or an alternative input device as those with some vision tended to prefer the touch screen.

Further research is indicated as to how this device supports those with reduced manual dexterity and those with reduced mobility (particularly in wheelchairs) but in general the device does seem to provide a beneficial addition to a touchscreen.

References

Day, P.N., Chandler, E., A. Colley, M. Carlisle, C. Riley, C. Rohan, S. Tyler, 2012, The universal navigator: a proposed accessible alternative to touchscreens for self-service. In: M. Anderson (ed.) *Contemporary Ergonomics and Human Factors 2012*, (Taylor and Francis, London), 31–38

Meuter, M., Ostrom, A., Bitner, M. & Roundtree, R., 2003, The Influence of Technology Anxiety on Consumer Use and Experiences with Self-Service Technologeies. *Journal of Business Research*, 56(11), 899–906

Penn, Schoen and Berland, 2004, *Elo TouchSystems In Touch Survey*, from http://www.elotouch.co.uk/AboutElo/PressReleases/040617.asp

RNIB, 2009. RNIB Group current five year plan 2009 to 2014. http://www.rnib.org.uk/aboutus/organisation/thefuture/Pages/rnib_current_plan.aspx. Last updated 16th July 2012

UKONS, 2010, *Older People's Day 2010 Statistical Bulletin*, UK Office for National Statistics, from http://www.ons.gov.uk/ons/rel/mortality-ageing/focus-on-older-people/older-people-s-day-2010/focus-on-older-people.pdf

Zaidi, 2008, *Features and challenges of population ageing*, from http://www.euro.centre.org/data/1204800003_27721.pdf

AGE, TECHNOLOGY PRIOR EXPERIENCE AND EASE OF USE: WHO'S DOING WHAT?

Mike Bradley[1], Joy Goodman-Deane[1], Sam Waller[1], Raji Tenneti[2],
Pat Langdon[1] & P. John Clarkson[1]

[1]*Engineering Design Centre, University of Cambridge, UK*
[2]*University of Western Australia, Australia*

Designers often assume that their users will have some digital technological prior experience. We examined these levels of prior experience by surveying frequency and ease of technology use with a range of technology products. 362 people participated as part of a UK nationwide larger survey of people's capabilities and characteristics to inform product design. We found that frequency and self-reported ease of use are indeed correlated for all of the products. Furthermore, both frequency and ease of use declined significantly with age for most of the products. In fact, 29% of the over 65s had never or rarely used any of the products, except for digital TV. We conclude that interfaces need to be designed carefully to avoid implicit assumptions about users' previous technology use.

Introduction

Venkatesh et al (2003) through their Unified Theory of Acceptance and Use of Technology (UTAUT), showed that technology use and acceptance is dependent on other factors such as performance expectancy, effort expectancy (related to perceptions of ease of use), social influence, facilitating conditions as well as age, gender, experience and voluntariness of use. Of these factors, designers of digital devices (and in particular interaction designers) often rely on users having some prior experience with technology, and often they expect a significant amount of such experience. For example, they may assume knowledge of common interface controls, symbols or paradigms.

However, not all users have wide technology experience, particularly among older age groups. For example, Morris et al (2007) found that computer and internet use decline sharply with age among the over 50s, with corresponding decline in the use of many other common technologies. Czaja et al (2006) also found that people over 60 are less likely than younger people to use technology in general, although O'Brien et al (2012) found that the lower usage was primarily limited to computer-based technologies. Other studies have found that technology experience does have an impact on product interaction. Blackler (2006) and Langdon et al (2008) found that prior exposure to products with similar features helped participants to use products more quickly and intuitively.

This paper examines this issue by comparing frequency of use and self-reported ease of use across a range of common technologies. It presents data from a survey of 362 people in the UK aged between 16 and 92, allowing a comparison of technology experience across a wide range of ages.

Method

The data in this paper is taken from a survey of people's capabilities and characteristics related to product use, conducted in 2010 (Tenneti et al, in progress). The survey was carried out as a pilot, testing the methods and materials in preparation for a full national survey. However, there were 362 participants, with the sample taken to represent the general adult population living in private households in England and Wales. It can therefore provide useful data and enable preliminary conclusions. The sample was 53.6% female and the age distribution was: 16–34 (23%), 35–49 (29%), 50–64 (24%) and over 65 (23%).

The results reported in this paper are pure sample data. They have not been weighted, e.g. by age and gender, since part of the purpose of this paper is to explore the differences between such demographic groups.

Procedure

The survey was conducted face-to-face in participants' homes. It comprised a series of modules examining a wide range of areas, including vision, hearing, dexterity, mobility, reach, cognitive function, technology/product use, psychological resources, and anthropometrics. The study included both self-report questions and performance tests to assess respondents' capabilities.

This paper focuses on the results from the technology use module. In this module, participants were first asked how often they performed various technology-related tasks, such as making a phone call on a mobile telephone. A full list of the tasks is shown in Figure 1. For each task, they could choose a response from: Frequently, Occasionally, Rarely and Never. Some respondents declined to answer and some said that they did not know. This only affected three or fewer participants per question.

Participants were then asked how easy or difficult they found it to do each of those tasks. They could choose responses from: Very easy, Easy, Neither easy nor difficult, Difficult, Very difficult or Impossible. They were not asked this question for a particular task if they had reported that they had never done that task, e.g. they were not asked how easy they find making a call on a mobile phone if they had never made a mobile phone call.

The questionnaire asked about technology-related tasks rather than technology products, in order to give more emphasis on technology experience. This is particularly relevant for products such as mobile phones which often have many functions with varying degrees of complexity. The tasks were chosen to cover a

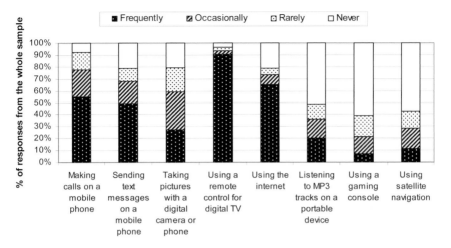

Figure 1.   The reported frequency of various technology-related tasks.

range of common interaction patterns, such as button pressing on remote controls and using Internet navigation features.

Results

Overall results

Figure 1 shows the results for how often participants said that they performed various technology-related tasks. The frequencies vary greatly between different tasks. Most participants had reported that they used digital TV remote controls at least occasionally, while more than half of the sample reported they had never used a gaming console or a satellite navigation system. The proportion of the UK sample who had never used a mobile phone to make a call was 7.8%, 21.1% had never sent a text message, and 21.4% had never used the internet. However it is worth noting with the self-reported frequency of use data, that interpretations of usage described 'frequently', 'occasionally' and 'rarely' may be influenced by the perception of a 'normal' frequency of use of the task. For example, daily use of digital camera may be considered frequent, whereas daily use of a mobile phone may be reported less than frequently.

Relationship between frequency and ease of use

Frequency of use and ease of use were significantly correlated for all of the technologies as shown in Table 1. It should be noted that participants were only asked about the ease of a task if they said they had ever done it. Thus, if instead of self-reporting ease of use actual usage performance on a technology had been measured, those with no prior experience could have been expected to perform the worst. If frequency and ease of use are closely connected, we would expect the biggest impact to be on those who have never used the technology.

Table 1.   Correlations between frequency and ease of use.

Technology task	Correlation between frequency and ease of use
Calling on a mobile phone	$\rho = .435$, p $< .001$
Sending text message on a mobile phone	$\rho = .419$, p $< .001$
Taking pictures with a digital camera, or your phone	$\rho = .340$, p $< .001$
Using a remote control for digital TV	$\rho = .229$, p $< .001$
Using the internet	$\rho = .529$, p $< .001$
Listening to MP3 tracks on a portable device	$\rho = .447$, p $< .001$
Using a gaming console	$\rho = .482$, p $< .001$
Using satellite navigation	$\rho = .329$, p $< .001$

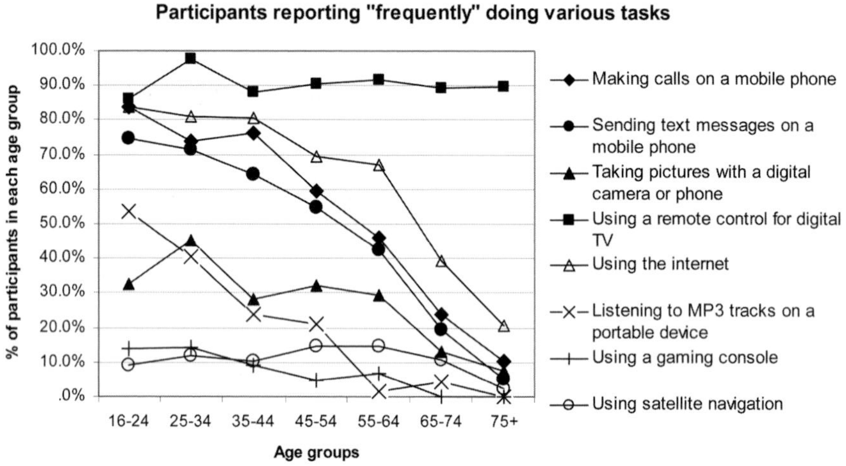

Figure 2.   High technology experience: participants reporting "frequently" doing various technology-related tasks, by age group.

Frequency of use with age

Figure 2 shows the percentage of each age category who said that they frequently carry out the technology-related tasks. For most digital technologies, reported use declines with increasing age. Digital TV remote control use however is high for all age groups and even in the over 75 age bracket, 89.7% of respondents report frequent use. At the other end of the scale, reported frequent use of satellite navigation systems are low for most age categories, peaking at 14.8% for the 55–64 age group and reported frequent use of gaming consoles is generally low, peaking at only 14.3% in the 25–34 age group.

Conversely, Figure 3 shows the percentage of each age category who reported never having done the tasks. This graph is particularly interesting because it indicates those who may have no experience of a technology at all, and of all the self-reported

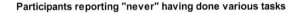

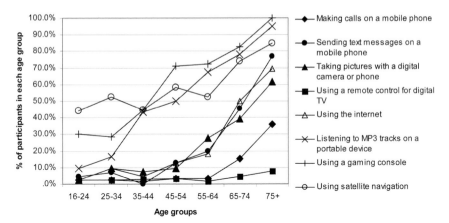

Figure 3.   No technology experience: Participants reporting "never" having done various technology-related tasks, by age group.

Table 2.   Proportions of age groups with very little technology experience.

Age group	% of age group who had never done any of the technology tasks (except use a digital TV remote control)	% of age group who had never *or rarely* done any of the technology tasks (except use a digital TV remote control)
16–34	2.4%	3.5%
35–49	0.9%	3.8%
50–64	2.3%	6.9%
65+	17.6%	29.4%

frequencies it is the most definitive. In the youngest age group, very few people had no experience with the technologies, except for satellite navigation and (to a lesser extent) games consoles. However, these figures rise with age, particularly over the age of 55. Over the half of those aged over 75 had never used a games console, an MP3 player, satellite navigation or the internet, nor sent a text message or taken digital photos. Lack of experience with MP3 players is particularly dependant on age. Even among those over 35, large numbers had never listened to MP3 tracks on a portable device.

Having no experience with a particular technology can affect product use. Having no experience with a wide range of technologies is likely to have an even greater impact. Table 2 shows the proportions of different age groups who had never done any of the technology tasks in the survey, except for using a digital TV remote control. The table also gives the proportions of whom had only rarely done any of the tasks, as this low level of technology use is also likely to have a great impact on ease of use. The numbers are fairly low except in among those aged over 65, where 17.6% had never used any of the technologies, except digital TV remote controls, and almost a third had only rarely used them.

Discussion

Fewer than 80% of the total sample claimed to have ever used the internet, sent a text message or taken a digital photograph. Nearly a third of over 65's have never or rarely used any digital technology with the exception of a TV remote control. Over two-thirds of the over 75's had never used the internet or sent a text message. We have shown the correlations between the frequency of use of technology and the perceived ease of use of that technology, which should not come as a great surprise.

Previous work has highlighted that prior experience is a key factor in determining an individual's experience of the ease of use of technology devices. However, there is a huge gap between the prior technological experiences of those of working age and those in retirement, which may provide some insight into the techno-cultural gulf that exists between the designers and engineers who make the technology, and older people in the population who may wish to learn but don't necessarily find it easy to use the latest digital devices. There seems to be a vast opportunity to design digital technologies for those of limited digital technology experience, which do not presuppose detailed mental models of the function, nor interaction patterns which require specific learning events to unravel.

Conclusions and further work

Technology experience and perceptions of ease of use are clearly related. Given the differences in technology experiences between the mainly younger designers and engineers designing interfaces for technology products and services, there is a strong case for an inclusive design approach, to ensure that a wider range of technology experience is accommodated, and particularly for the design of products which are to be used by older age groups.

It is suggested that further work should examine in greater detail particular aspects of prior technology experiences, with regards to the effect on ease of use, for example in the roles of icons and specific control interactions, to enable better understanding of the design implications.

References

Blackler, A. 2006. Intuitive interaction with complex artefacts. PhD Thesis. Queensland University of Technology: Australia

Czaja, S. J., Charness, N., Fisk, A. D., Hertzog, C., Nair, S. N., Rogers, W. A. and Sharit, J. 2006. Factors Predicting the Use of Technology: Findings From the Center for Research and Education on Aging and Technology Enhancement (CREATE). *Psychol. Aging* 21(2): 333–352

Langdon, P., Lewis, T. and Clarkson, J. 2008. Measuring Prior Experience in Learned Product Interactions. In *Ergonomics 2008*, Nottingham, UK, Ergonomics Society

Morris, A., Goodman, J. and Brading, H. 2007. Internet use and non-use: views of older users. Universal Access in the Information Society 6(1): 43–57

O'Brien, M. A., Rogers, W. A. and Fisk, A. D. 2012. Understanding age and technology experience differences in use of prior knowledge for everyday technology interactions. ACM Transactions on Accessible Computing 4(2)

Tenneti, R., Langdon, P., Waller, S., Goodman-Deane, J., Ruggeri, K., Clarkson, P.J., and Huppert, F.A., in progress. Design and delivery of a national pilot survey of capabilities. Submitted to *Applied Ergonomics*

Venkatesh, V., Morris, M. G., Davis, G. B. and Davis, F. D. 2003. User acceptance of information technology: toward a unified view. MIS Quarterly 27 (3): 425–478.

YOUNGER OLDER CONSUMERS OF ASSISTIVE TECHNOLOGY PRODUCTS

Andrée Woodcock[1], Gill Ward[1], Sujata Ray[2], Nikki Holliday[1], Louise Prothero[1], Jane Osmond[1] & Simon Fielden[1]

[1]*Coventry University, UK*
[2]*Age UK*

There are over 21 million people aged over 50 years in the UK – a third of the total population – with 14 million people being aged over 60. By 2083, it is estimated that 1 in 3 people in the UK will be aged over 60. Assistive Living Technology products, such as telehealth and telecare and services are being developed to meet the needs of this consumer group. However, consumers have been slow to recognise the potential of these products. In this paper, a range of factors are presented which may act as barriers or enablers to purchase.

Introduction

Successful ageing has been defined as 'growing old with good health, strength and vitality' (Rowe and Kahn, 1998). Psychological theories focus on 'optimization' (Baltes and Baltes, 1990) and the discovery of strengths that enhance quality of life. Assistive technologies can aid by providing equipment or environments that enable maintenance of quality of life. Information and Communication Technology applications can provide new ways to help older people retain independence (EC, 2008). Uptake and use of these products has not been extensive or sustained. Purchase and usage are influenced by factors, such as design and usability, individual and societal attitudes towards assisted living technologies, changes in health and social care provision. As most research has focused on older people, there is a need to discover what the barriers and enablers may be to purchase by younger consumers (50 years plus).

Research suggests that in order for technology to be used, it needs to be perceived as relevant to the everyday lives of older people. Recognizing that assistive technology can be of help (i.e. that it can bridge the gap between changing ability and the everyday activities that people want to do) is a key stage in obtaining and making successful use of Assistive Living Technologies (ALTs), such as motion and fall detectors). In general, people want to manage for themselves, regardless of living arrangements. Following this, the quality of the product or service itself becomes key; efficiency, reliability, simplicity and safety are important factors. Other issues relevant to the uptake of ALTs are shown in Table 1.

Table 1.   Factors relevant to the uptake of ALTs.

Factors	Examples
User	Age, gender, diagnosis and ability, education, socio economic group, values, goals and preferences, culture, needs, goals of the carers, psychological readiness
Technology	Aesthetic appeal, comfort, cost, reliability, durability, ease of use and effectiveness in meeting a need
Environment	Type of home the user lives in, outdoor places or other environments where assistive products might be used
Training	How well the user, carer or family is trained or prepared in using the assistive technology

Consumerism, retail models and statutory versus private purchase

The past fifteen years have witnessed a series of moves towards a retail model of purchase of assistive living products in England. This has been concurrent with fundamental changes in the funding of social care. Since 1997, certain user groups have been able to receive 'direct payments' in lieu of services, which they can use to pay for their care. Direct payments are now available in the form of a 'personal budget' – an amount of funding based on the assessed needs of the person. This budget can, of course, be used to pay for assistive technology. In 2001, the Department of Health published the '*Guide to Integrating Community Equipment Services*' recommending that provision of assistive technology was integrated between health and social care to provide a less disjointed service to users. Winchcombe (2008) highlighted that users received little benefit from the new approach and did not access 'community equipment' that might be of use. A subsequent initiative, 'Transforming Community Equipment Services' (DoH, 2010), gave prescriptions to users for 'simple' items, allowing choice from a range of products and the ability to 'top up' their personal budget for alternative items, to increase user choice and control. In 2008, the retail model was rolled out across England. However only a limited number of local authorities have adopted it, and reasons for the slow uptake of the consumer market, especially for young older people are limited.

The 'halfway' retail model of high street delivery normalizes the provision of assistive living products and reduces stigma (Ricability, 2009). Many service users like being able to 'buy products in a shop in the ordinary way'. There may also be age differences in whether people prefer to be supplied with products by care services or whether they prefer to choose their own.

Trust in the information or advice given has also been recognized as being important (DoH, 2010). In line with the prevalent medical models of disability (Oliver, 1996), trust is still placed in care professionals such as occupational therapists. Consumers may lack trust in private retailers' independence. Charities report concern that retailers and sole traders, under pressure to increase profits, pressurize users to buy products. For high street suppliers of assistive living products, it seems that retailer accreditation may be important to ensure that 'individuals don't have advantage taken of them' (Ricability, 2009).

The retail market is potentially very large compared and is yet to be fully realized. In 2008, local authority expenditure on assistive living products was approximately £177 million, compared to £244 million spent by private purchasers (Market and Business Development, 2009; WHICH?, 2009). Aspects of private supply need to be considered, e.g., the broad range of equipment versus the ability for retail outlets to offer comprehensive choice (de Jonge, 2008), the move from block supply contracts with large deliveries to few locations, to smaller deliveries to a larger number of locations. The private spend on ALTs is likely to increase and public spend decrease. Demographic change means more people will become consumers of assistive living technologies.

The COMODAL project

The aim of the three year, TSB funded, COMODAL – **CO**nsumer **MOD**els for **A**ssisted **L**iving – project is to develop a consumer market for assisted living technologies (ALTs) for younger older people (i.e. those approaching retirement and older age in the 50–70 year age range). This paper reports on research conducted in Year 1 to understand the needs of younger older people as both users and purchasers of ALTs and to investigate the barriers and enablers to ALT adoption for this group. This information will be used to design effective solutions to overcome these barriers and promote factors that will enable the ALT market to flourish through consumer led business models. In order to address this, the younger older age group was broken down into four segments.

Younger Older Personal Consumers	50 to 70 year old purchasers of ALTs for own use
Younger Older Prospective Consumers	50 to 70 year olds who neither buy nor use ALTs
Younger Older Carer Consumers	50 to 70 year old 'decision gatekeepers' involved in purchasing decisions for older ALT end users
Younger Older Non Purchasing User/Consumer	50 to 70 year olds who use ALTs but refer to other 'decision gatekeepers'

Methods

- **Market analysis** to categorize available products, identify the main suppliers and summarize their propositions, including economy, standard or premium products, routes to market through direct, third party or business-to business routes, and access to professional advice.
- **Product review** to understand trends in product development and how well current products meet the needs of consumers in their fifties and sixties.

- **Consumer street survey** with over 500 users to help understand the importance of the different barriers and enablers affecting the purchase and use of ALTs by younger older people.
- **Consumer focus groups** to explore the barriers and enablers using a more in-depth approach to gain a better understanding of attitudes towards the purchase and use of ALTs. Four focus groups were carried out across the UK with representatives from each of the groups mentioned above.

Results

Market review

The review confirmed that the ALT market is not fully realized. It is distinguished by its low levels of public awareness, high need for expert, trusted advice and VAT complications. However, there are examples of where good inclusive design has resulted in the mainstreaming of specialist products, such as the Oxo range of 'Good Grips' kitchen utensils. With more people becoming aware of ALT products and how they can be used to maintain their quality of life, mainstream outlets will start to stock them. Other drivers include recognition amongst mainstream manufacturers of an ageing customer base, saturation of other markets, lower entry barriers to the market, changing societal attitudes to self-help and self-provision, wider recognition of the opportunities and challenges of an ageing population.

Product review

Interviews with designers and developers confirmed awareness of younger older people (YOP) as potential consumers with distinguishing characteristics. The YOP market would be best served through better design of mainstream products, rather than specific items targeted at older people. Younger older people do not want to be stigmatized or singled out as in need of 'assistive' products or 'equipment', so products which are discretely tailorable are well received.

ALT products have tended to focus on the disability rather than the person and fail to fulfill higher level needs, wishes and aspirations. Unsurprisingly this makes products less attractive and desirable. It might also make them less functional as addressing just one disability may mean failure to recognize other needs. Other companies have seen an opportunity to market existing products as assistive technologies, with little effort made to redesign them to a different usage context. Products which are suitable for medical and office environments, in most cases, do not suit the home environment. The review highlighted factors which need to be given attention in the design and delivery of ALTs:

1. Many ALTs especially in telehealth and telecare services take control away from the user; for example remote monitoring of behaviour.
2. Products tend to focus on lower rather than higher order needs (Maslow, 1943).
3. People in later life value higher a higher level of service, including after sales care and advice.

4. Adherence to standards is weak – many products do not have ISO or CEN authentication.
5. Discretion and privacy – younger older consumers do not want products that characterize them as disabled or in need of assistance.
6. Pleasure and experiential design – few products take into account the whole experience of the user, such as its attractiveness, the meaning of using the product, and the feelings that arise during its use (Jordan, 2000).
7. Products need to be adaptable or tailorable to cater for a range of needs or disabilities and be adaptable to an individual's needs over time.
8. Fitting into peoples' homes – many recommended products are rejected because they *'did not want their homes looking like hospital clinics'*.
9. Economic feasibility – the affordability of the product to its target market, products with built-in obsolescence.
10. Products should be developed based on a clear understanding of user needs.

Consumer street survey

500 consumers (390 aged 50–70, 110 carers aged 30–49 or 71+) were asked to rank the importance and influence of the barriers and enablers to purchase and use of ALT products and services and the wider factors affecting purchasing decision making. The results highlighted how the consumer market has been held back by the dominance of the statutory provision of ALTs through health and social care services. Key results are summarized:

- The **top three barriers** having the strongest influence for the 50–70 age groups were cost, knowing how to choose what to buy and a lack of awareness the existence of useful products.
- **Top three enablers** were beliefs that a product could make a difference, would be affordable and worth it and make life safer at home.
- **Younger carers** appeared more sensitive to issues such as how the person they care for feels about using ALTs and their quality of life.
- **54%** said that they would buy a product or service to help with daily living.
- **28%** referred to simply *struggling on*, particularly those who had not bought or used ALTs. This indicates the importance of raising awareness of ALTs.
- **Social economic groups** (based on MRS (2006) definitions), differed in their approach to purchase of ALTs. AB consumers were more likely to take advice from healthcare professionals. Shopping via the internet was more important for consumers in AB and CI groups. Friends and family are more important influences on the purchase decision than the internet for C2 and DE consumers. These trends are important for market segmentation and provide useful insights for industry in terms of how they promote their products.
- **Purchasers of ALTs** did so with lack of knowledge and did not know where to seek advice.
- The cautious behaviour and attitudes of inexperienced consumers and carers may be significant for future developments. Healthcare professionals will remain

important influence and gatekeeper, but their future role, particularly at the less complex end of the product/service spectrum will diminish as the market place changes and becomes more mainstream.

- Few people had purchased an electronic ALT and even fewer telecare or telehealth products (8%). The popularity and acceptance of this product group amongst YOPs will grow faster than for any other ALT.

Consumer focus groups

Again participants revealed a lack of awareness of and information about ALTs. Some ALT products were viewed as stigmatising and not meeting functional needs. Additional worries related to quality and value for money.

A number of enablers to the purchase process were identified with rental or "try before you buy" services and more information increasing confidence. The normalisation of assistive technology and its increasing prevalence were seen as important enablers to purchase. Participants felt that the use of certain products by older age groups (e.g. mobility scooters) would become more acceptable. However, they were also keen to point out that such products could be helpful for all age groups. Mainstreaming would propagate awareness of products and their benefits as more people would be using the products (thereby improving acceptability in the public eye), and economies of scale would reduce cost. Positive images of those using assistive technology and the affirmative portrayal of disabled and older people in soap operas were important.

Although there was much agreement regarding many enablers and barriers; distinct differences between did emerge. **Prospective Consumers** discussed benefits of ALTs for their parents, not for younger people; they were most influenced by the stigma associated with ALTs with zimmer frames and shopping trolleys being distinguished as particularly stigmatising. This may be overridden if the product fulfilled an essential need. **Personal Consumers** felt many had low awareness of ATs, and that they were lucky to be part of good social networks which enabled them to find out about ATs. Generally, those with good social networks had higher levels of awareness of assistive technology. **Carer Consumers** commented most on the limitations and functionality of AT, as they have seen those they care for struggle to use various products. They were aware of the cost of AT, and felt that with a growing market, AT would become cheaper in the future. The Carers were similar to the Personal Consumers in their awareness of the usefulness of social networks in raising AT awareness. **Non-Purchasing Users** were the only group who thought AT should be provided by statutory services, possibly because most had been supplied with AT by their Local Authority, and had little experience or awareness of purchasing privately.

Electronic Assistive Technology (EAT) produced views ranging from "terrifying" to "fantastic". Participants had seen blood pressure monitors, fall detectors and products aiding memory loss or dementia in retail outlets. They felt that if EAT was to meet its potential it would need to be supported by education and information.

At present EAT was considered too complicated, with concerns about accuracy, increasing anxiety, ease of use and reduced social contact.

Conclusions

The barriers and enablers relevant for the YOP consumer market can be grouped into four categories: 1) Feeling a need or a desire to make a purchase, 2) Taking action/routes to purchase, 3) Choosing between options/design issues, 4) Using the product/customer service. Almost all the barriers found in existing studies were true for younger older people. However, they showed less concern about sharing personal data (e.g. with health services), privacy and control of monitoring services. Only 8% had purchased telehealth and telecare equipment and none talked spontaneously about technology "services" that might be able to help them. This indicates that the consumer market here is still in its infancy.

Lack of information was a key barrier to uptake. Recognizing that assistive technology can be of help (i.e. that it can bridge the gap between changing ability and the everyday activities that people want to do) is a key stage in the process of obtaining it and making successful use of it. As a general rule, people wanted to manage themselves and were open to ALTs supporting them to do so. However, the Prospective Consumers did not see the relevance of ALTs to their lives and could not imagine themselves using it. All groups except for the Non-Purchasing Users were open to purchasing ALTs as private consumers. This supports the scope and opportunity to develop the private consumer ALT marketplace.

A key finding is the lack of high quality research confirming the effectiveness of assisted living technologies. It is widely believed that such technologies can improve the lives of older people and provide better cost-effectiveness in terms of health and social care; this has not yet been verified. A lack of awareness/information about ALTs was identified as the largest barrier to uptake and use of ALTs. The market review uncovered inconsistencies in pricing and low levels of competition in traditional segments. All leading manufacturers sold their products to a wide range of outlets resulting in wide price variations.

An area that clearly influenced what people chose to buy was the stigma attached to it. The product review showed that designers and product developers are only now beginning to recognize the market of younger older people as consumers or potential consumers and customers. Transgenerational design provides an opportunity to develop nonstigmatizing, aspirational products which do not single out those who are aging or have a disability.

An industry workshop supported the need for market-changing products, a focus on solutions and services (rather than just technologies), the need for ALTs that were attractive, functional, discreet, simple and easy to use. **Industrial partners** revealed that they had a poor definition and understanding of the needs of potential ALT users and the structure of the market.**Consumers** revealed a clear preference for greater levels of hands-on experience through demonstrations, "try before you buy", money back guarantees, rental options and lease hire. User confidence could be built by increased reliability, maintenance and servicing, hearing about other customers' experiences, better information provision and advice. A positive purchase/retail experience that offers outstanding customer service was also seen as desirable.

As the population ages and those between 50–70 years old become the 'new' older generation, behaviours and demands will change, creating new market opportunities for innovation and adoption. This research provides insights for new and existing suppliers and retailers of ALT about consumer preferences. The evidence indicates an opportunity to open and develop the market so that all consumers have access to and can benefit from assisted living solutions, thereby easing the challenges presented by the natural process of ageing.

References

Age UK. 2012, *Later Life in the United Kingdom*

Baltes, P.B. and Baltes, M.M. (eds), *Successful Aging: Perspectives from the Behavioural Sciences*, Cambridge, MA: University Press

de Jonge, D. 2008, "The Retail Model may Bring Uncertain Future." *International Journal of Therapy & Rehabilitation* 15 (3), 119–120

Department of Health. 2010, *Transforming Community Equipment Services: Customers Questions and Answers*. London: Department of Health

European Commission. 2008, *Active Ageing and Independent Living Services: The Role of Information and Communication Technology*. Luxembourg: Office for Official Publications of the European Communities

Jordan, P.W. 2000, *Designing Pleasurable Products*. London: Taylor & Francis

Market and Business Development. 2009, *The UK Domiciliary Care Market Development*. Market and Business Development Ltd: Manchester

Market Research Society, *Occupation Groupings: A Job Dictionary, 6th ed*, 2006

Maslow, A.H. 1943, "A theory of human motivation." *Psych. Rev*, 50(4), 370–96

Oliver, M. 1996, *Understanding Disability: From Theory to Practice*. MacMillan, Basingstoke

Ricability. 2009, *The Revolution in Equipment Supply and what it Means for Information*. London: Ricability

Rowe, J.H. and Kahn, R.L. 1998, *Successful aging*, New York: Dell

WHICH? 2009, *Briefing: Tools for Independent Living Forum*. London: WHICH

Winchcombe, M. 2008, "Developments in Assistive Technology." *Consumer Policy Review* 18 (5), 127–131

CONTRIBUTING TO UK INPUT TO NEW EUROPEAN INCLUSIVE DESIGN STANDARD

Anne Ferguson & Piera Johnson

BSI Standards Ltd., UK

Work will start on a new European Standard, related to inclusive design, during 2013. This session will provide an opportunity for accessibility experts to find out a little more about the existing British Standard and contribute ideas for content for the proposed new European standard. It will also be an opportunity to identify potential committee members for the UK Committee which will mirror the European work.

Context

BSI is the UK National Standards Body (NSB), responsible for balancing the needs of industry, consumers and end-users and government policy in the generation of consensus best practice codes, guidance and standards. Voluntary standards have been used by business and industry for over one hundred years to improve the effectiveness of markets, raise product quality and provide confidence for customers and suppliers.

As the NSB for the UK, BSI is also the gateway to European standardization (CEN & CENELEC) and international standardization (ISO and IEC).

In support of the European internal market, European standards must be adopted by all NSB members of CEN and CENELEC. In the UK, BSI will publish a European standard with the BS prefix in addition to the CEN or CENELEC identifier. The development of some standards is specifically requested ('mandated') by the European Commission (EC).

A multi-stakeholder committee at the national level determines the views to be taken forward into Europe. UK delegates from the BSI committee attend relevant European or international standards meetings to ensure the national view is incorporated in the development of the standard.

CEN-CENELEC joint working group on 'Design for All'

EC Mandate M/473, under Clause 4.1, requested the development of '*A new standard (or other deliverable)*' that '*describes how the goods manufacturing industry as well as public and private service entities in their processes can consider accessibility following Design for all approach with due consideration for assistive technologies and services that could help bridging the usage gap of the product or service'.*

The mandate referenced a British Standard BS 7000-6: 2005 as an example of the sort of document required. This UK document was developed with input from a number of ergonomists and disability specialists from organizations such as Helen Hamlyn Institute, the Engineering Design Unit at Cambridge, Reading University, RNIB and RNID. BS 7000-6 provides guidance on managing inclusive design at both project and organisational level, discussing all stages of the process from planning, through development and delivery and including evaluation. An Annex includes tools and techniques for managing inclusive design. Revision of the BS standard was held up by the proposed work being discussed in Europe.

The creation of a joint CEN-CENELEC working group has been proposed with a secretariat provided by the Dutch standards body, NEN and a Chair put forward by NSAI, the Irish standards body. There is no date set as yet for the first meeting of the European Working Group but April 2013 is likely to be a good time to collect UK views.

Providing input to the European work

This conference session follows several papers on accessibility. It is proposed that a short introduction to the proposed CEN-CENELEC work will be given, followed by a brief description of the existing British Standard. The remaining time will be available to gather comments to put forward to the UK committee to help form the UK view as well as to seek further membership suggestions for the UK committee.

References

BSI, BS 7000-6: 2005 *Design management systems. Managing inclusive design. Guide.* Retrieved 20 November 2012, from: https://bsol.bsigroup.com/en/Bsol-Item-Detail-Page/?pid=000000000030142267

European Commission Enterprise and Industry, EC Mandate M/473 *Standardisation mandate to CEN, CENELEC and ETSI to include "Design for All" in relevant standardisation initiatives.* Retrieved 20 November 2012, from: http://ec.europa.eu/enterprise/standards_policy/mandates/database/index.cfm?fuseaction=search.detail&id=461

TALKING TV FOR BLIND PEOPLE – SEE HOW IT WORKS!

Lori Di Bon-Conyers, Sabine Croxford & Cathy Rundle

*Innovation Unit, RNIB (Royal National Institute of Blind People),
Peterborough, UK*

Introduction

How do you use TV if you're blind? Why would you even want to – surely blind people listen to the radio?

Well, for most blind people TV is an important leisure activity but digital TV – whilst bringing benefits such as audio description – has meant that the user interface is more screen-based and hence visual which in turn means it's harder for blind and partially sighted people to use.

This workshop shows you how this is changing and how some TVs are becoming more accessible and easier to use for blind and partially sighted people – and you can try it all out for yourselves.

Accessible TV for blind and partially sighted people

Televisions are in all our homes and are something we use most days without thinking about it. However, with the introduction of the many channels and on-screen interfaces of digital TV blind and partially sighted people were suddenly faced with a massive change in the way they can use and interact with TV. Audio description of programmes brought about great leaps in accessibility enabling people to follow programmes more easily and increase independence. However, the use of on-screen programme guides, whilst useful to sighted people, are not accessible and not available to blind people. In addition, the proliferation of channels (and sometimes frequent changes of channel numbers) means that if you can't see the screen it's harder to remember the channel numbers you need and it's far too easy to get lost and miss programmes. Remote controls have also become more complex and harder to cater for this additional functionality.

What can be done about this? Good design of a remote control with tactile buttons, a clear layout and good colour contrast will help to some extent. However, in order to make a TV properly accessible to blind people you need to include text to speech (TTS) to speak the information on screen. RNIB has worked with industry to develop this. A TV with TTS uses a synthetic voice to tell the user what is being shown on screen – for example, the channel number you are typing in or the channel you have changed to and the programme that is on now.

The TV industry is itself now developing more accessible TVs by providing TVs with TTS and clear visual interfaces. In addition, different ways are being developed to control the TV, consider voice and gesture input or apps on smart phones connected to the TV via a home Wifi network.

RNIB will demonstrate some of these technologies including a fully talking set top box and a talking digital TV. We'll show you how they work and talk about the experiences of users from our extensive usability testing and research into this area. We'll have a look at some apps – good and bad points – and highlight the research conclusions we've found for voice and gesture control of TV.

There'll be plenty of opportunity for people to try out the technology for themselves and have a go at using a TV or an app when you can't see it.

We'll highlight some of the visual elements that help make a TV or app accessible and people can try out different technology wearing "sim specs" (glasses that simulate different eye conditions) to see how the visual design can help or hinder people with partial sight.

People will be able to find out what makes good TTS in a TV or app and the things that are useful to think about when developing a product or software that speaks.

INNOVATION AND CREATIVITY

ADULTS' AND CHILDRENS' REACTIONS TO TECHNOLOGICAL INNOVATIONS

Martin C. Maguire

Design School, Loughborough University, UK

This paper reviews the findings of four surveys that studied the reactions of both adults and children to new ideas for technological development. These included the use of TV for new applications in the home, enhancements to make work easier in the kitchen, and the use of computers versus traditional methods for activities such as shopping and learning. The studies show some common themes – that convenience, familiarity, practicality, feasibility and relevance strongly influence participants' support for particular applications. People do not necessarily see technology as the preferred means of carrying out tasks but seem to consider 'high and low tech' solutions on their merits.

Introduction

Technological innovation leading to new technology products and services is one indicator of a society's progress. Yet it is not a simple consequence that all consumers will wish to make use of them. For instance, by the first quarter of 2012, it was reported that 8.12 million adults in the UK (16.1 per cent) had never used the Internet (Office of National Statistics, 2012). This paper was stimulated by a need to understand how people in general react to new technologies and what aspects of new developments might encourage or discourage their take up. Mack and Sharples (2009) found in a study of mobile phones that while usability is an important attribute of successful products and interfaces, other aspects are equally if not more important including: features, aesthetics, and cost. Product appearance also plays an important role in consumer choice (Creusen and Schoormans, 2004) so it is interesting to study how people feel about product concepts that they cannot see so need to visualise. Previous evidence of research with young people aged between five and seventeen revealed that they could evaluate the cons as well as the pros of consumer technology (Maguire, 2001).

Four survey studies are reported: one with adults, one with older adults and two with children. Each survey asked respondents to consider a list of technological ideas or innovations and choose which ones they would want or find useful. Some general findings from across the studies are reported. Ethical criteria were satisfied by conducting all the studies as anonymous surveys. The two surveys involving school children were administered by the teacher as part of a lesson.

Study 1: Adult's reactions to new ways of using TV

The use of TV as a common interactive device in the home is seen as a way to provide new services and functions for consumers, particularly those who may be less confident with computers or do not wish to own a computer. In 2008 a survey was conducted which presented 60 adult consumers 16 possible TV innovations or functions which participants could rate as 'useful', 'quite useful' or 'not useful'. The participants were selected from the Design School's participant database to give a cross section of people of different ages, genders and backgrounds but generally having a critical interest in new technology. The group included 36 males and 24 females with an age profile of 18–30 years (6 participants), 31–45 years (20 participants), 46–60 years (18 participants) and 61+ (16 participants). Figure 1 shows preferences for each technology in order of popularity.

It can be seen that while some functions received a relatively high number of 'useful' ratings, participants tended to be circumspect and most functions received a larger number of only 'quite useful' responses. A chi-squared test was carried out to see which items were significantly different from the average pattern shown in the bottom row of the chart across all sixteen functions (with a p value < 0.05).

The ability to link the TV to a security camera to see who was at the front door was seen as significantly more positive (p = 0.02) than average. This was possibly due to people's concerns about security. Example quotes from respondents were: *"I'm very security conscious"*, *"it makes you feel more safe"*, *"as a lone female I often feel vulnerable"*, *"I live on a corner property and a security wire would make me*

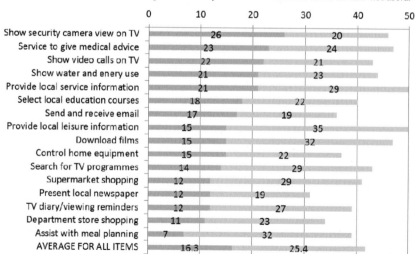

Figure 1.   Ratings of usefulness of TV technology (n = 60).

feel safer". One participant was actually implementing a similar system in his own home. Another raised a practical issue: *"It needs to be automatic. The doorbell or someone coming into camera view should supercede the TV programme. Otherwise you'd be continually pressing buttons"*.

The use of TV to assist with meal planning by displaying recipes was only regarded as 'useful' by 7 participants so significantly less than average (p = 0.02). Comments made were: *"I can plan my own meals thank you"*, *"I love cooking and have loads of ideas myself"*, *"I'm already a pretty resourceful cook"* and *"only as a play thing as there are already lots of places that cover this"*. These comments give the impression that this was redundant functionality as there were lots of other sources such as cookery books and people's own knowledge. However some comments indicated a lack of understanding of what the function was about, thinking that the TV service would actually help to prepare meals: *"useful if unable to prepare and cook one's own meals"* and *"would be an excuse to be lazy."*

Using TV as the medium for a local newspaper was also seen more negatively than the average response pattern (p = 0.01). Comments made by different participants were: *"I enjoy sitting in comfort to browse a newspaper and it is probably easier to read than pressing lots of buttons. Paper can be recycled as cardboard etc., if bleach is not used"*, *"I am old fashioned. I like to go to the newsagent to buy and then read it in bed/garden/bathroom, etc. "*, *"I like to pick up and put down such material when it's convenient to me, not look at TV"*, *"Would not find it comfortable reading a lot of text on a screen"*. However some positive comments were made: *"Modern newspapers are very unwieldy. Broadsheets too big (arms ache). Tabloid and compacts have too many pages and fall apart"* and *"No additions to the piles of paper one accrues with newspapers and magazines."*

The use of TV to provide local service information was a further idea with responses significantly different from the average view (p = 0.01) in that a high number of people regarded such a facility as 'quite useful' (35 participants) rather than 'useful' or 'not useful'. Comments were generally positive but several people qualified them: *"brilliant if kept up-to-date; many organisation's web pages are a waste of time because they are not updated frequently enough"* and *"only if you have no PC."* However others were unequivocally positive towards the idea: *"best use of this technology so far, I like this idea a lot"*, *"saves filing details on paper or searching directories"*, *"I could see myself using this information"*, *"might save the council money too – no glossy literature – more trees saved."* Some were negative: *"prefer to use computer"*, *"I keep things like this in my diary"*, *"already available via Internet-only, occasionally used"*.

It appears that where people can visualise a particular service and how it would work such as door camera linked to a TV and a newspaper on TV, they felt confident to accept or reject the idea. For a facility such as meal planning which was seen as redundant or where the service was not clearly understood, respondents were more clearly negative. For a service like local leisure information, people could visualise it and felt fairly positive towards it but some considered it was more appropriate if the same information was accessed via a computer or kept in a personal file.

Table 1.   Children preferences for advanced TV functions (n = 27).

TV technology	Yes	No
Choose music to listen to	19	8
Info for homework	15	12
See who is at door	14	13
Speak to friend or grandparent	12	15
Send message to teacher	10	17
Phone and see on TV	8	19
Buy present through TV	7	20
Read book on the TV	6	21

Study 2: Children's reactions to potential new TV functions

This study, which was carried out with a class of seven and eight year old children in 2010 explored reactions to possible new TV functions. This age group was chosen as it was felt they had reached the stage of being knowledgeable about technology but not too influenced by pre-conceived ideas or peer pressure. The survey was similar in nature to Study 1, with adults, but was reduced to include just 8 possible functions. Children simply had to indicate whether they would like to have or use each function on the TV or not. The results are shown in Table 1. Using a chi-squared test to compare the average ratings of all the functions (Yes = 11, No = 16) it was found that the function of choosing music on the TV to listen to was significantly more preferred than the average for all 8 functions (p = 0.003) while reading a book on the TV was seen as significantly less preferred (p = 0.04).

The children were asked to indicate what else they would like to do through the TV. Three commented that they did not think 'you could do anything else on the TV' and without further knowledge they did not seem able to imagine possibilities for other functions. However 11 children did have ideas which included: *"look on the TV and see all my class"*, *"watch TV from the past"*, *"buy army vehicles"*, *"send messages"*, *"type in where you want to see inside Mars"*, *"make an invention"*, *"paint"* and *"dance and see me and my friend Mollie dancing with me"*. Influenced by current TV facilities two suggested they would want to play games on the TV.

Study 3: Survey of new kitchen technology

The Open University and Loughborough University conducted a project funded by the ESRC New Dynamics of Ageing Programme to study people's lives in relation to the kitchen. The 'Transitions in Kitchen Living' (TiKL) project which ran from 2009 to 2011, examined older people's experiences of the kitchen living in a variety of accommodation including houses, bungalows, flats and sheltered apartments.

The TiKL project enabled the research team to study problems that older people face in the kitchen and to identify practical solutions they used to overcome them.

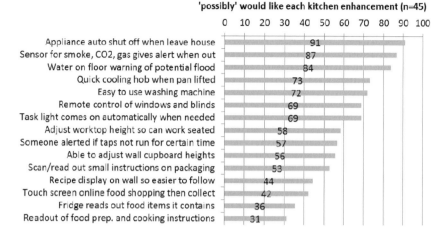

Figure 2.   Preferences for advanced technology in the kitchen (n = 45).

A natural extension of this was to review the possibilities for using more advanced technology to make kitchen life easier. A number of ideas were generated based on the 'ambient kitchen' where technologies are integrated into the kitchen environment to provide support to meet people's needs whether physical or cognitive. A survey was conducted with a group of older people to obtain their reactions to ideas and concepts behind the smart or 'techno kitchen'. There were 45 respondents including 30 females and 15 males, aged from 60 to over 90 years. Questions consisted of a list of possible technological enhancements. Each respondent was asked to state whether they would themselves like to have each facility in their own kitchen or not.

The results in Figure 2 show that there is enthusiasm for certain types of technological support. Most interest was expressed in ideas that could increase safety in the kitchen or address immediate problems such as: automatically turning off the cooker or iron when leaving the house, alerts for water or gas leak or CO_2 detection when away, and a quick cooling oven hob. All received over 70% support. Related comments were: *"it is easy to go out and leave it on"*, *"a very good safety measure"*, *"have been burnt [by the hob] in the past"*. However one person thought that a sophisticated sensor system would be *"too much hassle setting up"* while a quick cooling hob was considered by another to be *"an expensive feature"*.

Although reasonably well received, one person thought that a reader for food preparation instruction was *"too scary!"* Other innovations, although helping with well-known problems, were seen as too technically complicated, impractical or more suitable for people with more severe disabilities. For the smart fridge one person said *"you would have to tell the fridge what is in it"*. For reading out cooking instructions another said: *"I think this would be annoying and could get out of step with the person's actions"* while a further comment was that it was *"too complicated to copy all personal recipes to digital"*. Some thought that with the motorised wall cupboard there was the risk of damaging items on the worktop below.

Advanced technological support seems to have the potential for making kitchens easier to use but more needs to be done to make consumers more aware of how they work or of the fact that they are currently available. Clearly an understanding of older people's views is important if such developments are to be useful and acceptable to kitchen users in the future.

Study 4: Children's preferences for technology v traditional methods

Looking at a broader range of technologies, a class of 28 seven and eight year old children (14 boys and 14 girls) were surveyed to find out how they would prefer to perform a range of everyday tasks using either a traditional method or a more technological approach. The aim was to investigate whether they could see benefits in traditional methods as opposed to advanced technology. As for Study 2, children of this age were chosen as it was thought that they were knowledgeable and open minded about technology. Five main questions were asked:

- If you wanted to contact friend in the evening or at the weekend, how would you like to do this? The options were to: 'speak on phone', 'text them' or 'email them'.
- Would you rather read a book on a computer tablet or a 'real' book?
- Would you prefer to buy something using a computer or going out to a shop?
- Do you prefer to learn by listening to a teacher or using a computer?
- If you wanted to find some information would you prefer to look in a book or go on Internet?

The results are summarised in Table 2. Two children only completed half the survey each so all scores are out of 27. Regarding speaking to a friend at the weekend, there was a preference to speak directly using an audio phone call (chosen by 15 children) rather than using text (chosen by 11) or email (chosen by 1). Reasons for using a simple phone call was that it was *"quicker and more straightforward"*, *"you can 'face time'"* (on an Apple phone), *"I don't have their email address"* and

Table 2.   Children preferences for achieving tasks (n = 28).

Task	Number of children choosing each option		
Contacting a friend	Phone = 15	Text = 11	Email = 1
Reading a book	Paper book = 17	Computer tablet = 10	
Buying something	Go to a shop = 18	Computer/online = 9	
Method of learning	Teacher = 16	Computer = 11	
Looking up information	Internet = 18	Reference book = 9	

"you can hear their voice". Text was seen as preferable because *"it was shorter"* and *"I think it's a faster, simple way of talking"*. Only one person preferred email because *"you don't use so much money"*.

A physical paper book was preferred by the majority because *"you don't have to scroll down"* and *"you can't find any stories on a tablet"*. However a tablet was had the advantage that *"it is lighter than a book"* and *"you can put it in big writing"*.

Eighteen children preferred to go to a real shop because *"I enjoy the exercise and you know what you are buying"* and *"you can see things better and in more detail"*. Online shopping was favoured by 9 children because *"in the shop they might be sold out"* and *"you don't have to travel a long way"*. Comparing listening to a teacher for learning and using a computer, the majority, 16 versus 11, preferred the human teacher. This was because *"you learn more"*, *"teachers always know the right answers to sums"*, *"if you forget you can ask her again"* and *"she wants you to know what she knows"*. Computers on the other hand *"probably know a bit more things than teachers"* and with them *"you get different facts about the subject"*.

In order to look up something, only 9 preferred to look in a book while 18 would go on the Internet. The children were asked when a book might be a better option. Answers given were often practical i.e. books were better if you were in a library, a car, on holiday, when the Internet is down and because computers can *"strain your eyes"*. Other comments were that a book would be better when *"you don't have a computer"*, *"your Mum is on the computer"* and *"you cannot find what you are looking for on the computer"*.

A sixth question in the survey was: "If you could invent a new gadget what would you like it to do?" Responses included *"something to do everything for you that you dislike doing"*, *"a house that moves"*, *"a torch with a purple light which could set things up before you even know it!"*, *"a skate board that can read your mind"*, something to *"make ice cream and chocolate"*, *"turn your book pages"* and *"help you ride your bike and make you feel better"*.

Conclusions

The four studies have explored people's preferences for technological innovations. Convenience plays a part so the ease of looking up information on the Internet rather than in a book is attractive to many people and children. Familiarity is a further driver so that simply holding small instructions on packaging to the light or using a magnifying glass may seem preferable to using an unfamiliar device such as a scanner with speech synthesis. Looking across the studies, other themes seem to emerge. Respondents tend to be sceptical about developments that do not seem feasible or which they have little understanding of how they would work. This was true in the kitchen study with new innovations such as the smart fridge whereas the use of a door camera linked to the TV screen was perhaps more realistic and therefore more acceptable. Similarly children perhaps felt that in terms of novel uses for TV, listening to music and helping with homework seemed the most practical. Although technological developments might seem new and exciting, respondents have a practical and realistic approach to them. For example it was said that the ability to send email through the TV would lack access to photos and files to attach to messages, while email privacy if a shared room would be a limitation. As was found in the kitchen and children's survey, traditional methods of achieving a task were still valued such as using a recipe book rather than having instructions read out, going out to a physical shop rather than online and learning from a teacher rather than a computer. Respondents felt most positively towards ideas that really addressed activities important to them or an issue of concern such as kitchen safety or home security. Participants were also aware of the need for careful adaptation of an innovation such as establishing a new etiquette for video phone calls and the need to reformat a newspaper presentation on TV. As always, technological innovations that address genuine user needs are likely to be the most successful.

References

Creusen, M.E.H., Schoormans, J.P.L. 2005. The Different Roles of Product Appearance in Consumer Choice. *The Journal of Product Information Management* 22: 63–81

Mack, Z. and Sharples, S. 2009. The importance of usability in product choice: A mobile phone case study. *Ergonomics* 52(12): 1514–1528

Maguire, M. 2001. Young peoples' views and perceptions of information and communications technology. In M.A. Hanson (ed.) *Contemporary Ergonomics 2001*, (Taylor and Francis, London), 231–236

Maguire, M., Nicolle, C., Marshall, R., Sims, R., Lawton, C., Peace, S., Percival, J. 2011. A Study of User Needs for the 'Techno Kitchen', In: C. Stephanidis, ed., Posters, Part II, *Human-Computer Interaction International Conference*, 2011, CCIS 174, 66–71. Heidelberg: Springer

Office of National Statistics, 2012. "Statistical bulletin: Internet Access Quarterly Update, 2012 Q1", from: http://www.ons.gov.uk/ons

UNDERSTANDING HUMAN COGNITION

BUILDING A HUMAN CAPABILITY DECISION ENGINE

K. Tara Smith

HFE Solutions Ltd as part of the TeamREACH Research Consortium, Dunfermline, UK

This paper presents an overview of the work currently being carried out for Dstl that produced a "model of models" to compare the accuracy and completeness of decisions being made in relation to the introduction of technological or systematic interventions. The work described here formed only a small part of an overall programme looking at the assessment of mission effectiveness.

This work involved a review of the current state of knowledge of the human component of decision-making to identify causal links between elements and culminated in the development of an integrated conceptual model. This conceptual model is currently being further developed into a mathematical representation of the identified components of the decision-making process.

Introduction

This paper is based on work that was conducted for the Defence Science and Technology Laboratory (Dstl), who funded a multi-phase research project into the impact of information technology on the human component of mission effectiveness. Throughout the course of this programme, a significant amount of work has been undertaken to construct predictive models of the principle human elements affected by the introduction of new information technology on the battlefield. These models are based on best practice and covered the two primary areas of workload and situational awareness.

This work has led to a number of innovative models: most noticeably Blizzard3, which was reported on as "Predictive Operational Performance (PrOPer) Model" at the Ergonomics Society 60th Anniversary Conference 2009 [Smith & Mistry 2009].

This paper reports on the initial phases of the development of a "model of models" and specifically on the Decision Engine, which we have developed to allow us to glue the vast array of disparate models together. The next phase of this work will be the creation of a mathematical representation of the relationships identified.

Baseline understanding

The following subsections present an overview of some of the components that have gone into the construction of the Decision Engine.

Cognitive workload

When considering predictions and estimates of workload models, it is important to understand the context of it in relation to the user's mental model of the situation. The following statements should always be borne in mind when considering workload:

- People build cognitive representations of the Past, the Present and the Future world that affects them.
- These models are fuzzy.
- People use their model as a basis for their decisions.
- Human Science has failed to produce a coherent representation of these models.
- However there is a coherent representation of the cognitive workload involved in using those models.

Cognitive workload has a critical effect on overall operational performance. Examination of the detailed drivers to individual and team workload provides critical insight into design and strategic decisions. As all tasks have a component of concurrent and continuous monitoring, the assessment needs to represent those concurrencies. Each task domain has its own workload tempo that allows the user to time slice appropriately within that domain. An activity can be represented as a set of generic task components. Detailed workload can only be accurately considered in the context of a particular scenario. The sensitivity of the task to workload can be examined by looking at variations in team make-up, training and experience levels, etc.

The Blizzard3 model builds on a number of established and emerging theories, including Hopkins' Integrated Skills Theory (unpublished) and Tuckman's Group Development Model [Tuckman 1965]. It utilises data capture and presentation methods, such as GOTA (Goals, Objectives, Tasks and Activities) [Smith 1999] and ETLX (Extended Task Load Index – developed from NASA-TLX [Hart & Staveland 1988]). A number of accepted and validated models have been combined within this workload model: the key models within this model are Wickens Multiple Resource Theory [Wickens 1984], McCracken & Aldrich [Aldrich & McCracken 1984] and Dstl's STORM model (classified) which allows the consideration of team and cultural dynamics.

Although this summary is a representation of the Blizzard3 model, all of these statements are also true for any overarching model that would assess the workload component of mission effectiveness.

Situational Awareness (SA)

The research programme has adopted Endsley's definition of SA, "the perception of elements in the environment within a volume of time and space, the comprehension of their meaning, and the projection of their status in the near future" [Endsley 1995].

From the experiments conducted during the research programme there is a clear indication that an individual aims to gain an acceptable level of SA prior to making a command decision: this is referred to as the critical SA threshold. The amount of SA needed to reach this critical SA threshold is dependent on the style and experience of the individual.

From the findings of the research programme, the cycle of actively gaining additional SA was always initiated by a trigger event, such as a single report that posed a significant risk of significant change to the individual's SA picture at SA Level 2 (comprehension) and/or 3 (prediction): this is referred to as the SA trigger. If this SA trigger is late or wrong then it has a direct consequence to an individual's ability to reach an effective decision point (EDP). There is an additional threshold for the SA trigger in that if the data SA is inadequate the SA trigger does not initiate the proactive SA gathering cycle by the individual.

Effective decision point

The critical measure of operational performance in the context of SA is the time to effective decision point (EDP). This time is also affected by the level of cognitive workload required to maintain the individual's background SA and achieve his critical SA threshold. This trend was demonstrated during experiments conducted as part of the programme, but was not proven as the sample size was too small.

Another aspect that was examined during the programme was the relationship between the media by which information was transferred and presented and the cognitive workload. It was clearly identified that during times of concentrated activity, single-mode information exchange was a limiting factor. Therefore the balance between the different information channels (primarily visual and auditory, but could also include haptic) has a direct effect on workload and thus the time to EDP.

Additionally, the importance of having SA is not constant: i.e. it is more critical to the outcome to have better SA for non-standard situations, such as points of high information traffic, extraneous activity and unforeseeable events. These points/stressors are likely to cause additional workload on the individuals and therefore affect their SA and the time to EDP.

Although it was not proven during the experiments (due to the small sample size), there is a strong indication that an individual will trade workload, and to some extent SA, to allow a faster decision to be made.

Additional elements

As indicated above, the areas of workload and situational awareness have been well grounded throughout the research programme. However further work has questioned the boundaries that have been assumed in the modelling to date.

A literature review was carried out focusing on the relationships between a list of critical factors (identified from a conceptual map of the problem space) in relation to the human factors aspects of the decision process and specifically the use of technology to enhance the information flow to and from an individual. The terms used to search for relevant literature, papers and other previous research included:

- Usability, both cognitive and physical and "fightability"
- Cognitive and physical burden, workload and fatigue
- Extreme environments
- Training and experience
- Confidence in decision making
- Situational Awareness (SA) and information assimilation rate
- Operational performance

The key and critical findings from this review fell into the following categories:

- Data collection and measurement methods: to identify the criteria for assessing an element
- Technological interventions: an overview of the types of interventions assessed to indicate the context in which the collection methods were used
- Criteria type: the aspects measured during research experiments
- Findings: a selection of key, unique and critical results from the research, which were used to inform the resultant model.

There was a strong degree of coherence across the identified research, which was also in line with both the methodology and the findings of the research programme to date. Over 150 papers were identified as relevant and around 50 provided critical and/or unique inputs. Using the number of papers indicating a relationship between nodes and issues as a crude indication of the level of the strength and complexity of the relationship, it can be argued that the following nodes and issues have a strong and complex relationship:

- Decision-making with confidence, training and experience, information presentation style and information presentation media
- SA with cognitive workload, operational performance and information presentation media
- Information presentation style with decision-making and confidence
- Information presentation media with decision-making, confidence, SA, cognitive workload and operational performance

The research indicates that the key impactors on decision-making and confidence are the style and media of the information presentation together with the training and experience of the individual.

Causal relationships

Figure 1 – Causal Relationships from Literature, illustrates the causal relationships identified from the review. There was a strong link between SA, confidence and mission effectiveness reported in the literature, therefore although there was no direct causal relationship identified, this link is still represented in the figure below.

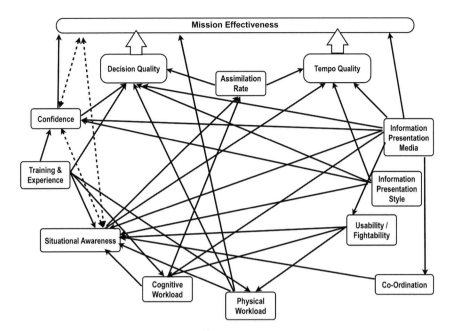

Figure 1.   Causal Relationships from Literature.

One issue was that the various different pieces of research utilised different conceptual models to frame their research within: Figure 2 – Adjusted Causal Relationship
presents a modified causal map, to provide a clearer picture and to reflect the conceptual model adopted for this work. This was achieved through applying two rule
sets: 1) creating a structured set of definitions, allowing sub-distinctions within
one category and equating all of the terms used in the academic research to these
definitions; 2) where the relationship identified above can be argued to be acting
through a chain of other elements, the direct link has been removed and the strength
of the indirect linkage has been increased. This has resulted in a clearer relational
diagram, while maintaining the subtleties of some of the research findings.

The Decision Engine

The major components (linked by the thicker lines) in Figure 2 all touch on workload,
SA or Signal Detection Theory [Wickens 2002]. These were therefore considered
to be the primary links of the conceptual model, with the other elements being
sub-components: i.e. these form the central core modules of the conceptual model.

This modified conceptual structure is presented in Figure 3 – Cognitive Decision
Engine, showing the conceptual relationships between the modules. The rates of
perception, comprehension, projection and decision (the quality and speed of the
information being processed by the individual) are directly impacted upon by SA
and workload. The other identified relationships are presented as impactors on

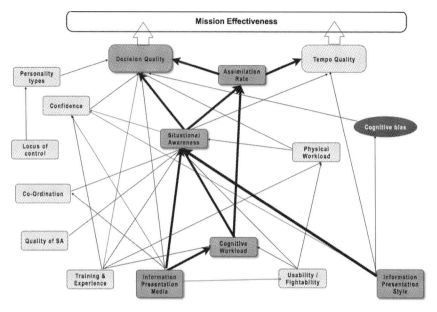

Figure 2.   Adjusted Causal Relationship.

these core modules. Put simply it integrates the Wickens [Wickens 1984] workload model with Endsley's [Endsley 1995, Endsley et al 1998] Situational Awareness model with Signal Detection Theory [Wickens 2002], in a way that allows other identified relationships to be integrated.

Mathematical representation

The next phase of this work is to represent these causal relationships mathematically so that the model can be used to compare the impact of the introduction of technological or systematic interventions on the decision-making process. This mathematical representation is currently under development.

Conclusion

To sum up, wherever you have an individual who is making complex decisions based on information they are receiving from multiple sources in a high pressure/fast paced environment, this conceptual model helps an analyst consider the potential advantages/disadvantages of modifying factors that can affect that overall decision-making process and allows the comparison of disparate alternatives. For example, it allows a comparison between modifying training, changing the team structure and introducing new equipment in a coherent way so that the different alternatives can be judged on the same criteria: i.e. their effect on the decision-making process.

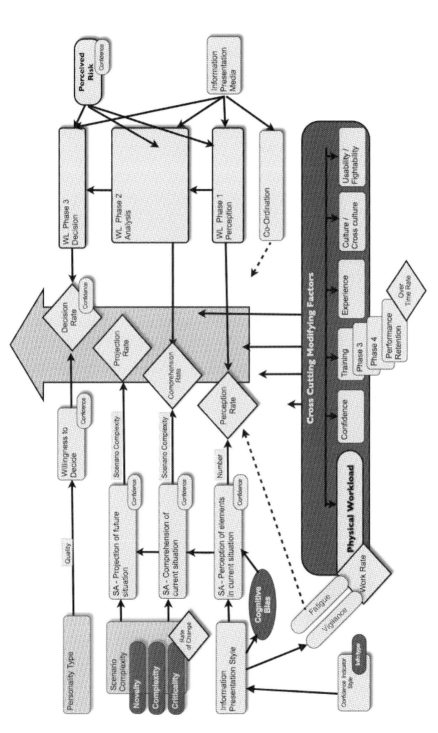

Figure 3.   Cognitive Decision Engine.

References

Aldrich, T.B. and McCracken, J.H. 1984. *"A computer analysis to predict crew workload during LHX Scout-Attack Missions"* US Army Research Institute Field Unit, Fort Rucker, Alabama

Endsley, M. R. 1995 *"Toward a theory of situation awareness in dynamic systems"* in Human Factors 37(1), 32–64

Endsley, M.R., Farley, T.C., Jones, W.M., Midkiff, A.H. and Hansman, R.J. 1998 *"Situation Awareness Information Requirements For Commercial Airline Pilots"* International Center for Air Transportation

Hart, S.G., & Staveland, L.E. 1988 *"Development of NASA-TLX (Task Load Index): Results of empirical and theoretical research"* In P.A. Hancock & N. Meshkati (Eds.), Human mental workload (pp. 139–183). (Amsterdam: North Holland)

Smith K T, 1999 *"A Framework for Modelling Goals and Objectives"* Ergonomics Society Conference 1999

Smith K T, Mistry, B. 2009 *"Predictive Operational Performance (PrOPer) Model"* Ergonomics Society 60th Anniversary Conference 2009

Tuckman, Bruce 1965. *"Developmental sequence in small groups"*. Psychological Bulletin 63

Wickens, C.D. 1984. *"Processing resources in attention"*, in R. Parasuraman & D.R. Davies (Eds.), Varieties of attention, (pp. 63–102). (New York: Academic Press)

Wickens, T D. 2002 *"Elementary Signal Detection Theory"*. (New York: Oxford University Press)

WAITING FOR WARNING: DRIVER SITUATION AWARENESS AT RURAL RAIL LEVEL CROSSINGS

Paul M. Salmon[1,2], Vanessa Beanland[2], Michael G. Lenné[2],
Ashleigh J. Filtness[2] & Neville A. Stanton[3]

[1] *University of the Sunshine Coast, Maroochydore, QLD, Australia*
[2] *Monash University Accident Research Centre, Melbourne, VIC, Australia*
[3] *University of Southampton, Highfield, Southampton, UK*

Driver behaviour at rail level crossings represents a key area for further research. This paper describes an on-road study comparing novice and experienced driver situation awareness at rural rail level crossings. Participants provided verbal protocols while driving a pre-determined rural route incorporating ten rail level crossings. Driver situation awareness was assessed using a network analysis approach. The analysis revealed key differences between novice and experienced drivers' situation awareness. In particular, the novice drivers seemed to be more reliant on rail level crossing warnings and their situation awareness was less focussed on the environment outside of the rail level crossing. In closing, the implications for rail level crossing safety are discussed.

Introduction

Collisions between vehicles and trains at Rail Level Crossings (RLX) remain a persistent road and rail safety problem across the world. In Australia, between 2000 and 2009, there were 695 collisions between road vehicles and trains at RLXs, resulting in 97 fatalities (Independent Transport Safety Regulator, 2011). Within Australia, a large proportion of RLXs are located in rural areas. Although the rural driving environment is less complex in nature than urban locations, the low frequency of trains creates expectancy issues whereby drivers may not expect to encounter trains at rural RLXs (e.g. Salmon et al, In Press). This issue is often compounded by the fact that the majority of rural RLXs are *passive* and do not have *active* warnings such as boom gates, which provide a highly salient cue when a train is approaching and are known to achieve the best safety performance (e.g. Saccomanno, Park & Fu, 2007). The problem is exemplified by the Kerang tragedy in which a loaded semi-trailer truck collided with a passenger train at a passive RLX in northern Victoria, Australia, killing 11 train passengers. A recent analysis of the incident suggested that the truck driver's Situation Awareness (SA) was diminished due to his schema-driven expectancy that a train would not be passing through the RLX (Salmon et al, In Press).

SA, which comprises drivers' schema-driven interaction with the world and resultant understanding of the current situation, therefore has a key role to play in RLX safety. Despite this, there has been little research investigating road user SA at RLXs, especially in the rural context. This paper presents some of the findings from a recent on-road study designed to investigate novice and experienced driver SA at rural RLXs.

Situation awareness at rail level crossings

It has been estimated that unintentional non-compliance at RLXs, whereby drivers fail to detect warnings or comprehend their meaning and enter the RLX as a train is approaching, account for almost half of all RLX crashes in Australia (ATSB, 2002). Poor driver SA is likely to be a key factor in unintentional non-compliance (Salmon et al, In Press). The development of SA (e.g. seeking and retrieving information from the world) and the content of SA (i.e. information and knowledge) are driven by schema. Schema direct exploration in the world (i.e. sampling of the environment) and shape how information from the world is interpreted, which in turn directs behaviour, which in turn modifies schema and so on (e.g. Neisser, 1976). Schema-related failures, including activation of wrong schema, failure to active appropriate schema, and the faulty triggering of active schema (Norman, 1981), potentially play a role in these cases where well intentioned drivers unintentionally do not comply with RLX warnings. For example, when encountering an RLX when a train is approaching, drivers with the wrong schema activated (e.g. schema for the RLX *without* a train approaching) may not direct attention to RLX warnings, or even if they do, may not perceive the warnings, and ultimately the oncoming train.

Driving experience, in particular experience of RLXs both with and without trains present, is therefore a key factor shaping driver SA and performance when negotiating RLXs. This becomes even more important in the rural context where encounters with trains at RLXs are typically low. Previous investigations into experience and drivers' attentional strategies (e.g. visual search) have demonstrated that novice drivers are vulnerable to driving errors associated with an inadequate visual search of the road environment (Underwood et al, 2002) and also that their visual search patterns may not differentiate between different road environments (Crundall and Underwood, 1998). In the RLX context, this suggests that novice drivers may not direct sufficient attention to approaching trains or RLX warnings and that they may not discriminate between RLXs with and without active warnings. The resulting diminished level of SA may result in delayed recognition of approaching trains is or in extreme cases no recognition at all.

In the context of a series of on-road studies examining driver behaviour at RLXs, the present analysis was undertaken to investigate novice and experienced driver SA at rural RLXs. The on-road study involved drivers driving a pre-determined route in regional Victoria incorporating 10 rural RLXs. Based on previous research which shows that drivers of differing experience levels achieve different levels of SA (e.g. Salmon et al, In Press), the hypothesis was that novice and experienced driver SA at RLXs would differ. Beyond this, a broader aim was to explore the nature of these differences and their implications for unintentional non-compliance at RLXs.

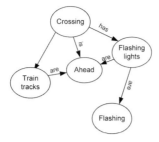

Figure 1.   Example network showing relationships between concepts.

Assessing driver situation awareness at rail level crossings

A network analysis procedure was used to describe and assess driver SA. This involves the construction of 'SA networks' to describe the information or concepts underlying awareness and the relationships between them (Walker et al, 2011). The SA networks are constructed using data derived from the Verbal Protocol Analysis (VPA) method, which involves participants 'thinking aloud' as they drive, with the resulting verbal transcript subjected to content analysis procedures to identify keywords or concepts and the relationships between them, which produces an SA network. For example, an extract of an SA network is presented in Figure 1. This shows the concepts 'crossing', 'train tracks', 'ahead', 'flashing lights' and 'flashing' along with the relationships between them; for example the 'crossing' *has* 'flashing lights', which *are* 'flashing', the 'train tracks' *are* 'ahead' etc.

Method

Design

The study was an on-road study whereby participants drove an instrumented vehicle around a pre-defined rural route incorporating ten RLXs.

Participants

Twenty-two drivers (10 males, 12 females) took part in the study. Participants were sorted into an experienced or novice driver group. The experienced driver group ($n = 11$, $M_{age} = 45.1$ years) had an average of 27.3 years solo driving experience ($SD = 7.6$). The novice driver group ($n = 11$, $M_{age} = 19.3$ years) had an average of 1.6 years solo driving experience ($SD = 0.3$). Participants were recruited through local newspapers, notice boards, community groups and word of mouth. All participants regularly drove in the study area.

Materials

The study route was approximately 30 km long, situated in and around Greater Bendigo, Victoria, Australia. The route encompassed a range of road types, including city streets, residential and suburban streets, highways, unmarked roads, gravel and dirt roads. Speed limits varied along the route, ranging between 40 km/h to 100 km/h. The route included ten RLXs. Six were active RLXs (Five had flashing lights and boom gates, one had flashing lights only), and four were passive RLXs (three with stop sign only, one with give way sign only). Participants drove the route using the On-Road Test Vehicle (ORTeV), which is a 2004 Holden Calais equipped to record vehicle and road scene data. A Dictaphone was used to record participant verbal protocols. The SA network construction process was undertaken using the Leximancer™ content analysis software.

Procedure

Upon completion of an informed consent form and demographic questionnaire, participants were briefed on the research and its aims, which were expressed broadly as a study of everyday driving. Participants were then given a short VPA training session, following which they were taken to the ORTeV and told to establish themselves in a comfortable driving position. Two observers were present in the vehicle throughout the drive. Participants completed a short practice drive whilst providing a concurrent verbal protocol. At the end of the practice route, participants were informed that the test had begun and that data collection had commenced. On-route, the observer located in the front passenger seat provided directions. Participants provided verbal protocols continuously throughout the drive. Participants' verbal protocols were transcribed post drive using Microsoft Word. For data reduction purposes, extracts of each participant's verbal transcript covering the approach to, and negotiation of, each RLX were taken from the overall transcripts based on set points located in the road environment. The transcripts were then used to build SA networks.

Results

The SA networks were analysed quantitatively, using network analysis metrics, and qualitatively, via content analysis. Due to space constraints, a sub-set of the qualitative analysis is presented in this paper. Two seperate network construction procedures were used. The first involved the use of the Leximancer™ software tool to auto-create overall SA networks for the experienced and novice driver groups at each of the four RLX types (booms, lights only, stop sign, give way). The second involved a manual network construction process in order to construct individual SA networks for each participant at each RLX. The second analysis was undertaken as the individual participant transcripts for each RLX were not large enough to support Leximancer™ analyses.

Leximancer analysis

Leximancer™ automates the SA network construction procedure by processing verbal transcript data through five stages: conversion of raw text data, concept identification, thesaurus learning, concept location, and mapping (i.e. creation of network). This led to the creation of eight SA networks (overall experienced and

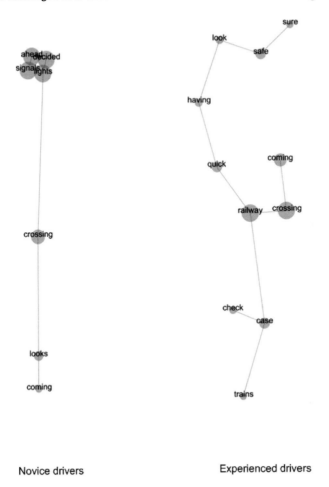

Novice drivers   Experienced drivers

Figure 2. Experienced/Novice drivers overall SA networks (lights only RLX).

novice driver SA networks for each RLX type). Example overall SA networks for each group at the 'lights only' RLX are presented in Figure 2.

Although the full Leximancer™ analysis is not presented, it is worth discussing the key points. The networks demonstrated that SA is different both across RLX types and also across the novice and experienced drivers. Notable differences across the novice and experienced driver networks were that the novice driver networks for all four RLX types contained concepts related to active RLX warnings (e.g. 'lights', 'booms'), even at the *passive* RLXs that have no *active* warnings (e.g. they had stop and give way sign warnings only). Further, the novices had fewer concepts related to trains and checking for them. For example, at the lights only crossing, the novice networks contained no concepts related to trains, the train line, or checking, whereas the experienced driver networks had both 'train' and 'trains' concepts, 'check', 'look', and 'sure' concepts, and 'railway'. Finally, at the RLXs with boom

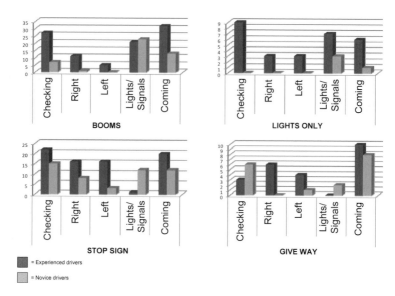

Figure 3.   Frequency count of key checking concepts at each RLX type.

gate warnings, the novices had more concepts relating to the RLX and RLX infrastructure, whereas experienced drivers also had concepts related to scanning and checking the environment e.g. 'clear', 'look', 'ahead'.

Manual network construction

Ten SA networks were manually constructed for each participant. To summarise this analysis, frequency counts of the concepts within the networks were undertaken. Similar to the Leximancer™ analysis, this revealed that a key difference between the experienced and novice driver SA networks was the presence of concepts related to drivers checking the road and rail environment on approach to each RLX. This is summarised in Figure 3, which presents frequency counts for the concepts 'check/checking', 'left', 'right', 'lights/signals' and 'coming' across all participants at each RLX type. The experienced driver networks had more 'check/checking' concepts at booms, lights only and stop RLXs and also more 'left' and 'right' concepts across all four RLX types (as in checking left and right for trains). Notably, at active RLXs (boom gates and lights only) the novice drivers had very few 'check/checking' and 'left' and 'right' concepts. At the passive RLXs (e.g. stop and give way), the novice drivers had more 'lights/signals' concepts. Finally, the experienced drivers had more 'coming' concepts across all four RLX types. Taken together, these findings suggest that experienced drivers' SA was underpinned more by checking the environment for trains, whereas novice driver SA was more focussed on information derived specifically from the RLX warnings. Further, the presence of 'lights/signals' concepts at passive RLXs in the novice driver networks is indicative of an expectation or desire to see RLX warnings regardless of RLX type.

Discussion

The aim of this paper was to present a snapshot of the results derived from an on-road study comparing novice and experienced driver SA at rural RLXs. Based on the analysis presented, there are some clear and important differences in novice and experienced driver SA whilst negotiating rural RLXs. First, as part of developing SA at the RLXs, it is apparent that the experienced drivers tested reported more checking behaviours, in particular checks of the railway tracks to the left and right of the roadway for approaching trains. This indicates that experienced driver schema directs checking behaviours outside of the road environment directly ahead that incorporate approaching trains, regardless of RLX type. Second, novice drivers appear to have some expectancy for RLX warning devices at all RLXs regardless of type, and engage less in checking for trains as a result. This is evidenced by the fact that they had less concepts related to checking across all RLXs, had concepts related to RLX warnings at passive RLXs, and had less concepts related to checking and trains at active RLXs. The findings thus suggest that novice driver SA at rural RLXs is focussed on RLX warnings, regardless of RLX type, indicating that they expect to be warned in the event of an approaching train. This could be problematic at passive RLXs (which do not provide an active warning) where novice drivers might approach expecting warnings but do not receive them. Although in most cases this will lead only to a delayed recognition of the approaching train, in extreme cases where the RLX environment makes the approaching train inconspicuous (as in Kerang e.g. through trees, sun glare, shadowing), recognition of the train may not occur until the driver is in close proximity to the RLX.

These findings point to the importance of experience, expectancy and schema at rural RLXs, and provide a strong case for their further examination. The ways in which RLX design can support SA across road users, regardless of experience, is another avenue requiring further exploration. Although exposure to RLXs with and without trains present is required for building appropriate schema in drivers, RLX design also has a key role to play. Low cost warnings that are as effective as boom gates in alerting drivers to the presence of a train are required in order to eradicate the problem of unintentional non-compliance at rural RLXs. Moreover, active warnings (e.g. flashing light signage) provided early along the approach to RLXs are likely to be beneficial. The author team are currently engaged in a major program of research involving the design and testing of such low cost warnings.

One limitation of the study is that the analysis presented is based entirely on participant verbal protocols. Additional data, such as eye fixations, would confirm some of the verbal protocol content (e.g. participants reporting that they are checking for trains) and enhance the validity of the findings. Studies currently being undertaken by the authors involve the use of verbal protocols and an eye tracking system together. Another issue lies in the lack of reliability and validity evidence associated with the use of VPA and Leximancer™ to assess SA. However, previous road user SA studies have employed this approach (e.g. Walker, 2011) and Grech et al (2002) found similar outputs when comparing Leximancer™ and manual analyses of SA in maritime accidents. That said, further studies examining the reliability and validity of Leximancer™ for SA assessments are recommended.

Acknowledgements

This project is funded by an Australian Research Council Linkage Grant (LP100200387) to Monash University in partnership with the Victorian Rail Track Corporation, Transport Safety Victoria, Public Transport Victoria, Transport Accident Commission, Roads Corporation (VicRoads) and V/Line Passenger. We gratefully acknowledge the support of the project partners and community participants. Without their valuable input and commitment, this research would not be possible. Dr Paul Salmon's contribution to this research and article is funded through the Australian National Health and Medical Research Council post-doctoral training fellowship scheme.

References

ATSB Transport Safety Statistics Unit. (2002). Level crossing accidents: fatal crashes at level crossings. In: Paper Presented at the 7th International Symposium on Railroad-highway Grade Crossing Research and Safety, Melbourne, Victoria, Australia.

Crundall D. E., & Underwood G, (1998). The effects of experience and processing demands on visual information acquisition in drivers. *Ergonomics*, 41, pp. 448–458.

Grech, M., Horberry, T., & Smith, A. (2002). Human error in maritime operations: Analyses of accident reports using the Leximancer tool. HFES Annual Meeting, USA, 46, pp. 1718–1722.

Independent Transport Safety Regulator. (2011). Level Crossing Accidents in Australia. ITSR, Sydney.

Neisser, U. (1976). Cognition and reality: Principles and implications of cognitive psychology. Freeman, San Francisco.

Norman, D. A. (1981). Categorization of action slips. *Psychological Review*, 88, pp. 1–15.

Saccomanno, F. F., Park, P. Y.-J., & Fu, L. (2007). Estimating countermeasure effects for reducing collisions at highway-railway grade crossings. *Accident Analysis & Prevention*, 39, 406–416.

Salmon, P. M., Read, G., Stanton, N. A., & Lenné, M. G. (In Press). The Crash at Kerang: Investigating systemic and psychological factors leading to unintentional non-compliance at rail level crossings. *Accident Analysis and Prevention*. Accepted for publication 25th September 2012.

Underwood, G., Chapman, P., Bowden, K., & Crundall, D. (2002). Visual search while driving: skill and awareness during inspection of the scene. *Transportation Research Part F*, pp. 87–97.

Walker, G. H., Stanton, N. A., & Salmon, P. M. (2011). Cognitive compatibility of motorcyclists and car drivers. *Accident Analysis and Prevention*. 43:3, pp. 878–888.

A QUICK METHOD OF ASSESSING SITUATION AWARENESS IN AIR TRAFFIC CONTROL

Jim Nixon[1] & Andrew Lowrey[2]

[1]*BAE Systems, Advanced Technology Centre, Bristol, UK*
[2]*NATS, Safety Policy & Assurance, Fareham, UK*

The Controller Picture Scale is a tool which has been used to assess controller situation awareness (SA) in a variety of contexts, including the introduction of new technology into air traffic management and the effects of airspace changes. Four dimensions relating to controller tasks are rated by the controller following an activity. These ratings give the practitioner a method of communicating SA meaningfully to non-specialists and can form the basis of debriefing discussions with controllers. Practitioner experience with the scale has shown its utility and good user acceptance by controllers.

Introduction

The Controller Picture Scale is a tool that has been used to assess controller situation awareness (SA) in both area and terminal operations. Controllers must develop a mental picture of traffic from a variety of remote, two-dimensional displays. To successfully de-conflict aircraft controllers must also be able to understand future states based on aircraft trajectory and behaviour. SA is a key construct in this domain, underlying safe and effective air traffic management. Controllers rate four controlling activities (Figure 1) and these judgements can then be used by the practitioner in conjunction with other human factors measures to gain an insight into controller SA.

Feeling of being behind	1	2	3	4	5	6	7	Comfortably ahead of the game
Poor understanding of the traffic situation	1	2	3	4	5	6	7	Full understanding of the traffic situation
Lost control of the RT	1	2	3	4	5	6	7	In full control of the RT
Aircraft call you and you have to hunt to recall them	1	2	3	4	5	6	7	Aware of aircraft coming into your sector before they call you

Figure 1.   The Controller Picture Scales.

The scale

Scale dimensions have been derived through short workshops with controllers. The dimensions selected represent recurring themes articulated in the workshops. The scale dimensions broadly align with Endsley's model of SA (Endsley, 1995), representing controller comprehension of the current situation and their awareness of near-future states. A ranking exercise conducted with controllers showed that no single dimension was more influential than any other overall. A seven-point scale was selected to give a higher degree of granularity than a five or three point scale. Other scales may also work well including, for example, visual analogue scales. Radiotelephony is abbreviated to RT in the scale since it is a common abbreviation used to describe communication between the pilot and controller.

Using the picture scale

Controllers rate the four dimensions immediately following any event requiring measurement of SA. Events have included simulation or immediately following a period of live operation. The scale has been used to assess the impact on SA associated with the introduction of new technology, airspace changes and SA change in response to changes in job and task.

Advantages of the scale include the speed of administration and relation to actual activities performed by controllers rather than needing to understand SA at a conceptual level, often required when using other scales such as China Lakes (see Gawron, 2000). The scale is simple, making analysis and interpretation as straightforward as possible. The scale connects with controller tasks and so has good user comprehension and acceptance. So far, the scales show a degree of independence, improving diagnosticity over single-rating scales which have been used previously.

The scale has been used as a basis for discussion during debriefs and reporting relative levels of SA. Currently the scale is subject to ongoing validation in different ATC environments, the numerical properties of the scale being further investigated in order to develop effective thresholds and success criteria.

References

Endsley, M. R. 1995, Toward a theory of situation awareness in dynamic systems. Human Factors 37 (1), pp. 32–64.

Gawron, V. J. 2000, Human performance measures handbook. Mahwah, NJ: Lawrence Erlbaum.

FUNDAMENTAL ISSUES

FROM CREATION TO COMPLIANCE: DO'S AND DON'TS OF NEGOTIATING REQUIREMENTS WITH DEVELOPERS

John R. Wilson

Human Factors Research Group Faculty, University of Nottingham and Ergonomics Team, Network Rail, UK

Requirements identification, generation, verification and representation is increasingly a hot topic, in part because of the use of scenarios – for good or ill – by the user experience community, but largely due to the much increased involvement of E/HF in large infrastructure projects. This paper is based on practical experience, built around a series of do's, don'ts, cautions and tricks of the trade.

Introduction

Generating, representing, negotiating and employing user requirements and ergonomics/human factors (E/HF) requirements is a hot topic. It is difficult to overstate the vital role that user E/HF requirements play in systems development. To ignore them, misunderstand them or identify them carelessly invites potential for systems failure and wasted resources of time, people and costs. For example the UK FIREControl IT project was flawed from the outset because it "... did not have the support of the majority of those essential to its success – its users [p. 5] ... there was no single, authoritative owner of the user requirements ..." [p. 6] (National Audit Office, 2011). Other "spectacular" failures highlight the importance of requirements capture, the leading reason for failures being that "project teams did not know how adequately to generate user information requirements" (Lindgaard et al, 2006, p. 50).

This paper summarises organizational and logistic issues faced in working on E/HF requirements. It is not a full guide to E/HF requirements, whether molecular (listed or tabulated) (e.g. FAA, 2003), or graphical and other scenarios (e.g. Alexander and Maiden, 2004). There are many guides to doing this (e.g. Alexander and Beus-Dukic, 2009; Robertson and Robertson, 2006). The purpose is to provide a succinct set of do's and don'ts, warnings and cautions, and tricks of the trade, drawing from a Network Rail guidance document produced by the author, his work over many years in several areas of E/HF, and other colleagues in systems engineering and software requirements. Particularly the paper addresses requirements in light of widespread extension of E/HF in large, contractually complex infrastructure projects and, at the other end of the spectrum, the rise in the user experience community and contributions to user involvement in web and similar design.

Overview of the process to develop ergonomics requirements

A straightforward process of developing ergonomics requirements will follow five key stages. After understanding the project or system goals and constraints the ergonomics/human factors *scope* must be set, highlighting what human factors are important. Second, a *human machine systems analysis* is carried out – assessing and understanding more about the key human factors in the current system and work domain (baseline) and identifying stakeholders and the critical factors for the future system. Third, *E/HF requirements* are generated and delivered in an E/HF Requirements Document, based on user needs, user requirements, and E/HF performance requirement. Fourth, optionally, *E/HF design requirements* for the system are defined – what are the people-centred design priorities, what form must the human machine system design take and how will ergonomics be evaluated? Fifth, all requirements must be *quality assured and validated*, assessed continually and updated. For smaller and less complex system designs, the process may only comprise certain stages and a sub-set of analysis techniques and sources of information. For larger systems there will be more iteration loops, but even then not all the techniques possible are used. All inputs, activities and outputs outlined within this process are **indicative not prescriptive** for how ergonomists will approach developing E/HF requirements in systems design so as to enhance the human factors.

Basis for the paper – adventures in requirements over the years

This paper draws upon experience of requirements in several contexts, some turning out successfully, others less so – which makes them better learning exercises: a) Participatory (workplace) design, where participants were their own designers, developing requirements for themselves, and expressing these in the form of sketched scenarios, simple checklists, and mid-process rapid reconfigurable Virtual Reality (VR) mock ups; b) Developing small scale VR models using storyboarding, as part of the Virtual Environment Development Systems (VEDS), with early storyboards used to define presentation quality, and walkthrough and manipulation interactivity; c) Bridging user-developer gaps for collaborative engineering VR/Augmented Reality (AR) tools, using in part a Living Labs approach, video based scenarios enabling potential users to visualise what they might expect, and followed by a set of tabulated scenarios and detailed design-related requirements; d) Agreeing a Human Factors Integration Plan (HFIP) meaning subsequent requirements were developed to support compliance checking and contractual arrangements; e) Rail E/HF projects providing understanding of rail traffic control functions and roles and identifying design requirements, but in danger of slipping over into providing detailed design guidance at too early a stage; f) Development of a requirements development process based on functional analysis for use in rail engineering and maintenance, starting with visual scenario analysis and with subsequent use of functional analysis requirements generation.

Hard-learned lessons, dos and don'ts, cautions and hints

Oranges not the only fruit: what type of E/HF requirement for what?

There are different classes of E/HF requirement, all with their place in supporting systems development. However, ergonomists, the clients, engineers, developers and business managers they work with, must all be quite clear about which sort of requirement is being produced and for what purpose. Classes and levels of ergonomics requirement include: *User requirements* – directly from user needs and stakeholder analyses; *Ergonomics/human factors requirements* – broadened and modified from user requirements to embrace the wider socio-technical system; *Ergonomics process requirements* – how E/HF should be integrated; *Usability requirements* – generally for human-machine interaction and addressing effectiveness, efficiency, utility, accessibility, satisfaction; *"Look & feel" requirements*; *Guidelines and standards requirements*; *Human impacts requirements* – user health, satisfaction, comfort; *Human-machine system performance requirements* – what people must be capable (and supported) to do, what the human-machine system must achieve; *HF sustainability requirements* – environment, participation, health and safety, work systems adaptability; *Implementation requirements* – user learning and training, documentation; *Derived requirements* – implied by or extended from higher level requirements.

Compliance v creativity

The contribution of ergonomists via requirements can be directly into the development life cycle, where the support is eventually into creativity and design. Classic examples of this are where the ergonomists work with the developers of a new web site, handheld computer interface or stand-alone control interface. On the other hand, in large projects and system developments, the requirements will be more concerned with what the suppliers must do to meet E/HF and end user needs; such contractually-oriented requirements will be more to do with HOW they will get to suitable designs, and checking compliance, than with the content of any design.

Process or content

The distinction above carries over into the distinction between process and content-based requirements. The former are strongly connected to a Human Factors Integration Plan and define what must be done in terms of the human factors programme as a whole. The latter are more concerned with individual aspects of design content – the human-machine interface, the workplace, roles and so on.

Size matters – fitness for purpose

Processes and methods employed in requirements generation will depend upon the needs, constraints and resources of the project. The effort committed and degree of sophistication and detail of the E/HF requirements work should be appropriate to project size, complexity and importance. Small product design projects generate fewer more detailed requirements using quick and simple methods; larger system projects have a much greater number, developed using more detailed analyses but often in less detail but with more levels of prioritisation. The requirements process must have fitness for purpose, with an appropriate degree of parsimony.

Don't be greedy – cuttings coats according to our cloth

It can be tempting to try to include within the E/HF requirements everything relevant to people and human performance. However, all agreed requirements will have to be accounted for by developers and suppliers, perhaps governed by contractual terms, and any proposed system component will have to be assessed or tested against requirements subsequently. It does not help the credibility of E/HF if assessment and agreement is long or difficult to carry out for requirements which could be optional. Not everything will be needed or even advisable for every project.

Do not be drawn into redundant debate over functional/non-functional

Whatever happens do not get involved in a redundant debate over the vexed question of functional and non-functional requirements. The view of many designers, engineers and business analysts is that functional requirements are what the system must do to achieve its goals; identified restrictions and constraints on the functional requirements or on the system are then defined as non-functional requirements. In this view non-functional requirements include usability, reliability, capacity, efficiency, security, maintainability, legal and political constraints. However, this is an engineering-oriented view of non-functional requirements which only makes sense if the system is conceived as comprising only the integrated hardware and software components. The human factors view of "system" is that it is the human-machine system and people are thus an integrated part of the system and not a constraint or restriction on it; many requirements to do with people will then be functional ones. Usability embraces functionality – if the system does not provide, or the user cannot access, the basic functionality to complete their tasks then the system is not usable for the purpose intended. It is better that E/HF requirements are delivered free of any labels to do with being "functional" or "non-functional".

E/HF requirements are not necessarily user requirements

The next guideline may seem to be counter-intuitive, but **E/HF requirements** are not necessarily the same as **user requirements**. The latter are elicited or emerge (sometimes unfiltered) from expressed user needs and subsequent analyses, often focused on the human-machine interaction and on immediate operator support. The former are an extended version of user requirements, including systems aspects perhaps not considered by the users and frequently cover a broader scope of the complete socio-technical system. Also E/HF requirements will be an edited version based on assessment of whether user requirements are appropriate, feasible and deliverable before translation into E/HF requirements, cutting those not feasible or within scope. Some degree of second guessing what the users should have is involved but a two stage process is usually vital to allow the system development to have a chance of meeting specifications and to allow useful evaluation to take place.

Which user are we talking about?

One clear difference between some developers and ergonomists is if former see user requirements as expressing what the budget holder wants, any contract meaning that the customer cannot demand anything outside them. In fact, "the customer" may be quite separate from real end users and from what end users really do. Client (cost centre) users may concentrate on project costs to the exclusion of trying to satisfy end user needs. A thorough stakeholder analysis can allow the needs of all stakeholders – operators, maintainers, managers, engineers, support functions, customers and bystanders – to be expressed within requirements. There is evidence that an ergonomist can act as a mediator between different "users" and developers, in one interpretation acting as the "lawyer" of the end users (Nies and Pelayo, 2010).

Bridging the imagination gap

Some end users may not have experiences with, nor be able to imagine, new technologies or work processes, whereas others may have a fanciful view of what new systems can do. Careful facilitation is needed to encourage some users to move beyond being locked up in "what I do now" and to focus others on potentially feasible solutions rather than flights of fancy built around Hollywood hype.

Users are not paragons of virtue ...

Many tracts on ergonomics stress the virtues of knowing and working with end users as if they are the redeeming angels of design. However be honest – they are not always easy (or even likeable!) to work with. Not all end users are fully knowledgeable about their current systems, work processes or the problems and flaws in these, and may be cynical, lazy or unable to express needs clearly.

... But they may have good reasons

Users may be reluctant to participate in requirements gathering, because of previous poor experiences or with good reason. Reluctance by end users to consider changes to how they do things now is often ascribed by managers or developers as being due to general resistance to change or technophobia, but can be due to rational fears of job losses or deskilling; opposition can in fact can be practical and well-founded, as users with expert knowledge of how things really happen, and of the work-arounds they have to use now, can often see technology or organisation system flaws and thus potential systems failures, far better than developers and managers.

Not set in tablets of stone – win the war not every battle

"But surely a requirement means it IS required" – a plaintiff cry from a junior colleague battling with developers. Whilst the very word "requirement" implies that all must be met, it is preferable for all but the most minimal set of requirements that some priorities are allocated, to enable project compliance and minimise subsequent negotiation time between developers, ergonomists and clients. Pragmatism, compromise and social skills of negotiation are required; would we really argue that a particular shade of red preferred on a certain display is at the same level of criticality as ensuring that staff levels can allow emergencies to be managed? On the other hand great care is needed, and for any requirement not deemed as of highest priority the consequences of not meeting it should be traced through and assessed. The process of gaining agreement in a continual process of verification, quality assurance (Boness et al, 2011; Rexfelt and Rosenblad, 2006) can be very difficult and the skills to navigate the minefield are much prized. The wise ergonomist keeps bargaining chips in reserve, requirements they can let go – after an appropriate period of apparent reluctance – to ensure that the important ones are accepted. The ergonomists should continually test their assumptions: how important and relevant is each requirement, what time, effort and financial cost will be needed, what are the consequences if it is not met? Along the way some cherished requirements may have to go; give way gracefully and live to fight the bigger battles ahead.

The won't haves

Although "won't have" requirements are often ignored, or at least not returned to in any detail, they can be almost as important as the "Must haves". They can reflect end user or business views that such a system attribute or component is to be avoided, because of concern with over-functionality, complexity, instability, non-supported platforms, or known unworkable routines (Hoorn et al, 2007).

Over-ruling the user

A special case of not being able to have everything is when the expressed desires of end user participants conflict with what the E/HF specialist believes to be good practice. Persuasion of the participant group needs managing with great care.

Life cycle effects

Requirements evolve over the system life cycle, broadly from business to technological requirements. E/HF requirements are strongly related to the system life cycle in two ways. First, different issues are concerned with different stages in the life cycle – development, evaluation, implementation, maintenance and repair, and decommissioning including recycling. Typical input is into development related requirements but with great interest in environmental sustainability and design for maintainability in light of error and failure rates in maintenance, service and repair. Second, developing E/HF requirements in many ways parallels the development of the system itself, becoming more detailed, specific, sophisticated and richer.

Paper trails and audits – "where did this come from?"

As projects progress their form and content will change, features, components and even sub-systems may be modified or omitted; as a consequence the human-system interactions will change. Updates and changes to the ergonomics requirements will have to be made throughout systems development for several reasons: new evidence or early prototype testing may show requirements to be wrong, inappropriate or not complete; it may become clear that not all can be met by project managers, analysts, engineers and suppliers; changed ideas in system development will lead to the need for new requirements or omission of redundant ones. When requirements are removed, modified or added, the rationale for this and any trade-offs involved – e.g. where performance reliability has been prioritised over use of capacity – must be noted. Caution must be exercised in updating – changed user needs and ergonomics requirements cannot continue *ad infinitum,* and design coherence lost through too many updates (e.g. Lindgaard et al, 2006, pp. 48–490).

Use all sources to generate requirements

All manner of innovative E/HF methods exist but essentially come down to: "Seeing" – direct observation of what people (and equipment) actually do in practice in the workplace; "Asking" – self-report from the job holders or users through questionnaires, interviews, workshops, diaries, incident records and structured exercises; "Simplifying" – modeling and simulation; "Thinking" – expert analysis often in mixed function groups; and "Borrowing" – use of existing data and knowledge in archives or knowledge stores.

Beware recycling

A special case of "borrowing" is requirements re-use – from related projects, similar earlier systems etc. This seems efficient but is not always easy or convenient, risking losing context, structure and traceability details (Alexander and Beus-Dukic, 2009).

Horses for courses: Lists or pictures?

E/HF requirements can be represented in a number of ways: computationally, molecular (textual – lists or tables), or scenarios (stories, storyboards, use cases) in text, graphics, pictures, videos, multi-media or VR models etc. Molecular requirements are more useful for large systems and projects, better allow for cross-referencing and provide far easier checking on compliance and audit trails and updating, but can be time consuming and risk the designers not seeing the wood for the trees. Scenarios are useful for smaller less complex products and the early stages of large projects, lend themselves to needs for creativity, and are motivational for participatory design, but risk not seeing the trees for the wood.

Requirements for design or not

The decision about when and how far to develop design related requirements is critical, with no easy or universal answer. Most views on system development say that design ideas should be left to the developers, as including them in E/HF requirements will constrain development and transfer risk from developers to clients and ergonomists. E/HF design requirements can be very useful where: parts of the system are already in the detailed design stage; there are known constraints from context and environment of use or special user needs; and the requirements team has previous knowledge of proven design specifications. Often design of the user interface cannot be held up until all requirements are captured, and an evolutionary process can allow a start to be made with adaptation as requirements become clearer.

> "trying to write product specifications without making any assumptions about design is frankly silly . . . trying to specify . . . capabilities without using words that admit a solution has been chosen." Alexander and Beus-Dukic (2009, p. 349)

Nonetheless it is best that ergonomists avoid producing design-specific requirements in early or early-middle stages of design, but employ them carefully and selectively in later detailed design.

References

Alexander, I and Beus-Dukic, L, 2009, Discovering Requirements: How to Specify Products and Services. West Sussex, UK: John Wiley and Sons

Alexander, I and Maiden, N, 2004, Scenarios, Stories, Use Cases: Through the Systems Development Life Cycle. Chichester: John Wiley and Sons

Boness, K, Finkelstein, A and Harrison, R, 2011 (in press), A method for assessing confidence in requirements analysis. Information and Software Technology, doi 10.1016/j.infsof.2011.05.003

Federal Aviation and Administration, 2003, Guidelines for Human Factors Requirements Development. Office of the Chief Scientific and Technical Advisor for Human Factors, Report AAR-100

Hoorn, J F, Konijn, E A, van Vliet, H and van der veer, G, 2007, Requirements change: Fears dictate the must haves: desires the won't haves. The Journal of Systems and Software, 80, 328–355

Lindgaard, G, Dillon, R, Trbovich, P, White, R, Fernandes, G, Lundahl, S and Pinnamaneni, A, 2006, User needs analysis and requirements engineering: Theory and practice. Interacting with Computers, 18, 47–70

National Audit Office, 2011, The failure of the FIREControl project. Report by the Comptroller and Auditor General, HC 1272. London: The Statioery Office, July

Nies, J and Pelayo, S, 2010, From users' involvement to users' needs understanding: A case study. International Journal of Medical Informatics, 79, e76–e82

Rexfelt, O and Rosenblad, E, 2006, The progress of user requirements through a software development project. Int J of Industrial Ergonomics, 36, 73–81

Robertson, S and Robertson, J, 2006, Mastering the Requirements process (2nd edition). Boston MA: Addison-Wesley.

WHAT MAKES SCIENTIFIC JOURNAL ARTICLES APPEAL TO HUMAN FACTORS PROFESSIONALS?

Amy Z.Q. Chung[1], Ann Williamson[1] & Steven T. Shorrock[1,2]

[1]*The University of New South Wales, Sydney, Australia*
[2]*Eurocontrol, Brétigny sur Orge, France*

The research-practice gap is of concern as it is believed that HF/E research may not be making an impact on practice in the 'real world'. A potential issue is what researchers and practitioners perceive as important in HF/E journal articles as a primary means of conveying research findings to practitioners. This study looked at the characteristics that make scientific journal articles appeal to HF/E researchers and practitioners using a web-based survey. The results compared the views of professionals from a related discipline, psychology. HF/E professionals placed importance on the practical significance of journal articles and were more similar to each other than to psychologists. HF/E researchers and practitioners were more similar than expected in judgements of important attributes of articles.

Introduction

Many disciplines have identified the existence of a research-practice gap. This gap is characterised by practitioners not implementing research findings and researchers not addressing questions relevant to practitioners (Cascio & Aguinis, 2008; McNicol, 2004). The problem has been highlighted in disciplines including industrial, work and organisational psychology, library and information science, and healthcare. The gap is of concern as it has potential implications for the adequacy of research and the implementation of research findings into practice.

A schism between researchers and practitioners has been identified in Human Factors and Ergonomics (HF/E) since the 1960's (Meister, 1999; Waterson & Sell, 2006). A recent international survey of HF/E professionals found that respondents from academic/research institutions rated HF/E journals as significantly more useful compared to participants working in applied settings (Chung and Shorrock, 2011). The main barriers to research application in practice included lack of access to journals or time to read them, and the view that the research was not relevant to practice. It seems that the research-practice gap may be at least partly due to differences in perceptions of the value and importance of HF/E research publications between researchers and practitioners.

A study by Sternberg and Gordeeva (1996) took up the question of what makes a journal article have impact. A survey of 252 psychologists required them to rate the importance of 45 characteristics of articles in determining whether or not journal articles have impact. The results highlighted six factors underlying judgements of an article's importance: *Quality of Presentation, Theoretical Significance, Practical Significance, Substantive Interest, Methodological Interest,* and *Value for Future Research*. The study participants were recruited from the American Psychological Society, which requires doctoral level admission and emphasises research. This almost certainly influenced their perceptions of what is important. For HF/E, it is important to know whether and how researchers and practitioners differ in perceptions of the importance of research and journal articles. Therefore, the aim of this study was to determine the beliefs of HF/E professionals (researchers and practitioners) about the attributes that make scientific journal articles have an impact in the field. This study also aimed to extend Sternberg and Gordeeva's (1996) research into the HF/E discipline. We hypothesised that those who work predominantly in research would have different views to those working mainly in practice.

Method

Questionnaire development and format

Following pre-testing, the questionnaire was hosted on www.surveymonkey.com. The questionnaire was open to HF/E professionals in both research and practice, and comprised a welcome/information page followed by 15 questions in two sections: Section 1) About you, and Section 2) Attributes of HF/Ergonomics journal articles.

In Section 2, respondents were asked to rate the importance of 28 statements presented in random order. Each statement represented a possible attribute of a journal article that may have an impact in the field of HF/E, i.e. an article that influences research or practice, or both. Respondents were asked to consider each attribute from the perspective of the discipline and profession, rather than from their own personal perspective. Each attribute was rated on a six-point scale from 'Not at all important' (1) to 'Extremely important' (6). The statements were selected from each of the six factors from Sternberg and Gordeeva's (1996) Principle-Components Analysis based on their loadings of greater than .55, with the exception of one statement in the *Substantive Interest* factor with a loading of .50, which was kept because there were only 3 items in this factor. This resulted in 28 statements being selected from the original list of 45 from Sternberg and Gordeeva's (1996) study, with references to psychology modified to apply more closely to HF/E.

Procedure and respondents

No universal list of HF/E professionals is available for this type of research so it was not possible to determine a total population size. In order to recruit as many HF/E professionals as possible, recruitment methods included contacting:

- HF/E Societies. The Presidents and Secretariats of each of the 47 International Ergonomics Association (IEA) Federated Societies were contacted in May 2011 to request that they invite their members to complete the survey. Responses were received from 27 out of the 47 IEA Federated Societies.

- LinkedIn Groups. A notice, "Survey of HF/Ergonomics Professionals: Appeal for Participation", was posted to members of 19 groups associated with HF/E on the professional networking website www.linkedin.com. Additionally, 203 members of these groups were sent the invitation individually via LinkedIn email.
- Professional contacts. 988 professional contacts (including 375 contacts from a previous study and 613 corresponding authors from the journals: *Human Factors*, *Ergonomics*, and *Applied Ergonomics*) were asked by email to complete the survey and to forward the invitation to their own HF/E colleagues.

Respondents could choose to be anonymous, or leave their email address to receive the results. There was no material incentive to complete the questionnaire.

Results

In total, 309 usable questionnaires were submitted. The study aimed to compare researchers and practitioners, but many respondents did both, so the analysis included respondents who spent at least 25% of their time in research or practice and the percentage of time spent in research was greater than that spent in practice (vice versa for practitioners). This resulted in exclusion of 49 respondents, plus another 10 were excluded because they were not currently employed or failed to specify details of work time. The study sample therefore included 250 respondents, with 104 classified as researchers and 146 classified as practitioners.

Demographics

The respondents represented 38 countries. Estimated work time over the last 12 months as an active HF/E professional was split on average as: research (30%), application (46%), education (15%), and other (10%). Most had postgraduate qualifications. For researchers, 66% were at doctorate level and 29% at postgraduate certificate, diploma or masters level. For practitioners, 31% were at doctorate level and 56% at postgraduate certificate, diploma or masters level.

Attributes of HF/Ergonomics journal articles

Table 1 present the top and bottom ten attribute ratings of HF/E professionals versus the psychologists in Sternberg and Gordeeva's (1996) study. The ranks for psychologists in Table 1 represent the order of the 28 items selected in the current study rather than the original rank. The HF/E groups differed on the top two ranked attributes: *a clearly stated and conceptualized problem* and *well written, structured and organised* for researchers and *containing useful implications for practice* and *results of major practical significance* for practitioners. Overall, however, HF/E researchers and practitioners rated the attributes similarly. The groups shared nine out of the top 10 attributes and seven of the bottom 10 attributes.

Table 1. Rank, mean and SD for the top and bottom ten attribute of HF/E professionals vs. Sternberg & Gordeeva's psychologists.

Attribute	HF Researchers	HF Practitioners	Psychologists
The problem is clearly stated and well-conceptualised	1. 4.94 (0.92)	4. 4.86 (0.98)	5. 4.33 (1.33)
Well-written, well-structured, and well-organised	2. 4.89 (0.89)	6. 4.78 (0.99)	7. 4.28 (1.26)
Results of analysis are presented clearly and discussed carefully …	3. 4.84 (0.94)	3. 4.91 (0.86)	4. 4.43 (1.25)
Topic is interesting and important	4. 4.76 (1.01)	5. 4.79 (0.95)	6. 4.29 (1.17)
Has a logical flow and organisation of ideas	5. 4.73 (0.91)	8. 4.69 (1.03)	13. 4.05 (1.29)
The writing is succinct and internally consistent	6. 4.60 (1.04)	9. 4.61 (0.98)	11. 4.13 (1.31)
Contains useful implications for professional practice	7=. 4.58 (1.17)	1. 5.04 (0.98)	27. 2.87 (1.40)
Results are of major practical significance	7=. 4.58 (1.16)	2. 5.02 (0.95)	19. 3.74 (1.45)
Hypotheses are clearly stated and testable	9. 4.39 (1.16)	10. 4.56 (1.11)	10. 4.14 (1.30)
Captures reader's interest	10. 4.34 (1.22)	11=. 4.40 (1.15)	18. 3.79 (1.34)
Includes concrete examples	18. 4.14 (1.22)	7. 4.73 (0.99)	28. 2.32 (1.26)
Presents a useful new theory or theoretical framework	19. 4.10 (1.21)	20=. 3.98 (1.25)	2. 4.79 (1.18)
Is clearly understandable to a broad cross-section of HF/E professionals	20. 4.05 (1.17)	14. 4.26 (1.21)	25. 3.40 (1.39)
Presents a new and useful test or technique	21. 4.04 (1.23)	16. 4.18 (1.09)	21. 3.70 (1.21)
Presented results are of major theoretical significance	22. 4.03 (1.18)	24. 3.82 (1.12)	1. 4.91 (1.11)
Demonstrates a useful experimental paradigm	23. 3.78 (1.14)	26. 3.69 (1.11)	22. 3.68 (1.13)
Integrates many different areas of data …	24. 3.74 (1.29)	22. 3.97 (1.22)	3. 4.56 (1.14)
Appears at the right moment, when people are ready to hear the message	25. 3.61 (1.35)	25. 3.76 (1.38)	14. 4.03 (1.54)
Debunks an existing theory or way of thinking	26. 3.54 (1.39)	27. 3.44 (1.46)	17. 3.88 (1.39)
Is applicable to work in many areas of HF/E	27. 3.52 (1.32)	18. 4.06 (1.38)	24. 3.46 (1.49)
Presents an experiment with a particularly clever paradigm …	28. 3.46 (1.40)	28. 3.17 (1.29)	23. 3.48 (1.18)

In contrast, the rankings between HF/E professionals and psychologists in Sternberg and Gordeeva's (1996) study were rather different. The three attributes ranked highest by psychologists, relating to *presenting a new theory*, *paper of major theoretical importance* and *integration of different areas of data* were in the bottom 10 of the HF/E researcher rankings. The psychologists shared only five of the attributes ranked in the top 10 by HF/E professionals. Overall, in contrast to both HF/E groups, psychologists rated attributes from the *Theoretical Significance* factor to be much more important than attributes from the *Practical Significance* factor.

Following Sternberg and Gordeeva (1996), a Principle-Components Analysis with varimax rotation was performed on the 28 attributes. Inspection of the scree plot revealed a clear break after four components. This four-factor solution explained 52.1% of variance and included 23 items, each loading above .5 on one of the four factors while not loading similarly on another factor. Table 2 shows the results of the final solution along with a comparison with the Sternberg and Gordeeva factor structure. The two factor structures were very similar, except that three factors from the original psychologists structure, *Theoretical Significance*, *Methodological Interest*, and *Value for Future Research* grouped into one factor for the HF/E professionals, titled *Research and Methodological Significance*.

The 28 attributes in the current study clustered under similar factors as in Sternberg and Gordeeva's (1996) study. The first factor was interpreted as *Research and Methodological Significance* as it contained items relating to theory and methods, and the value of research. The second and third factors were interpreted as *Quality of Presentation* and *Practical Significance* respectively. The items in these two factors corresponded exactly with factors of the same names in Sternberg and Gordeeva's (1996) factor analysis. The fourth factor was interpreted as *Interest and Impact* because it contains 3 items relating to capturing reader's attention and making an impact. All factors contained at least 3 items and the internal consistency across items in each factor (alpha) was high for all factors (from 0.67 to 0.85).

A logistic regression was conducted to determine whether any of these factors distinguished researchers and practitioners. Table 3 shows that Factor 1 and Factor 3 significantly distinguished researchers and practitioners.

The odds ratio for Factor 1: *Research and Methodological Significance* was significantly greater than 1, indicating that researchers were more likely than practitioners to rate that factor as important. On the other hand, Factor 3: *Practical Significance* had an odds ratio lower than 1, indicating that practitioners were more likely to rate this factor as important compared to researchers. The results show that researchers were 1.5 times more likely to rate attributes related to *Research and Methodological Significance* as important, and about half as less likely to rate attributes related to *Practical Significance* as important. The two groups did not differ on their ratings of the *Quality of Presentation*, and *Interest and Impact*.

Table 2. Results of the final Varimax-rotated principle components for the 23 attributes, the factor name, original factors from Sternberg and Gordeeva's (1996) study, variance explained and alpha (α) for each factor.

S&G Factor	Item	Loading
	Factor 1: Research and Methodological Significance (variance = 17%, α = 0.85)	
T	Debunks an existing theory or way of thinking	0.69
M	Presents an experiment with a particularly clever paradigm …	0.68
T	Presented results are of major theoretical significance	0.66
T	Presents a useful new theory or theoretical framework	0.65
M	Demonstrates a useful experimental paradigm	0.63
T	Provides a better explanation of existing phenomena	0.61
V	Contains useful implications for a scholarly understanding of the field	0.61
T	Integrates many different areas of data …	0.61
V	Contains useful recommendations for further research …	0.57
	Factor 2: Quality of Presentation (variance = 14%, α = 0.83)	
Q	Results of analysis are presented clearly and discussed carefully …	0.76
Q	The problem is clearly stated and well-conceptualised	0.75
Q	Has a logical flow and organisation of ideas	0.69
Q	The writing is succinct and internally consistent	0.63
Q	Well-written, well-structured, and well-organised	0.62
Q	Tone is unbiased and impartial	0.61
Q	Hypotheses are clearly stated and testable	0.54
	Factor 3: Practical Significance (variance = 11%, α = 0.74)	
P	Results are of major practical significance	0.79
P	Contains useful implications for professional practice	0.72
P	Includes concrete examples	0.62
V	Contains useful recommendations for changes or modifications in accepted theoretical constructs	0.50
	Factor 4: Interest and Impact (variance = 10%, α = 0.67)	
S	Captures reader's interest	0.82
Q	Starts and ends strongly, attracting attention and interest from the first paragraph and ending with clear take-home message	0.66
S	Appears at the right moment …	0.51

Discussion

The results revealed that HF/E researchers and practitioners have similar views on what attributes are important to them in a journal article. Both groups agreed that the quality of presentation (i.e. clear and well-structured writing) is crucial. This accords with one of the top suggestions for researchers in Chung and Shorrock's (2011) study, that researchers should adopt a reader-friendly writing style using

Table 3.   Logistic regression model of the four factors.

Factor	Coefficient	Standard error	Odds ratio	95% confidence interval	p value
1. Research and Methodological Significance	0.40	0.16	1.49	1.10, 2.02	0.010
2. Quality of Presentation	0.15	0.15	1.16	0.86, 1.56	0.335
3. Practical Significance	−0.83	0.17	0.44	0.31, 0.60	0.000
4. Interest and Impact	0.00	0.15	1.00	0.74, 1.35	0.998

*Hosmer-Lemeshow test $\chi^2_{(8)} = 9.08, p = 0.34$.

plain language and less scientific jargon to ensure that professionals within and outside of HF/E and their industry can understand the research findings. The current study indicates that researchers also rate these attributes highly, suggesting that the lack of appreciation of this problem is not a reason for the research-practice gap.

Overall, HF/E researchers and practitioners rated attributes related to theoretical significance (i.e. relating to new and existing theory) and methodological interest (i.e. relating to the experimental paradigm) relatively low in importance for impact in the field. Despite this, researchers still rated attributes relating to research and methodological significance higher in importance than did practitioners. In contrast, as might be expected, practitioners rated attributes related to practical significance as more important than did researchers. This is consistent with the finding of Chung & Shorrock's (2011) study where practitioners felt that the lack of consideration of the application of research findings was a barrier to the translation of research into practice. Again, however, the same two practice-related attributes were rated in the top 10 attributes for HF/E researchers and practitioners.

There appear to be clear differences between what HF/E professionals and psychologists view as determinants of impact for journal articles. In Sternberg and Gordeeva's (1996) study, psychologists rated theoretical significance more highly than the HF/E researchers in the current study. Thus despite the criticisms about the lack of practicality and applicability of HF/E research over the past 50 years, HF/E researchers do appear to value the practical significance of journal articles. However, a lack of theoretical focus in HF/E compared to research in psychology may restrict continuous improvements in the theories that should be the basis for practical applications. Well-written journal articles are universally appreciated by researchers and practitioners across disciplines. Overall, however, researchers and practitioners in the HF/E discipline are more alike in their views about impact, than researchers are across the HF/E and psychology disciplines.

Key findings

- HF/E researchers and practitioners have similar views on what attributes are important to them in a journal article.

- HF/E researchers and practitioners rated attributes related to theoretical significance and methodological interest as relatively low in importance.
- Researchers rated attributes relating to research and methodological significance as relatively higher in importance. Practitioners rated attributes related to practical significance as relatively higher in importance.
- HF/E researchers interest in theoretical significance appears to be relatively low when compared to psychologists.
- Well-written journal articles are universally appreciated across disciplines, regardless of professional orientation towards research or practice.

References

Cascio, W. F., & Aguinis, H. (2008). Research in industrial and organizational psychology from 1963 to 2007: Changes, choices, and trends. *Journal of Applied Psychology, 93*(5), 1062–1081.

Chung, A. Z. Q., & Shorrock, S. T. (2011). The research-practice relationship in ergonomics and human factors – Surveying and bridging the gap. *Ergonomics, 54*(5), 413–429.

McNicol, S. (2004). Is research an untapped resource in the library and information profession? *Science Journal of Librarianship and Information, 36*(3), 119–126.

Meister, D. (1999). *The history of human factors and ergonomics.* Mahwah, NJ: Lawrence Erlbaum Associates.

Sternberg, R. J., & Gordeeva, T. (1996). The anatomy of impact: What makes an article influential? *Psychological Science, 7*(2), 69–75.

ERGONOMICS EDUCATION AND STUDENTS' TENDENCY TO USE RESEARCH METHODS

Ilgim Eroglu, Abdusselam Selami Cifter & Kerem Ozcan

Department of Industrial Product Design, Mimar Sinan Fine Arts University, Turkey

Ergonomics lectures have been a stable element of industrial design education in Turkey. In this paper, effectiveness of education and possible improvements to education methods are researched. The aim was to identify how students use ergonomics evaluation methods during their design projects, as well as to understand whether it could be linked with the effectiveness of their ergonomics education. For this intention, an online questionnaire survey was conducted with 79 students from various universities in Turkey. The results were discussed and evaluated for further studies.

Introduction

Ergonomics is a vital element to be considered during Industrial Design education. Students getting education in this discipline frequently require utilizing ergonomics methods in their design projects. Most of the Turkish Universities giving education in this field provide ergonomics and human factors modules in both graduate and postgraduate levels. However, it is an ambiguity regarding how effectively students implement ergonomics during their education.

It is frequently thought that designers tend to built projects for themselves rather than potential users (Margolin, 1997; Kates & Clarkson, 2003; Kates & Clarkson, 2004). In fact, Pheasant (2003) states that, one of the common mistakes designers make is to think as "The design is satisfactory for me – it will therefore, be satisfactory for everybody else". Therefore, designers require using data that are linked with ergonomics, as it "... addresses the psychological aspects of how people interact with products, such as user perception, cognition, memory, reasoning and emotion" (Rodgers & Milton, 2011). This is why ergonomics is a part of industrial design education in Turkish universities, as well as other universities worldwide.

The aim of this study was to evaluate ergonomics and human factors education of industrial design departments in Turkey, combined with evaluation of their tendencies to use pre-defined research methods. This research is meant to be basis for a further study, in order to develop new approaches to education, where the aim is to improve students' knowledge and capability to use it.

In our study, we held a survey among 79 students from industrial design departments of various universities, including; Anadolu University, Bahçeşehir University,

Doğuş University, East Mediterranean University, Gazi University, Istanbul Technical University, Işık University, Kadir Has University, Marmara University, Mimar Sinan Fine Arts University, Middle East Technical University.

Regarding the aim of this research, this survey focused on the following topics:

- To what degree were the ergonomics modules found satisfying in their point of view?
- What are the main reasons for their satisfaction/dissatisfaction?
- How do they think their projects are assessed in terms of ergonomics?
- How do they test and develop their works to address ergonomics issues?
- How do they find the necessary information about ergonomics?

Methodology

A structured self-administered online questionnaire survey was used as the main method of this study (Robson, 2002). This method was chosen because the intention was to send the questionnaire to potential participants easily and more anonymously. Also this method enabled the researchers to find possible participants from various universities from different cities.

During the preparation of the questionnaire, SurveyMonkey online questionnaire tool was used. This is an effective tool because it facilitates the data collection and also provides basic analysis options. Where the researchers required more in depth analysis options, SPSS Statistical Analysis software was used. Mainly descriptive analysis techniques were used.

Two main question types were used in the questionnaire: i.e., open-ended questions and Likert scale questions. Likert scale questions were used to measure students' tendency to use certain research methods. For this purpose, a five-point scale was used in order to allow the neutrality to these questions. On the other hand, open-ended questions were used where the intention was to seek more in-depth information. We tried to adjust questionnaire application time to approximately 15 minutes. Therefore limited number of open-ended questions were included, also these questions were optional.

Results

The study was examined in two main parts. In the first part, the participants were asked about their thoughts on ergonomics education. They answered questions about their level of satisfaction and its reasons, if they reach published data easily and whether they think their evaluators (instructors) pay enough attention to ergonomics. On the other hand second part was about understanding of how design students seek ergonomics data during a project. They also answered questions whether they carry out an evaluation study by using ergonomics research methods when they reach to a certain stage at their design projects.

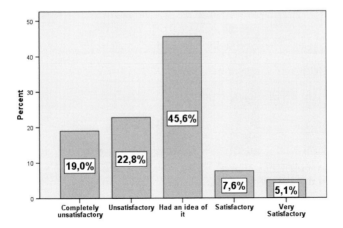

Figure 1. How satisfactory was your ergonomics and human factors module?

Evaluation of ergonomics education

Students were asked about their opinion to what extend is ergonomics taken into account when projects are evaluated by using a five-point Likert scale question. The mean value for this criterion is 3.8, which is above the average tendency. None of the 79 student chose the answer "never" for this question. The most common answer was "Usually", with a percentage of 40.5%.

It was also important to understand whether design students in Turkey were able to reach published data on ergonomics easily. A 'five-point' Likert scale question was used for this purpose as well. Their answers have a mean value of 2.7, which may be pointing out the limitation of published resources on ergonomics in Turkey.

Students were also asked if they were satisfied with their ergonomics education. The mean value for this question was 2.6, which is below average. The most common answer is "Had an idea of it" (which was ranked 3) with a percentage of 45.6%. This could be linked with the findings about difficulty in reaching published resources again.

Regarding the subsequent open-ended part of the question, several reasons reflecting students dissatisfaction, were expressed by 43 out of the 79 participants. Their answers were reviewed and coded into five categories as follows: (1) instructor was incapable, (2) insufficient applied studies, (3) syllabus was not suitable for industrial design, (4) course duration was short, and (5) syllabus was found poor. The results are presented for each of the category in Figure 2.

As can be seen from the results below, "insufficient applied studies" was the most mentioned category. Most of the respondents indicated that the education they received was mainly theoretical. Their general idea was, for industrial design

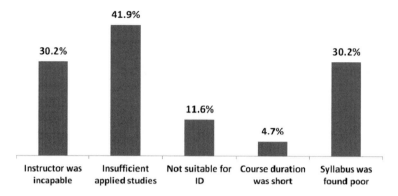

Figure 2.   The reasons reflecting students' dissatisfaction of the ergonomics modules they took.

education, ergonomics should be thought through hands-on applications and case studies. Some of their comments are as follows:

"It could be better. It should include more applied studies, and if there is any, more importance should be given to ergonomics."

"Our workshop studies were insufficient. Mostly we had visual studies."

This was followed by the categories of 'instructor was incapable' and 'syllabus was not suitable for industrial design' which share the same percentage (30.2% each). Criticisms on instructors varied from their lack of knowledge to their lack of attendance. On the other hand, participants who believed the course was not suitable for industrial design, expressed problems with respect to the lack of connection between ergonomics data given and their relevance to actual design problems. They also argued the knowledge provided was being more suitable for engineering purposes. Some of their comments were:

"I think neither students nor the instructor paid the necessary seriousness for this subject."

"Course was too theoretical. Industrial design and ergonomics concepts were not combined in the course."

From these questions, it can be inferred that students believe that their instructors take ergonomics into account, when evaluating their design projects. However, there are certain problems in ergonomics education itself. Students addressed some of the problems they feel about ergonomics education, which can be used as basis for improvement in further studies.

Students' tendency to use research methods before starting a project

Students were asked about what research methods they use to make an evaluation about ergonomics before starting a project. A five-point Likert scale question was

Figure 3.   Evaluation methods used in the early design phase.

prepared for this purpose; i.e. 1 means "I never use"; 5 means "I often use". A number of pre-determined research methods found in the relevant literature were given them to scale. The predetermined research methods frequently used by designers are as follows: the internet (Goodman et al., 2007), self (Goodman et al., 2007), comparisons (BSI, 2008), books and journals (Goodman et al., 2007), expert review (BSI, 2008), users (Goodman et al., 2007), other people (Goodman et al., 2007), contextual inquiry and observation (BSI, 2008), questionnaire (BSI, 2008).

The results are presented in Figure 3. A mean score over the average value '3' was considered as positive.

As seen in Figure 3, students' tendencies to use "Questionnaire", "Expert Review" and "Published Resources" received a score below average value, which was considered as the students do not utilise these methods frequently in their design projects. Other research methods were ranked all above the average value, with "the Internet" and "Contextual Inquiry/Observation" getting the highest mean values.

When listed from highest to the lowest ranked research methods, the results suggested that students' intention to use a certain method decreases if they have to contact other people to use that method. The only exception is "published resources", which ranks as the third least used method. However, as mentioned above, students also expressed their difficulty in finding published resources; which could be the reason for this finding.

Students tendency to use research methods during/at the end of the design project

Students were asked about what research methods they use to make an evaluation about their projects after it has reached a certain stage. A five-point Likert scale question was prepared for this purpose; i.e. 1 means "Never used"; 5 means "Always used". A number of pre-determined research methods found in the relevant literature

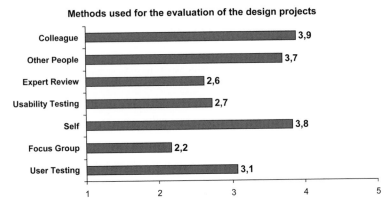

Figure 4.   Methods used during the late design phase.

were given them to scale. An "Other" option was also provided. The predetermined research methods are as follows: user tests (Stanton et al., 2005), focus group (Stanton et al., 2005), Self (Goodman et al., 2007), usability testing (BSI, 2008), expert review (BSI, 2008), users (Goodman et al., 2007), other people (family and friends), colleagues.

Results are presented in Figure 4. A mean score over the average value '3' was considered as positive.

"Colleagues", "Self" and "Other People" are ranked as the three mostly used research methods, where there is a clear gap between them and other research methods in terms of the mean values. "User Testing" method received a mean value of 3.1, which can be translated into an average tendency of use. On the other hand, the results suggested that "Expert Review", "Usability Testing" and "Focus Group" methods are not frequently utilised by design students in order to evaluate their design projects.

It can be inferred that after students reached a certain level at their design projects, they tend to use methods which are less structured and less formal. They prefer to discuss their projects with other people around them.

Discussion and conclusions

From this research, two main conclusions can be reached. Firstly, students feel their project instructors give enough importance on ergonomics when evaluating their design projects, but there are several issues related with ergonomics and human factors modules themselves. Criticisms were mostly on syllabuses, instructors and insufficient applied studies. Also they mentioned there was a lack of published resources.

About the second main conclusion; it is inferred that design students in Turkey do not tend to involve other people in their design projects if they require giving effort to contact them. Rather, they tend to do their studies within their own environment.

Another study which was done among students of Brunel University hints that there was a tendency among students to use qualitative data rather than quantitative data. The paper also states that the most popular research methods among students were found to be "Observation, Data Collection through literature search, Interview, and Questionnaire" (Dong & Nicholls, 2011). However, as "Users" and "Other People" are not usually used and "Questionnaire" is the least favourite method in our study, it may be claimed that the situation in Turkey is not about data being qualitative or quantitative. It may be estimated that students in Turkey prefer research methods that they can handle within their daily environment, as the tendency to use a research method decreases when unfamiliar people are involved at the process.

These two main conclusions may be providing hints for a following study to improve ergonomics education efficiency. The fact that students mentioned low tendency to speak with users/experts and seldom use of structured research methods, it was considered this could be linked with their inadequacy of ergonomics knowledge, as well as their problems in reaching ergonomics data. Mackie (2011) suggests that when students are provided with adequate ergonomics knowledge, their willingness to use their ergonomics skills in design projects increases. Therefore, it may be assumed that, students' tendency to use more structured research methods might be increased, if certain improvements are done in their ergonomics and human factors modules. However, it should be noted that this paper is the first step of a more in-depth study, and some of the possible linkages between mentioned outcomes will be sought with students in following studies.

Further studies

A following study will be carried out in an effort to suggest improvements to ergonomics and human factors modules with the participation of students. For this intention workshops are planned to shed light on why students prefer to work within their own environment, and the reasons of their avoidance of using ergonomics methods which require contacting with other people.

Results coming out of these studies may help to understand students' point of view on how to improve the efficiency of ergonomics and human factors modules. Intention would be to understand whether there is an increase in students' tendency to use research methods in their design projects when they are provided with the ergonomics knowledge.

References

BSI. (2008c). BS EN 62366:2008 – Medical Devices – Application of Usability Engineering to Medical Devices. British Standards Institute.

Dong, H. & Nicholls, S. 2011. How Design Students Use Ergonomics? In Andersen, M. (ed.) Contemporary Ergonomics and Human Factors 2011. (Taylor & Francis, Wiltshire), 113–120

Goodman, J., Langdon, P. & Clarkson, P.J. 2007, Formats for user data in inclusive design. In Proceeding UAHCI'07 Proceedings of the 4th international conference on Universal access in human computer interaction: coping with diversity, (Springer-Verlag Berlin, Heidelberg), 117–126

Keates, S. & Clarkson, J. 2003. Design Exclusion. In J. Clarkson, R. Coleman, S. Keates, & C. Lebbon, Inclusive Design: Design for the Whole Population. (London: Springer-Verlag London Ltd.) 88–102

Keates, S. & Clarkson, J. 2004. Countering Design Exclusion: An Introduction to Inclusive Design. (Springer-Verlag London Ltd.)

Margolin, V. 1997. Getting know the user. Design Studies, 18, 227–236.

Mackie, E. 2011. Ergonomics Teaching Within Industrial Design; An Evaluation of Evidence of Understanding. In Andersen, M. (ed.) Contemporary Ergonomics and Human Factors 2011. (Taylor & Francis, Wiltshire), 113–120

Pheasant, S. 2003. Bodyspace; Anthropometry, Ergonomics and Design of Work. (Taylor & Francis, London)

Rodgers, P. & Milton, A. 2011. Product Design. (Laurence King Publishing, London)

Stanton, N. A., Salmon, P. M., Walker, G. H., Baber, C. & Jenkins, D. P. 2005. Human Factors Methods: A Practical Guide for Engineering and Design. (Ashgate, Surrey, England)

POSTERS

EXPLORING THE BENEFITS OF BRIEF HEALTH PSYCHOLOGY INTERVENTIONS IN THE WORKPLACE

James I. Morgan

Psychology Research Group, Sheffield Hallam University, UK

Introduction

According to Labour Market Statistics, between the 2nd quarter of 2008 (the start of economic recession) and the 4th quarter of 2011, 2.68 million workers were laid off or made redundant in the UK (see CIPD, 2012). Whilst the implications of downsizing are explicitly apparent for the newly unemployed, there are also underlying consequences for those who remain in post (Armstrong-Stassen, 1994); 'downsize survivors'. Downsize survivors can experience a heightened sense of job threat, increased workload, and an overall increase in stress and anxiety (e.g., Armstrong-Stassen, 1994; Wiesenfeld et al., 2001). With British businesses losing £26 billion per year as a result of work-related stress (CIPD/MIND, 2011), maintaining employee well-being is of paramount importance. Failing to effectively manage the negative effects of downsizing may incur more cost and render the downsizing process counterproductive. Through the implementation of effective strategies to improve employee mental health UK businesses could save up to £8 billion per year (CIPD/MIND, 2011).

One potentially useful approach is to utilise the extensive health psychology research literature in order to develop and test interventions that can be easily applied in occupational settings, i.e., those that will not place any additional strain on employees or employers. The present study applies Steele's (1988) concept of self-affirmation to reduce stress levels in downsize survivors over both the short- and longer-term. The principle aim of the study was to test the effectiveness of a brief self-affirmation intervention adapted for use in a work setting.

Method

Sixty-six workers were recruited from a UK sixth form college during a period of organisational downsizing. Employees were randomly allocated to one of two conditions: one in which they were asked to create a self-affirmation implementation intention (S-AII; adapted from Armitage, Harris, & Arden, 2011), or a control. To create a S-AII workers were simply required to construct an 'if . . ., then . . .' plan (an implementation intention) to self-affirm (recall core beliefs or values) when feeling threatened or anxious at work. Feelings of anxiety and depression were measured before and after the intervention or control task (time one), and for some (twenty-eight workers) three weeks later (time two). Job satisfaction, self-efficacy, and self-esteem were also measured.

Results

There were statistically significant differences between the S-AII condition and the control. Self-affirmed workers reported an immediate reduction in anxiety. This reduction was evident 3 weeks later. Workers in the S-AII condition reported greater self-efficacy (a belief in one's own ability to succeed/prevail). There were no significant effects of S-AII's on any of the other outcomes.

Conclusion

The results of this small-scale study suggest that the integration of health psychology interventions, such as the S-AII, into existing organisational practices may be of benefit to the well-being of employees.

Acknowledgements

The author would like to thank Chris Broom for his role in data collection, and the HE college staff who kindly gave up their time to take part.

References

Armitage, C.J., Harris, P.R., and Arden, M.A. 2011, Evidence that self-affirmation reduces alcohol consumption: Randomized exploratory trial with a new, brief means of self-affirming, *Health Psychology, 30*, 633–641.

Armstrong-Stassen, M. 1994, Coping with transition: A study of layoff survivors, *Journal of Organizational Behavior, 15*, 597–622.

CIPD. 2012, Counting the cost of the jobs recession, Work Audit Issue 42, http://www.cipd.co.uk/binaries/5795workauditWEB.pdf (September 2012, date last accessed).

CIPD/MIND, 2011, Managing and supporting mental health at work: disclosure tools for managers, http://www.mind.org.uk/assets/0001/6314/Managing_and_supporting_MH_at_work.pdf (September 2012, date last accessed).

Steele, C. M. 1988, The psychology of self-affirmation: Sustaining the integrity of the self. In L. Berkowitz (ed.), *Advances in Experimental Social Psychology* (New York, NY: Academic Press), Vol. 21, pp. 261–302.

Wiesenfeld, B. M., Brockner, J., Petzall, B., Wolf, R., and Bailey, J. 2001, Stress and coping among layoff survivors: A self-affirmation analysis, *Anxiety, Stress, and Coping: An International Journal, 14*, 15–34.

USING TOUCHSCREEN DEVICES TO PROVIDE QUICKER, BETTER, AND CHEAPER REAL-TIME WORKLOAD MEASUREMENT

Selma Piric

Think Research Ltd. UK

Even though there are many methods available for measuring subjective cognitive workload, none of these methods provide the benefits if being portable, customisable, and freely available. Based on the ISA technique, the "iSA app", was developed to prove that workload assessment in ATM research can be all of the above. Bypassing traditional real-time workload measurement methods that require either expensive hardware or a team taking real-time measurements, the iSA app is freely available to everyone who needs quick and easy access to a subjective workload measurement device.

Overview of workload measures

Measuring workload has been an enduring effort in military and aviation domains. Elsewhere (EUROCONTROL, 2003) several cognitive workload measurement methods have been described and reviewed. These measures include the Modified Cooper-Harper scale; the SWAT measure; the NASA Task Load Index; and the Instantaneous Self-Assessment (ISA). These methods vary in, reliability, sensitivity, and also in intrusiveness if the measurement is taken real-time during operations.

Real time workload measurement in ATM

Measuring subjective real-time workload in Air Traffic Management (ATM) research and development simulations historically requires either dedicated hardware embedded into the R&D Platform or a team of analysts taking real-time measurements. These approaches are still the most widely used and are the main means of providing the measurement stated because there hasn't been an alternative. One option requires installing static, limited, hard-wired systems that are fixed in place (installation costs of ±£40k). The other option requires a resource intensive manual recording of workload measure during the research or trial session (costs of up to £15K per session).

Justifying the need for a move to mobile touchscreen devices

These previously mentioned options are expensive considerations when the user does not have an existing measurement process or system already available.

443

In addition, the existing measurement methods are fixed in place, not portable, and not usable in small places. Times have changed, flexible and powerful mobile touchscreen devices have the ability to be tailored to specific situations, and they are cheap, quick to update and portable. These systems can be made scalable from one measured position to many with limited cost implications.

The Instantaneous Self-Assessment (ISA) method is used extensively throughout ATM R&D projects as it is seen as the least intrusive of all subjective workload measures. This concept is simplistic in nature; it records a numerical value every x minutes that determines the perceived workload response. Therefore, this method of measurement lends itself to touchscreen devices, and for that reason it was chosen as the case study for further development. Developed by Think Research, the "iSA app" (lower case i), based on the ISA technique (capital I), provides flexibility in workload assessment in ATM research. This technique uses a uni-dimensional scale that uses five different ratings for perceived workload: from excessive, to under-utilized; which is subjectively given by the operator during the task. Figure 1 shows the screenshots of the app; the start screen, the settings screen, and asking the user in the third screen "*How do you rate your perceived workload at present?*"

The development of iSA cost Think Research around £500 and was developed to prove the following:

- **Portability**: The iSA app is portable, and can be used anywhere;
- **Always accessible**: The iSA app is non-invasive for engineering terms: it is always directly available and ready for use;
- **Configurable**: Questions are adjustable to what needs to be measured; the scale can be adjusted, as well as the input time limit;
- **Free**: With low development costs, Think Research has been able to make the app available for free via an online application distribution platform, in this case Apple's App Store.

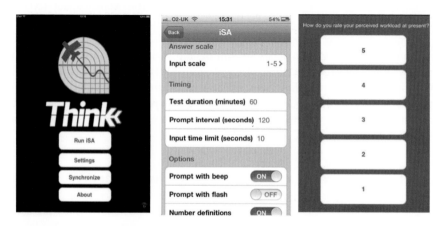

Figure 1.   iSA app screenshots: Start screen; Settings; and Input.

Data output

The iSA app logs the interval number, the score recorded by the user, the response time for the user to enter the response, and a time stamp of the actual time of data entry. The results can then be viewed and analyzed using a PC.

User experience, evaluation and V&V

The ISA technique is a well-used method for measuring subjective workload and therefore a V&V has not been carried out for the app. On iTunes there is a possibility to rate the application as a user after the download. One person has added comments in the review:

> *"Being a User Experience/Human Factors Engineer I'm probably squarely in the 'target audience' for this app so I was very interested to give it a try when I heard that this has been released.*
>
> *I've carried out a few studies using the ISA measurement scale and it's obvious that the authors know what they're talking about, as all of the configuration options and useful features are in the app (the app records non-responses for example, which can be a useful indication that the participant was just too busy to complete the rating scale at all).*
>
> *If the authors maintain this app then the ability to prevent the phone from dimming the screen or going to sleep where there is no response would be very handy.*
>
> *For engineers involved with Workload Analysis, Human Factors, or Operability studies in air traffic management or other industries this is likely to be a useful tool."*

Figure 2 shows the total amount of rating given by users and the score. The Figure indicated that 6 users (out of 10 reviewers) have rated it 5 stars, one user has rated the app 3 stars and 3 users have rated it a low one star.

Figure 2. iSA User ratings from iTunes

Conclusion

The ISA method is used extensively throughout ATM R&D projects as it is seen as the least intrusive of all subjective workload measures and therefore lends itself to touchscreen devices. Developed by Think Research, the "iSA app" (lower case i), based on the ISA technique (capital I), provides flexibility in workload assessment in ATM research. This app, available on iTunes is portable, always accessible, configurable, and free, unlike the traditional ISA measurement options.

Summary comments

Touchscreen devices provide a platform for ATM relevant tools such as the iSA app to be applied quickly and cheaply for research. Using the iSA app as a proof of concept, Think Research have taken an existing and proven measure and made it more accessible, user friendly, portable, and free; enabling over 500 users from 40 countries around the globe to use the ISA technique as has been shown in the amount of downloads from the iTunes store.

Reference

EUROCONTROL, Review of Workload Measurment, Analysis and Interpretation Methods, CARE-Integra-TRS-130-02-WP2, 2003.

DESIGNING AN EXPERT SYSTEM FOR RISK-MANAGEMENT TO SUPPORT OPERATORS' MENTAL MODEL

Sebastian Spundflasch, Johannes Herlemann & Heidi Krömker

Department of Media Production, Ilmenau University of Technology, Ilmenau, Germany

Introduction

The growing trend of centralization in German road-tunnel control results in an increasing number of tunnels and a higher amount of information that must be processed by individual operators. In order to support their main tasks – *monitoring*, *diagnosis* and *control* (VDI/VDE 3699, 2005) and in particular *risk management* – an expert system will be developed within the research project ESIMAS. To find out the requirements that the user interface of the expert systems must comply, we studied workflow and mental model of tunnel control operators. The results serve as basis for designing the interaction between operator and expert system.

Procedure

In a first analysis we wanted to find out, how risk management is being performed by different tunnel control centers within the 3 phases of process control [Figure 1]. 10 tunnel control centers, different in terms such as number of tunnels controlled, traffic volume, organizational structure and composition of its staff, were analyzed. Following the Contextual Design (Beyer & Holtzblatt, 1998) process, operators were observed at work and in-depth interviews were conducted with tunnel managers and operators. The focus was on the interdependences between operators, their tasks, their control system and their working environment.

Results

The analysis showed, that the main tasks control, monitoring and diagnosis can be observed in all analyzed control centers, but the extent to which the tasks are being

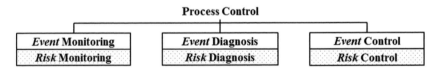

Figure 1.   Phases of Process Control according to (VDI/VDE 3699, 2005) expended by activities of Risk Management.

performed within subtask vary. With regard to risk management, a major difference between the centralized monitoring of several tunnels (multi tunnel control – MTC) and monitoring of a single tunnel (single tunnel control – STC) was observed. In STC, the operators mental model (Yin & Laberge, 2010) and thus the workflow concerning risk management, is much more precise than in MTC. A Hierarchical Task Analysis (Shepherd, 2001) of each control center yielded the result. In STC the process of monitoring includes monitoring for risks, carried out by proactive control behavior, e.g. the continuous monitoring of video images from inside the tunnel. The monitoring for risks is the prerequisite for activities in risk diagnosis and risk control. With an increasing number of tunnels to be monitored in MTC, a more and more reactive control behavior was observed. The operators just react to manually or automatically detected *events*. A monitoring for *risks* does not take place. Not seeing the *risk,* the operator is not able to do preventive tasks. This can cause events which could have been prevented.

Next steps

The results showed that there is a need for support in monitoring for risks especially in MTC. The next step will be the user interface design of the expert system. It should support the operators by providing information they need to perform effective risk management. The required Information will be extracted from the analysis of STC operators' mental model and their performed subtasks in risk management. The activity of monitoring will be extended by an automatic detection of risks. The activities of diagnosis and control are to be improved by selective information for an efficient assessment of risks and decision making. The ultimate goal of the expert system is to represent selective data that enables the operators to perform the differentiated subtasks in MTC as well as in STC.

References

VDI/VDE 3699 Part 2. 2005, *Process control using display screens – Prinicples,* (Beuth Verlag, Berlin).
Beyer, H. and Holtzblatt, K. 1998, *Contextual Design. Defining Customer-Centered Systems*, (Morgan Kaufmann Publishers, San Francisco).
Yin, S. and Laberge J. 2010, *How Process Control Operators Derive, Update, and Apply Mental Models*. Human Factors and Ergonomics Society 54th Annual Meeting. San Francisco, CA.
Shepherd, A. 2001, *Hierarchical Task Analysis*, (Taylor & Francis, London).

Author index